Graph Theory and
Related Topics

Courtesy J. A. Bondy

W. T. Tutte

Graph Theory and Related Topics

*Proceedings of the conference
held in honour of Professor W. T. Tutte
on the occasion of his sixtieth birthday,
University of Waterloo, July 5–9, 1977*

Edited by J. A. BONDY   U. S. R. MURTY
DEPARTMENT OF COMBINATORICS AND OPTIMIZATION
FACULTY OF MATHEMATICS
UNIVERSITY OF WATERLOO
WATERLOO, ONTARIO

ACADEMIC PRESS   New York   San Francisco   London   1979
A Subsidiary of Harcourt Brace Jovanovich, Publishers

COPYRIGHT © 1979, BY ACADEMIC PRESS, INC.
ALL RIGHTS RESERVED.
NO PART OF THIS PUBLICATION MAY BE REPRODUCED OR
TRANSMITTED IN ANY FORM OR BY ANY MEANS, ELECTRONIC
OR MECHANICAL, INCLUDING PHOTOCOPY, RECORDING, OR ANY
INFORMATION STORAGE AND RETRIEVAL SYSTEM, WITHOUT
PERMISSION IN WRITING FROM THE PUBLISHER.

ACADEMIC PRESS, INC.
111 Fifth Avenue, New York, New York 10003

United Kingdom Edition published by
ACADEMIC PRESS, INC. (LONDON) LTD.
24/28 Oval Road, London NW1 7DX

Library of Congress Cataloging in Publication Data
Main entry under title:

Graph theory and related topics.

    A symposium honoring Professor W.T. Tutte,
held at the University of Waterloo, Ontario,
July 5-9, 1977.
    1. Graph theory--Congresses. 2. Tutte, W.T.
I. Bondy, John Adrian. II. Murty, U. S. R.
III. Tutte, W. T.
QA166.G73   511'.5   78-27025
ISBN 0-12-114350-3

PRINTED IN THE UNITED STATES OF AMERICA

79 80 81 82 83 84   9 8 7 6 5 4 3 2 1

To W. T. Tutte, on the occasion
of his sixtieth birthday

Contents

List of Participants and Contributors	xi
Preface	xvii

Early Reminiscences   xix
    CEDRIC A. B. SMITH

A Tribute   xxiii
    PAUL ERDÖS

A Note on Some of Professor Tutte's Mathematical Work   xxv
    C. St. J. A. NASH-WILLIAMS

Papers and Books by W. T. Tutte   xxix

Topological and Algebraic Methods in Graph Theory   1
    LÁSZLO LOVÁSZ

All the King's Horses (A Guide to Reconstruction)   15
    W. T. TUTTE

Hadwiger's Conjecture and Six-Chromatic Toroidal Graphs   35
    MICHAEL O. ALBERTSON and JOAN P. HUTCHINSON

Chromatic Polynomials and the Internal and External   41
Activities of Tutte
   RUTH A. BARI

Chromatic Equations   53
   ARTHUR BERNHART

Planar Colorings: A Theory   71
   FRANK R. BERNHART

On The Algebra of Graph Types   81
   NORMAN BIGGS

Matroids, Graphs, and 3-Connectivity   91
   ROBERT E. BIXBY and WILLIAM H. CUNNINGHAM

On the Mixed Achromatic Number and Other Functions of Graphs   105
   FRED BUCKLEY and A. J. HOFFMAN

On Tutte's Conjecture for Tangential 2-Blocks   121
   BISWA TOSH DATTA

Intersection and Distance Patterns   133
   M. DEZA and I. G. ROSENBERG

Regular Groups of Automorphisms of Cubic Graphs   145
   DRAGOMIR Ž. DJOKOVIČ and GARY L. MILLER

Problems and Results in Graph Theory   153
and Combinatorial Analysis
   PAUL ERDÖS

Strong Independence of Graphcopy Functions   165
   PAUL ERDÖS, LÁSZLÓ LOVÁSZ, and JOEL SPENCER

Squaring Rectangles and Squares 173
P. J. FEDERICO

Subsemigroups and Supergraphs 197
S. FOLDES and G. SABIDUSSI

There Are Finitely Many Kuratowski Graphs for the Projective Plane 201
HENRY H. GLOVER and J. PHILIP HUNEKE

On a Kuratowski Theorem for the Projective Plane 207
HENRY H. GLOVER, J. PHILIP HUNEKE, and CHIN SAN WANG

On F-Hamiltonian Graphs 219
ROLAND HÄGGKVIST

Rochromials and the Colourings of Circuits 233
DICK WICK HALL

A General Construction for Equidistant Permutation Arrays 247
KATHERINE HEINRICH, G. H. J. VAN REES, and W. D. WALLIS

J-Components, Bridges, and J-Fragments 253
ARTHUR M. HOBBS

Hamilton Cycles in Regular Two-Connected Graphs 261
BILL JACKSON

Representations of Matroids 267
CHRISTOPHER LANDAUER

On Some Generalizations of Partial Geometry 277
RENU LASKAR and J. A. THAS

A Graph-Theoretical Approach to Embedding $(r,1)$-Designs 289
  D. McCARTHY, N. M. SINGHI, and S. A. VANSTONE

Chromatic Enumeration for Triangulations 305
  J. D. McFALL

A Covering Problem in Binary Spaces of Finite Dimension 315
  R. C. MULLIN and R. G. STANTON

The Tutte Polynomial and Percolation 329
  J. G. OXLEY and D. J. A. WELSH

Sums of Circuits 341
  P. D. SEYMOUR

Unsolved Problems 357
  A Conjecture on Planar Graphs 357
  MICHAEL O. ALBERTSON and DAVID BERMAN

  Selected Open Problems in Graph Theory 358
  ANTON KOTZIG

  Some Unsolved Problems on One-Factorizations of Graphs 367
  P. D. SEYMOUR

  Imbalance of Trees 368
  B. SIMEONE

  The Four Color Problem for Locally Planar Graphs 369
  WALTER STROMQUIST

  A Dissection Problem and an Intertwining Problem 370
  PETER UNGAR

List of Participants and Contributors

Numbers in parentheses indicate the pages on which the authors' contributions begin.

H. L. ABBOTT, Department of Mathematics, University of Alberta, Edmonton, Alberta T6G 2E1

M. O. ALBERTSON (35, 357), Department of Mathematics, Smith College, Northampton, Massachusetts 01063

S. E. ANACKER, 2573 Muskingum Court, Columbus, Ohio 43210

K. APPEL, Department of Mathematics, University of Illinois, Urbana, Illinois 61801

R. A. BARI (357), Department of Mathematics, George Washington University, Washington, D.C. 20052

L. BATTEN, Department of Mathematics, University of Winnipeg, Winnipeg, Manitoba R3B 2E9

D. BERMAN (357), Department of Mathematics, University of New Orleans, New Orleans, Louisiana 70122

A. BERNHART (53), Department of Mathematics, University of Oklahoma, Norman, Oklahoma 73069

F. R. BERNHART (71), Department of Mathematics, Bloomsburg State College, Bloomsburg, Pennsylvania 17815

N. L. BIGGS (81), Department of Mathematics, Royal Holloway College, Egham, Surrey TW20 0EX, England

L. BILLERA, School of Operations Research, Cornell University, Ithaca, New York 14850

R. E. BIXBY (91), Department of Industrial Engineering and Management Science, Northwestern University, Evanston, Illinois 60201

G. S. BLOOM, Department of Computer Science, California State University, Fullerton, California 92634

F. BUCKLEY* (105), Department of Mathematics, Graduate Center, City University of New York, New York, New York 10036

J. W. BUTLER, Division of Mathematical Sciences, University of Iowa, Iowa City, Iowa 52242

J. CSIMA, Department of Mathematics, McMaster University, Hamilton, Ontario L8S 4K1

W. H. CUNNINGHAM, Department of Mathematics, Carleton University, Ottawa, Ontario K1S 5B6

B. T. DATTA (121), Department of Mathematics, The Ohio State University, Lima, Ohio 45804

A. K. DEWDNEY, Department of Mathematics, University of Western Ontario, London, Ontario N6A 5B9

M. DEZA (133), CNRS Paris, 3 rue de Duras, 75008 Paris, France

D. Ž. DJOKOVIC (145), Department of Pure Mathematics, University of Waterloo, Waterloo, Ontario N2L 3G1

R. DUKE, School of Mathematics, Georgia Institute of Technology, Atlanta, Georgia 30332

P. ERDÖS (153, 165), Mathematical Institute, Hungarian Academy of Sciences, Reáltanoda u. 13–15., Budapest V, Hungary

P. J. FEDERICO (173), 3634 Jocelyn Street, N. W., Washington, D.C. 20015

H. FLEISCHNER, Akademie der Wissenschaften, Fleischmarkt 20, 1 Stiege, A1010 Wien, Austria

J. FLORENCE, Department of Mathematics, University of Western Ontario, London, Ontario N6A 5B9

S. FOLDES (197), IRMA, Boîte Postale 53, 38041 Grenoble Cedex, France

G. FOURNIER, Department of Mathematics, Université de Sherbrooke, Sherbrooke, Québec J1K 2R1

R. FOURNIER, Department of Mathematics, Université de Sherbrooke, Sherbrooke, Québec J1K 2R1

M. L. GARDNER, Department of Mathematics, North Carolina State University, Raleigh, North Carolina 27606

P. M. GIBSON, Department of Mathematics, University of Alabama, Huntsville, Alabama 35807

F. R. GILES, School of Operations Research, Cornell University, Ithaca, New York 14850

H. GLOVER (201, 207), Department of Mathematics, The Ohio State University, Columbus, Ohio 43210

R. L. GRAHAM, Bell Laboratories, 600 Mountain Avenue, Murray Hill, New Jersey 07974

* Present address: Division of Mathematics and Science, St. John's University, Staten Island, New York 10301

LIST OF PARTICIPANTS AND CONTRIBUTORS xiii

D. GREENWELL, 7805 Cadillac Drive, Huntsville, Alabama 35802
J. L. GROSS, Department of Mathematics, Columbia University, New York, New York 10027
R. P. GUPTA, Department of Mathematics, The Ohio State University, Columbus, Ohio 43210
R. HÄGGKVIST* (219), Department of Mathematics, University of Ůmea, S-901 87 Ůmea, Sweden
W. HAKEN, Department of Mathematics, University of Illinois, Urbana, Illinois 61801
D. W. HALL (233), Department of Mathematics, State University of New York at Binghamton, Binghamton, New York 13903
D. HANSON, Department of Mathematics, University of Regina, Regina, Saskatchewan S4S 0A2
B. HARTNELL, Department of Computer Science, Brandon University, Brandon, Manitoba R7A 6A9
K. HEINRICH (247), Department of Mathematics, University of Newcastle, Newcastle, New South Wales 2308, Australia
A. P. HERON, 9586 Basket Ring Road, Columbia, Maryland 21045
A. M. HOBBS (253), Department of Mathematics, Texas A & M University, College Station, Texas 77843
A. J. HOFFMAN (105), IBM T. J. Watson Research Center, P.O. Box 218, Yorktown Heights, New York 10598
C. HUANG, 39 Glen Park Drive, Ottawa, Ontario K1B 3Y9
J. P. HUNEKE, Department of Mathematics, The Ohio State University, Columbus, Ohio 43210
J. P. HUTCHINSON (35), Department of Mathematics, Smith College, Northampton, Massachusetts 01063
W. JACKSON (261), Faculty of Mathematics, University of Waterloo, Waterloo, Ontario N2L 3G1
T. A. JENKYNS, Department of Mathematics, Brock University, St. Catherines, Ontario L2S 3A1
J. KAHN, Department of Mathematics, The Ohio State University, Columbus, Ohio 43210
R. A. KINGSLEY, Department of Mathematics, University of Winnipeg, Winnipeg, Manitoba, R3B 2E9
R. R. KORFHAGE, Department of Computer Science, Southern Methodist University, Dallas, Texas 75275
A. KOTZIG (358), 4850 Côte des Neiges #612, Montréal, Québec N3V 1G5
W. KUHN, St. Joseph's College, Philadelphia, Pennsylvania 19131

* Present address: Department of Combinatorics and Optimization, University of Waterloo, Waterloo, Ontario N2L 3G1

C. LANDAUER (267), Language/Systems Group, Pattern Analysis and Recognition Corporation, Rome, New York 13440
R. LASKAR (277), Department of Mathematical Sciences, Clemson University, Clemson, South Carolina 29631
A. LEHMAN, Department of Mathematics, University of Toronto, Toronto, Ontario M5S 1A1
B. LOERINC, Courant Institute of Mathematical Sciences, 251 Mercer Street, New York, New York 10012
J. Q. LONGYEAR, Department of Mathematics, Wayne State University, Detroit, Michigan 48202
L. LOVÁSZ (1,165), Bolyai Intezet, József Attila University, Aradi Vértanúk tere 1, 6720 Szeged, Hungary
D. McCARTHY (289), Department of Computer Science, University of Manitoba, Winnipeg, Manitoba R32 2N2
J. D. McFALL (305), Department of Combinatorics and Optimization, University of Waterloo, Waterloo, Ontario N21 3G1
E. MENDELSOHN, Department of Mathematics, University of Toronto, Toronto, Ontario M5S 1A1
M. MILGRAM, 11213 Monticello Avenue, Silver Spring, Maryland 20902
G. L. MILLER (145), Department of Mathematics, Massachusetts Institute of Technology, Cambridge, Massachusetts 02139
G. J. MINTY, Department of Mathematics, Indiana University, Bloomington, Indiana 47401
R. C. MULLIN (315), Faculty of Mathematics, University of Waterloo, Waterloo, Ontario N2L 3G1
C. St. J. A. NASH-WILLIAMS, Department of Mathematics, University of Reading, Whiteknights, Reading, Berkshire RG6 2AX, England
J. G. OXLEY* (329), Merton College, Oxford OX1 4JD, England.
C. M. PAREEK, Department of Mathematics, Kuwait University, P.O. Box 5969, Kuwait
D. K. RAY-CHAUDHURI, Department of Mathematics, The Ohio State University, Columbus, Ohio 43210
K. B. REID, Department of Mathematics, Louisiana State University, Baton Rouge, Lousiana 70803
G. N. ROBERTSON, Department of Mathematics, The Ohio State University, Columbus, Ohio 43210
I. G. ROSENBERG (133), Centre de Recherches Mathématiques, Université de Montréal, C.P. 6128, Montréal, Québec H3C 3J7
B. ROTHSCHILD, Department of Mathematics, University of California, Los Angeles, Los Angeles, California 90024

* Present address: Department of Mathematics, I.A.S. Australian National University, Canberra, Australia

G. SABIDUSSI (197), Department of Mathematics, Université de Montréal, C.P. 6128, Montréal, Québec H3C 3J7

F. SCHAFFER, Department of Mathematics, University of Lethbridge, Lethbridge, Alberta T1K 3M4

J. SELFRIDGE, Department of Mathematics, Northern Illinois University, Dekalb, Illinois 60115

P. D. SEYMOUR (341, 367), Department of Combinatorics and Optimization, Faculty of Mathematics, University of Waterloo, Waterloo, Ontario N2L 3G1

Y. SHIMAMOTO, Brookhaven National Laboratory, Upton, New York 11973

B. SIMEONE (368), Faculty of Mathematics, University of Waterloo, Waterloo, Ontario N2L 3G1

N. M. SINGHI (289), Department of Mathematics, School of Mathematics Tata Institute of Fundamental Research, Colaba, Bombay — 400005, India

J. SPENCER (165), Department of Mathematics, State University of New York at Stony Brook, Stony Brook, New York 11790

R. G. STANTON (315), Department of Computer Science, University of Manitoba, Winnipeg, Manitoba R32 2N2

R. STEINBERG, Faculty of Mathematics, University of Waterloo, Waterloo, Ontario N2L 3G1

W. R. STROMQUIST (369), 6001 Arlington Avenue, #720, Falls Church, Virginia 22044

R. SUEN, Department of Mathematics, University of Prince Edward Island, Charlottetown, Prince Edward Island C1A 4P3

J. A. THAS (277), Seminar of Higher Geometry, University of Ghent, Ghent, Belgium

S. TOIDA, Department of Systems Design, University of Waterloo, Waterloo, Ontario N2L 3G1

J. TOTTEN, 6138 Cedar Street, Halifax, Nova Scotia

W. T. TUTTE (15), Faculty of Mathematics, University of Waterloo, Waterloo, Ontario N2L 3G1

P. UNGAR (370), 109 Lea Road, Scarsdale, New York 10583

G. H. J. VAN REES (247), Department of Combinatorics and Optimization, University of Waterloo, Waterloo, Ontario N2L 3G1

S. A. VANSTONE (289), Department of Mathematics, St. Jerome's College, University of Waterloo, Waterloo, Ontario N2L 3G1

W. D. WALLIS (247), Department of Mathematics, University of Newcastle, Newcastle, New South Wales 2308, Australia.

C. S. WANG (207), Department of Mathematics, The Ohio State University, Columbus, Ohio 43210

D. J. A. WELSH (329), Merton College, Oxford OX1 4DJ, England
H. WILF, Department of Mathematics, University of Pennsylvania, Philadelphia, Pennsylvania 19174
L. WEINBERG, 11 Woodland Street, Tenafly, New Jersey 07670

Preface

The pioneering work of Professor W. T. Tutte has had a profound influence on the development of graph theory. We at Waterloo are fortunate indeed in having such a distinguished colleague. In appreciation of his contributions to mathematics and to the quality of academic life at the University, a conference was organized in his honour on the occasion of his sixtieth birthday. This volume is a record of the proceedings of that conference and reflects the many facets of Professor W. T. Tutte's work.

We are grateful to the National Research Council of Canada and the University of Waterloo, whose financial support made the conference possible. We should also like to thank Charles Haff, Heather Hergott, Kathy Knight, Ron Mullen, Joan Selwood, and Dan Younger for their help in organizing the conference.

J. A. Bondy
U. S. R. Murty

Early Reminiscences

The Combinatorial Conference at Waterloo has been organized to celebrate the sixtieth birthday of a distinguished mathematician, Professor W. T. Tutte. Now to have a sixtieth birthday is not usually too difficult—I have recently managed it without too much effort. But the story of how Tutte became a mathematician is rather less straightforward, and it may be worth explaining at least in broad outline.

An essential first step took place in 1935, when two first-year mathematics students happened to be walking down Broadhurst Gardens, in northwest London. They met, recognized each other by sight, and indulged in a highly intellectual conversation:

Student 1: What are you doing here?
Student 2: I live here (long pause). What are you doing here?
Student 1: I live here (even longer pause). I have some shopping to do. Good-bye.
Student 2: Good-bye.

The following term they met at Trinity College, Cambridge, and found that their respective names were Arthur Stone and Cedric Smith. Now, because of a misprint in the timetable, Cedric Smith had found himself, together with another freshman named R. Leonard Brooks, attending an advanced lecture (on almost periodic functions) on his very first day of lectures; and they became friends. So Cedric Smith introduced Leonard Brooks to Arthur Stone. And, to reciprocate, Leonard Brooks introduced us all to a chess-playing friend, a chemist named William T. Tutte.

Tutte looked with respect at three mathematicians, and asked us to find a semiexponential function, $S(x)$, that is, one satisfying $S(S(x)) = \exp x$. We tried our hands, without success, and told Tutte that we could not solve the problem. Indeed, no self-respecting mathematician could be expected to solve an impossible problem like that, or so we thought. Whereupon he replied, "But I have a solution." He explained, as follows.

Suppose that α is a number such that $S(\alpha) = \alpha$. Then clearly exp α = $S(S(\alpha)) = S(\alpha) = \alpha$. By solving (numerically) the equation exp α = α we find possible values for α, such as $0.318 + 1.337i$.

Now differentiate repeatedly the defining equation $S(S(x)) = \exp x$, to get a sequence of equations. Substitute $x = \alpha$ in these equations; we find in turn the values of the successive derivatives $S'(\alpha), S''(\alpha), \ldots$, and hence the Taylor series for $S(x)$ expanded around α.

After this, the real mathematicians looked at Tutte, the chemist, with considerable respect. (We could not then prove the Taylor series convergent. Several years later we found out that in fact it was easy to do so, using methods explained in E. Piccard's book, " Leçons sur Quelques Équations Fonctionelles.")

Stone now told us that while he was at school he had heard a talk by Mr. (later Professor) W. R. Dean, mentioning among other things a new problem: can a geometric square be divided into squares, all of unequal sizes? We set about this, soon discovering, as we thought, that it could all be reduced to graph theory. In fact, we were really talking about regular patroids. But perhaps we could be forgiven for not knowing that, since regular patroids were not discovered until 35 years later. Whenever we got stuck, and did not know what to do, Tutte provided the vital clue. (A lively account of the investigation is given in Martin Gardner's book, "More Mathematical Puzzles and Diversions.") Finally, several years later, Brooks invited the others to coffee after dinner. Unknown to him, Smith and Stone had been working hard in secret, and produced a square divided into equal squares. Unknown to them, Brooks had been working hard in secret, and produced a square divided into unequal squares. And when we met together, the squares were dramatically unveiled. Only Tutte was unsquared, which was a shame since he had contributed most of the ideas.

The next essential development was the unfortunate arrival of two depressions. An industrial depression left many unemployed, so Smith set off for camp where some of the unemployed of Cumbria were given a holiday —and so also did a Trinity College tutor and boy scout (later a professor) called Patrick Duff. A meteorological depression brought pouring rain, Patrick Duff's tent collapsed, and he went into a house to dry out. This gave him time to talk to Cedric Smith. "I'm worried about one of my pupils, named W. T. Tutte," he said. "It seems he's no good at chemistry." Smith looked surprised. "But he got a first class degree. And what is more, he's a mathematical genius." Patrick Duff looked surprised. "Do you think so? I will find out more." However, almost immediately war broke out, and before long Tutte went off to do some very secret work, even more secret than dividing squares into squares. So the whole question seemed to have lapsed.

But, one day toward the end of the war Smith was cycling down a Cambridge street when he saw Professor Duff on the sidewalk. "Stop! Stop!! STOP!!!" shouted Patrick Duff. Smith jammed on the brakes in alarm: what had gone wrong? "We've elected Tutte to a mathematical fellowship at Trinity College," said Duff, "but we don't know what he has done or where he lives. Have you got his address?" So, after 10 years, Tutte had become a genuine professional mathematician. And soon afterwards it was noticeable that graph theory and matroid theory were beginning to develop fast.

Tutte has now become a famous Canadian mathematician. At least, this is what is generally thought of him. A powerful voice in the contrary direction was that of B. Descartes, who wrote a letter in 1968 to Gian-Carlo Rota proving that none of Brooks, Smith, Stone, or Tutte really exists, and that their alleged paper "Dissection of rectangles into squares" showed by its title that they were merely a pseudonym for somebody called DORIS. [An English translation of this letter appeared in the *Journal of Recreational Mathematics* **2** (1969) 206–211.] As far as I know nobody has yet pointed out any fallacy in Descartes's arguments.

I hope this is not the end of the story. For me (if indeed I do exist despite Descartes) time has somehow turned in an almost full circle, after 40 years; for every time I go to work I pass Broadhurst Gardens. Alas I never see Arthur Stone there now (if indeed he also exists); but someday perhaps I will, and we will meet again with Bill Tutte and Leonard Brooks, and find some new and exciting problem to pursue. An interesting thought.

Cedric A. B. Smith

A Tribute

I first heard about Tutte in early September 1939 from Arthur Stone, who had just arrived from Cambridge to Princeton and told me of the Brooks–Smith–Stone–Tutte solution of the squaring the square problem.

But of the real powers I only learned later. T. Gallai and I as freshmen in 1930 took the course on graph theory by D. König—he mentioned the conjecture of Tait and the extension of Petersen's theorem on factorisation of graphs as important outstanding problems—we tried both unsuccessfully. As is well known, Tutte settled both problems—and many others. I hope that despite his enormous age he will settle many more problems before his distant cure.

Paul Erdös

A Note on Some of Professor Tutte's Mathematical Work

Combinatorial mathematicians sometimes express the feeling that almost any idea that they think of will be found, on further examination, to have been studied by Professor Tutte twenty years earlier. Such a sweeping generalization, even concerning Professor Tutte, may contain some element of exaggeration; but one can well see how this impression arose when one examines the range of his work and the many aspects of combinatorial mathematics on which it has had a major influence.

In consequence, I cannot possibly from my own limited knowledge describe even the highlights of all of this work, and perhaps only Professor Tutte himself can. Nevertheless, it was felt that somebody should say something about some of it in this volume, and that is all that I shall very briefly attempt to do.

Much of Professor Tutte's earliest work seems to have been motivated by the problem of dissecting a rectangle, whose sides have integral lengths, into squares of unequal sizes, whose sides also have integral lengths. This work was done largely in collaboration with R. L. Brooks, C. A. B. Smith, and A. H. Stone. It involved ingenious and elegant use of electric currents in graphs and seems to have links with many of the threads of Professor Tutte's later work on such themes as planar graphs, dichromatic polynomials, etc.; but perhaps only one of Professor Tutte's own inimitably clear and beautiful lectures, illustrated by diagrams on the blackboard, can adequately convey the flavour of all this.

A *1-factor*, or *perfect matching*, of a loopless graph G is a subset F of $E(G)$ such that each vertex of G is incident with exactly one edge in F. Petersen [119] proved, by a very complicated argument, that all bridgeless regular 3-valent finite graphs have 1-factors, and expressed the feeling that this theorem could probably be extended to bridgeless regular finite graphs of larger odd valency if one could face the complications involved. However,

König [118] pointed out that, despite the passage of 45 years, during which Petersen's proof had been substantially simplified, such a generalization had not been achieved. This indicates the importance of Professor Tutte's much greater achievement [8] in characterizing *all* finite graphs which have 1-factors. This theorem of [8] is clearly one of the basic theorems which might be expected to occupy a central place in any graph theory course or textbook: perhaps it is *the* canonical example of such a theorem. A generalization of the problem asks, for a given finite graph G and nonnegative integer-valued function f on $V(G)$, whether G has an *f-factor*, i.e., a subset F of $E(G)$ such that each vertex ξ is incident with exactly $f(\xi)$ elements of F (loops being counted twice). By an ingenious method of constructing one graph from another, Professor Tutte [20] has solved this problem by reducing it to the (already solved) problem of characterizing finite graphs with 1-factors. He has extended his work on 1-factors and f-factors to infinite but locally finite graphs [16, 17]; and it may now be ripe for generalization to all denumerably infinite graphs and perhaps to an even wider class of infinite graphs using ideas drawn from recent work of Damerell and Milner [114] and others on matchings in infinite bipartite graphs. In another direction, the theorem of [8] has led to the important work of Edmonds [115] on the "matching polyhedron" of a finite graph.

Another key theorem [32] of Professor Tutte's gives necessary and sufficient conditions for a finite graph to have n edge-disjoint spanning trees (or, equivalently, for it to be the union of n edge-disjoint connected spanning subgraphs). The proof used the remarkable idea of generalizing the problem to one in which we consider n graphs which all have the same set of edges but need not be related to one another in any other way and we imagine one of the n spanning trees to lie in each of these graphs. Some insight into why such an improbable-looking idea actually worked can be gained from the paper of Edmonds [116] in which the theorem of [32] is generalized to matroids (i.e., roughly speaking, sets endowed with a certain structure reminiscent of that of a finite subset of a vector space).

Professor Tutte himself has, of course, been one of the prime movers of matroid theory, in which, for example, one of his deep theorems characterizes graphic matroids (i.e., matroids related in a certain standard way to the family of circuits in some finite graph), using ideas reminiscent of Kuratowski's theorem which characterizes planar graphs. However, much of Professor Tutte's work on matroids has involved complicated technicalities, and I must leave the task of describing it to others who have studied it much more thoroughly than I have.

He has been one of the most influential developers of the theory of chromatic and dichromatic polynomials. (The *chromatic polynomial* $P_G(\lambda)$ of a finite graph G, first defined by Birkhoff [113], is a polynomial in an indeterminate λ such that, for each nonnegative integer n, $P_G(n)$ is the number of ways of colouring the vertices of G when n colours are available and no two adjacent vertices may receive the same colour. The *dichromatic polynomial* of G, introduced in [121], is a polynomial in two variables which embodies considerably more information about G.) Computer calculations made at Waterloo in about 1968–1969 (see [68]) showed that the chromatic polynomials of many 3-valent regular planar graphs tend to have a root near $\frac{1}{2}(3 + \sqrt{5})$, and an interesting theorem which seems to cast considerable light on this phenomenon was proved by Professor Tutte in [75]. There is in fact some question [122] as to how fully this theorem accounts for the empirical observation about roots near $\frac{1}{2}(3 + \sqrt{5})$; but this line of investigation has certainly been fruitful and has led to the idea that chromatic polynomials of many planar graphs may tend to behave in some particularly significant way at or near each of the points $\lambda = B_2, B_3, B_4, \ldots$, where B_n, the so-called *nth Beraha number*, is $2 + 2\cos(2\pi/n)$. Thus $\frac{1}{2}(3 + \sqrt{5}) = B_5$. The fact that $B_n \to 4$ as $n \to \infty$ might seem to have some connection with the four colour theorem, which asserts that 4 is not a root of the chromatic polynomial of any planar graph.

Professor Tutte has been mainly responsible for developing the theory of enumeration of planar graphs [37–41, 49, 66, 67, 69, 72, 82, 86–90, 93, 94, 100, 107, 111], which has close links with chromatic and dichromatic polynomials (see [82–90]). This work involves some highly innovative techniques of his own invention, requiring considerable manipulative dexterity in handling power series (whose coefficients count appropriate kinds of graphs) and the functions arising as their sums, as well as geometrical dexterity in extracting these power series from the graph-theoretic situation.

Much of this work was motivated in some degree by the four colour problem. When Appel and Haken eventually solved this problem, the reconstruction problem became regarded by many people as the most outstanding unsolved problem in graph theory; and Professor Tutte, nothing daunted, quickly showed [106] how this work on graph enumeration and chromatic and dichromatic polynomials could be adapted to make important and intriguing contributions to reconstruction theory. (The reconstruction problem asks whether a finite graph G with vertices $v_1, v_2, \ldots, v_n$ ($n \geq 3$) is determined up to isomorphism when we know its subgraphs $G - v_1, G - v_2, \ldots, G - v_n$ up to isomorphism, i.e. "can G be reconstructed from its vertex-deleted subgraphs?")

Harary [117] suggested that the "graphical diseases" (i.e., problems constituting the most persistent obsessions of graph theorists) were the four colour disease, the reconstruction disease, and the Hamilton disease. Having mentioned Professor Tutte's contributions to two of these diseases, we must not overlook his important influence on the third. His theorem

[22] that every 4-connected planar graph has a Hamilton circuit (generalizing a more restricted, but already difficult, theorem of Whitney [119]) is one of the deepest and most interesting contributions to Hamilton circuit theory; and, by contrast, he has also contributed [5, 28, 30, 85] to the development of elegant techniques for constructing nonobvious examples of planar graphs without Hamilton circuits.

As I suggested earlier, the main strands of Professor Tutte's wide-ranging work appear to be, in his own mind and in reality, fairly closely linked together in ways that it might require a much more ambitious survey to trace fully. In view of the breadth and depth of his contributions, the proceedings of this or any other combinatorial conference must almost surely contain a large proportion of material which has been substantially influenced by work of Professor Tutte.

REFERENCES

1–112. See list of Professor Tutte's publications.
113. G. D. Birkoff, A determinant formula for the number of ways of colouring a map, *Ann. of Math.* **14** (1912) 42–46.
114. R. M. Damerell and E. C. Milner, Necessary and sufficient conditions for transversals of countable set systems, *J. Combinatorial Theory Ser. A* **17** (1974) 350–374.
115. J. Edmonds, Maximum matching and a polyhedron with 0,1-vertices, *J. Res. Natl. Bur. Standards Sect. B* **69** (1965) 125–130.
116. J. Edmonds, On Lehman's Switching Game and a theorem of Tutte and Nash-Williams, *J. Res. Natl. Bur. Standards Sect. B* **69** (1965) 73–77.
117. F. Harary, The four color conjecture and other graphical diseases, in "Proof Techniques in Graph Theory, Proceedings of the Second Ann Arbor Graph Theory Conference, February 1968" (F. Harary, ed.), pp. 1–9. Academic Press, New York, 1969.
118. D. König, "Theorie der Endlichen und Unendlichen Graphen." Leipzig, 1936 (reprinted by Chelsea, Bronx, New York, 1950).
119. J. Petersen, Die Theorie der regularen Graphs, *Acta Math.* **15** (1891) 193–200.
120. H. Whitney, A theorem on graphs, *Ann. of Math.* **32** (1931) 378–390.
121. H. Whitney, The coloring of graphs, *Ann. of Math.* **33** (1932) 688–718.
122. D. R. Woodall, Zeros of chromatic polynomials, *in* "Combinatorial Surveys, Proceedings of the Sixth British Combinatorial Conference" (P. J. Cameron, ed.), pp. 199–223. Academic Press, New York, 1977.

C. St. J. A. Nash-Williams

Papers and Books by W. T. Tutte

1. (with G. B. B. M. Sutherland) Absorption of polymolecular films in the infra-red, *Nature (London)*, Oct. 21, 1939, p. 707.
2. (with W. C. Price) The absorption spectra of ethylene, deutero-ethylene and some alkyl-substituted ethylenes in the vacuum ultraviolet, *Proc. R. Soc. London Ser. A* **174** (1940) 207–220.
3. (with R. L. Brooks, C. A. B. Smith, and A. H. Stone) The dissection of rectangles into squares, *Duke Math. J.* **7** (1940) 312–340.
4. (with C. A. B. Smith) On unicursal paths in a network of degree 4, *Amer. Math. Monthly* **48** (1941) 233–237.
5. On Hamiltonian circuits, *J. London Math. Soc.* **21** (1946) 97–101.
6. A ring in graph theory, *Proc. Cambridge Philos. Soc.* **43** (1947) 26–40.
7. A family of cubical graphs, *Proc. Cambridge Philos Soc.* **43** (1947) 459–474.
8. The factorization of linear graphs, *J. London Math. Soc.* **22** (1947) 107–111.
9. (with R. L. Brooks, C. A. B. Smith, and A. H. Stone) A simple perfect square, *Ned. Akad. Wetensch. Proc. Ser. A* **50** (1947) 1300–1301.
10. The dissection of equilateral triangles into equilateral triangles, *Proc. Cambridge Philos. Soc.* **44** (1948) 463–482.
11. On the four-colour conjecture, *Proc. London Math. Soc.* **50** (1948) 137–149.
12. A note to a paper by C. J. Bouwkamp, *Ned. Akad. Wetensch. Proc. Ser. A* **51** (1948) 106–108.
13. On the imbedding of linear graphs in surfaces, *Proc. London Math. Soc.* **51** (1949) 474–483.
14. (with C. A. B. Smith) A class of self-dual maps, *Canad. J. Math.* **2** (1950) 179–196.
15. Squaring the square, *Canad. J. Math.* **2** (1950) 197–209.
16. The factorization of locally finite graphs, *Canad. J. Math.* **2** (1950) 44–49.
17. The factors of graphs, *Canad. J. Math.* **4** (1952) 314–328.
18. The 1-factors of oriented graphs, *Proc. Amer. Math. Soc.* **4** (1953) 922–931.
19. A contribution to the theory of chromatic polynomials, *Canad. J. Math.* **6** (1954) 80–91.
20. A short proof of the factor theorem for finite graphs, *Canad. J. Math.* **6** (1954) 347–353.
21. A class of Abelian groups, *Canad. J. Math.* **8** (1956) 13–28.
22. A theorem on planar graphs, *Trans. Amer. Math. Soc.* **82** (1956) 99–116.
23. The chords of the non-ruled quadric in $PG(3,3)$, *Canad. J. Math.* **10** (1958) 481–488.
24. A homotopy theorem for matroids, I., *Trans. Amer. Math. Soc.* **88** (1958) 144–160.
25. A homotopy theorem for matroids, II., *Trans. Amer. Math. Soc.* **88** (1958) 161–174.
26. Matroids and graphs, *Trans. Amer. Math. Soc.* **90** (1959) 527–552.
27. On the symmetry of cubic graphs, *Canad. J. Math.* **11** (1959) 621–624.
28. A non-Hamiltonian graph, *Canad. Math. Bull.* **3** (1960) 1–5.

29. An algorithm for determining whether a given binary matroid is graphic, *Proc. Amer. Math. Soc.* **11** (1960) 905–917.
30. A non-Hamiltonian planar graph, *Acta Math. Acad. Sci. Hung.* **11** (1960) 371–375.
31. Convex representations of graphs, *Proc. London Math. Soc.* **10** (1960) 304–320.
32. On the problem of decomposing a graph into *n* connected factors, *J. London Math. Soc.* **36** (1961) 221–230.
33. Symmetrical graphs and coloring problems, *Scr. Math.* **25** (1961) 305–316.
34. On the colourings of graphs, *Canad. Math. Bull.* **4** (1961) 157–160.
35. A theory of 3-connected graphs, *Ned. Akad. Wetensch. Proc. Ser. A* **64** (1961), 441–455.
36. Squaring the square, *in* "Second Scientific American Book of Mathematical Puzzles and Diversions" (M. Gardner, ed.), Chap. 17. Simon and Schuster, New York, 1961.
37. A census of planar triangulations, *Canad. J. Math.* **14** (1962) 21–38.
38. A census of Hamiltonian polygons, *Canad. J. Math.* **14** (1962) 402–417.
39. A census of slicings, *Canad. J. Math.* **14** (1962) 708–722.
40. A new branch of enumerative graph theory, *Bull. Amer. Math. Soc.* **68** (1962) 500–504.
41. A census of planar maps, *Canad. J. Math.* **15** (1963) 249–271.
42. How to draw a graph, *Proc. London Math. Soc.* **52** (1963) 743–767.
43. The non-biplanar character of the complete 9-graph, *Canad. Math. Bull.* **6** (1963) 319–330.
44. The thickness of a graph, *Ned. Akad. Wetensch. Proc. Ser. A* **66** (1963) 567–577.
45. From matrices to graphs, *Canad. J. Math.* **16** (1964) 108–127.
46. The number of planted trees with a given partition, *Amer. Math. Monthly* **71** (1964) 272–277.
47. (with F. Harary and G. Prins) The number of plane trees, *Ned. Akad. Wetensch. Proc. Ser. A* **67** (1964) 319–329.
48. (with F. Harary) The number of plane trees with a given partition, *Mathematika* **11** (1964) 99–101.
49. (with W. G. Brown) On the enumeration of rooted non-separable planar maps, *Canad. J. Math.* **16** (1964) 572–577.
50. The quest of the perfect square, *Amer. Math. Monthly* **72** (II) (1965) 29–35.
51. Lectures on matroids, *J. Res. Natl. Bur. Stand. B* **69** (1965) 1–47.
52. Menger's theorem for matroids, *J. Res. Natl. Bur. Stand. B* **69** (1965) 49–53.
53. (with P. Erdös and F. Harary) On the dimension of a graph, *Mathematika* **12** (1965) 118–122.
54. (with F. Harary) A dual form of Kuratowski's theorem, *Bull. Amer. Math. Soc.* **71** (1965) 168.
55. On the algebraic theory of graph-colorings, *J. Combinatorial Theory* **1** (1966) 15–50.
56. Connectivity in matroids, *Canad. J. Math.* **18** (1966) 1301–1324.
57. (with F. Harary) On the order of the group of a planar map, *J. Combinatorial Theory* **1** (1966) 394–395.
58. Introduction to the theory of matroids, Rand Corp. Report F-446-PR, 1966.
59. Squared Rectangles, *Proc. IBM Computing Symp. Combinatorial Problems* (1966) 3–9.
60. "Connectivity in Graphs." Univ. of Toronto Press, Toronto, 1966.
61. Antisymmetrical digraphs, *Canad. J. Math.* **19** (1967) 1101–1117.
62. (with F. Harary and R. M. Karp) A criterion for the planarity of the square of a graph, *J. Combinatorial Theory* **2** (1967) 395–405.
63. On dichromatic polynomials, *J. Combinatorial Theory* **2** (1967) 301–320.
64. On even matroids, *J. Res. Natl. Bur. Stand. B* **71** (1967) 213–214.
65. A geometrical version of the four color conjecture, *in* "Combinatorial Mathematics and Its Applications" (R. Rose and T. Dowling, eds.). Univ. of North Carolina Press, Chapel Hill, 1967.

PAPERS AND BOOKS BY W. T. TUTTE

66. On the enumeration of planar maps, *Bull. Amer. Math. Soc.* **74** (1968) 64–74.
67. Counting planar maps, *J. Recreational Math.* **1** (1968) 19–27.
68. (with G. Berman) The golden root of a chromatic polynomial, *J. Combinatorial Theory* **6** (1969) 301–302.
69. On the enumeration of four-coloured maps, *SIAM J. Appl. Math.* **17** (1969) 454–460.
70. Even and odd 4-colorings, *in* "Proof Techniques in Graph Theory" (F. Harary, ed.), pp. 161–169. Academic Press, New York, 1969.
71. Projective geometry and the four color problem, *in* "Recent Progress in Combinatorics" (W. T. Tutte, ed.), 199–207. Academic Press, New York, 1969.
72. On the enumeration of almost bicubic rooted maps, Rand Corp. Memorandum RM-5887-PR, February, 1969.
73. Toward a theory of crossing numbers, *J. Combinatorial Theory* **8** (1970) 45–53.
74. On the enumeration of two-coloured rooted and weighted plane trees, *Aequationes Math.* **4** (1970) 143–156.
75. On chromatic polynomials and the golden ratio, *J. Combinatorial Theory* **9** (1970) 289–296.
76. The golden ratio in the theory of chromatic polynomials, *Ann. New York Acad. Sci.* **175** (1970) 391–402.
77. More about chromatic polynomials and the golden ratio, *in* "Combinatorial Structures and Their Applications" (Richard Guy *et al.*, eds.), pp. 439–453. Gordon and Breach, New York, 1970.
78. Connectivity in matroids, *in* "Graph Theory and Its Applications" (B. Harris, ed.), pp. 113–119. Academic Press, New York, 1970.
79. On the 2-factors of bicubic graphs. *Discrete Math.* **1** (1971) 203–208.
80. "Introduction to the Theory of Matroids." American Elsevier, New York, 1971.
81. Wheels and whirls, *in* "Théorie des Matroïdes" (C. Bruter ed.), pp. 1–4. Springer-Verlag, Berlin and New York, 1971.
82. Dichromatic sums for rooted planar maps, *Proc. Symp. Pure Math.* **XIX** (*Combinatorics*) (1971) 235–245.
83. Three-connected planar maps, *Proc. Manitoba Conf. Numerical Math.* (1971) 43–52.
84. (with H. Whitney) Kempe chains and the four colour problem, *Utilitas Math.* **2** (1972) 241–281.
85. Non-Hamiltonian planar maps, *in* "Graph Theory and Computing" (Ronald C. Read, ed.), pp. 295–301. Academic Press, New York, 1972.
86. The use of numerical computations in the enumerative theory of planar maps, *Jeffery-Williams Lectures 1968-1972 Canad. Math. Congress* (1972) 73–89.
87. Chromatic sums for rooted planar triangulations: the cases $\lambda = 1$ and $\lambda = 2$, *Canad. J. Math.* **25** (1973) 426–447.
88. Chromatic sums for rooted planar triangulations II: the case $\lambda = \tau + 1$, *Canad. J. Math.* **25** (1973) 657–671.
89. Chromatic sums for rooted planar triangulations III: the case $\lambda = 3$, *Canad. J. Math.* **25** (1973) 780–790.
90. Chromatic sums for rooted planar triangulations IV: the case $\lambda = \infty$, *Canad. J. Math.* **25** (1973) 929–940.
91. What is a map? *in* "New Directions in the Theory of Graphs" (F. Harary, ed.), pp. 309–325. Academic Press, New York, 1973.
92. Duality and trinity, *Infinite and Finite Sets. Colloq. Math. Soc. János Bolyai* **10** (1973) 1459–1472.
93. Some polynomials associated with graphs, *Combinatorics Proc. British Combinatorial Conf.* (1973) 161–167.

94. Chromatic sums for rooted planar triangulations V: special equations, *Canad. J. Math.* **26** (1974) 893–907.
95. A theorem on spanning trees, *Quart. J. Math. Oxford* **25** (1974) 253–254.
96. Spanning subgraphs with specified valencies, *Discrete Math.* **9** (1974) 97–108.
97. Codichromatic graphs, *J. Combinatorial Theory B* **16** (1974) 168–174.
98. (with M. Gordon) On sums of determinants of intersection matrices of Petrie matrices, *Proc. Cambridge Philos. Soc.*, **75**, (1974) 155–163.
99. Map-coloring problems and chromatic polynomials, *Amer. Sci.* **62** (1974) 702–705.
100. On elementary calculus and the Good formula, *J. Combinatorial Theory B* **18** (1975) 97–137.
101. (with R. L. Brooks, C. A. B. Smith, and A. H. Stone) Leaky electricity and triangulated triangles, *Phillips Res.* **30** (1975) 205–219.
102. The dichromatic polynomial, *Congr. Numer. XV Proc. 5th British Combinatorial Conf.* (1975) 605–635.
103. Separation of vertices by a circuit, *Discrete Math.* **12** (1975) 173–184.
104. Hamiltonian circuits, *in* "Teorie Combinatorie," pp. 193–199. Accademia Nazionale dei Lincei, Rome, 1976.
105. The reconstruction problem in graph theory, *British Polymer. J.* (1977) 179–183.
106. All the King's horses, this volume, p. 15.
107. On a pair of functional equations of combinatorial interest, *Aequationes Math.* **17** (1978) 121–140.
108. The rotor effect with generalized electrical flows, *Ars Combinatoria* **1** (1976) 3–31.
109. (with C. St. J. A. Nash-Williams) More proofs of Menger's theorem, *J. Graph Theory* **1** (1977) 13–17.
110. Bridges and Hamiltonian circuits in planar graphs, *Aequationes Math.* **15** (1977) 1–33.
111. On the enumeration of convex polyhedra, *J. Combinatorial Theory B* (to appear).
112. The subgraph problem, *Ann. of Discrete Math.* **3** (1978) 289–295.

Graph Theory and
Related Topics

Topological and Algebraic Methods in Graph Theory

LÁSZLÓ LOVÁSZ

0

The aim of these notes is to give some applications of algebraic topology and linear algebra to graph theory. Most of the results I am going to mention have been published elsewhere and, in part, are due to other mathematicians. They are collected here because I feel that these or similar methods could be applied under other circumstances as well, and much further work could and should be done to determine the area of their applicability.

1

Graphs are just 1-dimensional simplicial complexes and this connection to topology has never been neglected. Problems of embedding graphs into surfaces as well as homology groups of graphs have been investigated very actively. We are, however, not concerned here with these connections. There are other ways to associate topological spaces (simplicial complexes) with graphs and these spaces may carry nontrivial information about purely graph-theoretical properties. In this paper we display two such constructions which help to solve graph-theoretical problems.

The following problem was raised by Frank [7] and Maurer [17].

Theorem *Let G be a k-connected graph, $\{v_1, \ldots, v_k\} \subseteq V(G)$, and $n_1, \ldots, n_k$ positive integers with $n_1 + \cdots + n_k = n = |V(G)|$. Then there exists a partition $\{V_1, \ldots, V_k\}$ of $V(G)$ such that*

(a) $v_i \in V_i$,
(b) $|V_i| = n_i$, and
(c) V_i spans a connected subgraph of G $(i = 1, \ldots, k)$.

The case $k = 2$ is quite easy. The case when $n_1, \ldots, n_{k-1} \leq 3$ was proved by Frank [7] and the case $k = 3$ by Milliken [18]. The general theorem was proved independently by Györi [9] and the author [11]. The proof of Györi is elementary but quite involved. In view of the remarks in the introduction, I sketch here my proof, which uses algebraic topology. Details have appeared in [11].

The idea of the proof is contained in the following result, which was obtained in connection with a problem of Bondy. Let G be a graph and fix a root $a \in V(G)$. Call two spanning trees T_1 and T_2 of G *adjacent* if $T_1 \cap T_2$ contains a tree on $n - 1$ points including a. So T_2 can be obtained from T_1 by replacing an endline by another endline.

Lemma 1 *If G is 2-connected then for any two spanning trees there is a chain of spanning trees connecting them, in which any two consecutive trees are adjacent.*

Let us see how the case $k = 2$ follows from this (it would be simpler to prove it directly, but this is the proof which generalizes).

Let us add a new point a and connect it to v_1 and v_2. For each spanning tree T of the resulting graph G', let $f(T)$ denote the number of points of T accessible from a along the line (a, v_1). Obviously, if T_1 and T_2 are adjacent then $|f(T_1) - f(T_2)| \leq 1$ (f is continuous). Now consider two spanning trees T_1 and T_2 of G' such that $f(T_1) = 0$ but $f(T_2) = n$. Connect them by a chain of spanning trees as in the lemma. By the "Bolzano–Weierstrass theorem" there will be a spanning tree T_0 with $f(T_0) = n_1$. Then the two components of $T_0 - a$ yield the partition of $V(G)$ as desired.

It may sound artificial to mention the Bolzano–Weierstrass theorem in a simple discrete argument like this, but this is what gives the idea of the generalization, knowing that various fixed point theorems of topology can be regarded as extensions of the Bolzano–Weierstrass theorem.

We shall discuss now the case $k = 3$. This already contains all the main points of the proof of the general case.

Let us formulate first the lemma as follows. Construct a graph $\tilde{G}$ whose vertices are the spanning trees of G' and "adjacency" is as introduced. Then Lemma 1 says that $\tilde{G}$ is connected if G is 2-connected.

Construct a 2-dimensional topological space as follows. Consider the triangles and quadrilaterals in G and span a 2-cell on each of them. Let K denote the resulting topological space.

Recall that a topological space is called *simply connected* if every closed curve can be contracted to a single point in it.

Lemma 2 *If G is 3-connected then K is simply connected.*

The proof of this lemma is technical and is omitted here. We show how it can be applied to get the theorem in the case $k = 3$.

So let G be a 3-connected graph, $v_1, v_2, v_3 \in V(G)$ and $n_1 + n_2 + n_3 = |V(G)|$. Take a new point a and connect it to v_1, v_2, and v_3. Consider the topological space K constructed for this new graph G'. For each spanning tree T of G', let $f_i(T)$ denote the number of points in T accessible from a along the edge (a, v_i) ($i = 1, 2$). Then the mapping

$$f: T \mapsto (f_1(T), f_2(T))$$

maps the vertices of K onto lattice points of the plane. Let us subdivide each quadrilateral 2-cell in K by a diagonal into two triangles; in this way we obtain a triangulation $\bar{K}$ of K. Extend f affinely to each such triangle so as to obtain a continuous mapping of K into the plane. Obviously, the image of K is contained in the triangle $\triangle = \{x \geq 0, y \geq 0, x + y \leq n\}$. We are going to show that the mapping is onto $\triangle$.

Let us pick three spanning trees T_1, T_2, T_3 first such that $f(T_1) = (n, 0)$, $f(T_2) = (0, n)$, $f(T_3) = (0, 0)$. Obviously, such trees exist. Next, by applying Lemma 1 to the graph $G' - (a, v_3)$, we select a polygon P_{12} in K connecting T_1 to T_2 and having $f_3(x) = 0$ at all points. Thus $f(P_{12})$ connects $(n, 0)$ to $(0, n)$ along the side of the triangle $\triangle$ with these endpoints. Let P_{23} and P_{31} be defined analogously.

By Lemma 2, $P_{12} + P_{23} + P_{31}$ can be contracted in K to a single point. Therefore, $f(P_{12}) + f(P_{23}) + f(P_{31})$ can be contracted in $f(K)$ to a single point. But "obviously" (or, rather, by applying the well-known fact that the boundary of a triangle cannot be contracted to a single point in the triangle with one interior point taken out), $f(K)$ must cover the whole triangle $\triangle$. So in particular the point (n_1, n_2) belongs to the image of K, and therefore it belongs to the image of a triangle of $\bar{K}$. But it is easy to see that this implies that (n_1, n_2) is the image of one of the vertices of $\bar{K}$; i.e., there exists a spanning tree T with

$$f_1(T) = n_1, \qquad f_2(T) = n_2.$$

The three components of $T - a$ now yield the desired partition of $V(G)$. ∎

2

A very important connection between topology and graph theory has been pointed out by Erdös and Hajnal [5]. Consider the graph $B_k(\varepsilon)$ whose vertices are the points of the k-dimensional unit sphere S^k, two of them being connected by an edge if and only if they are "almost antipodal," that is, their distance is greater than $2 - \varepsilon$ ($\varepsilon > 0$). Then this graph has chromatic number $k + 2$. It is easy to see that $B_k(\varepsilon)$ is $(k + 2)$-colorable. The fact that it cannot be colored with only $k + 1$ colors is highly nontrivial, and is equivalent to the following well-known theorem of algebraic topology:

Borsuk's Theorem *If S_k is represented as the union of $k + 1$ closed sets, one of these must contain antipodal points.*

We remark that the word "closed" could be replaced by the word "open."

The graphs $B_k(\varepsilon)$, also known as Borsuk's graphs, have many remarkable properties; among others if ε is small then they contain no short odd circuits. They are infinite but by the well-known de Bruijn–Erdös theorem, they contain finite subgraphs with the same chromatic number.

Kneser [10] posed the following problem. Define the graph $K(n, k)$ as the graph whose vertices are all the n-subsets of a $(2n + k)$-element set S, two being adjacent if and only if they are disjoint. It is easy to find a $(k + 2)$-coloration of this graph. Kneser conjectured that it is not $(k + 1)$-colorable. This also can be formulated as a Ramsey-type statement: If all n-subsets of a $(2n + k)$-element set are $(k + 1)$-colored, there always exist two disjoint n-subsets with the same color.

Erdös and Hajnal [5] noted that Kneser's graph $K(n, k)$ would have properties very similar to those of Borsuk's graph if the conjecture were true. For example, $K(n, k)$ contains no odd circuits shorter than $(2n + k)/k$. This analogy was pointed out to me by Simonovits [21].

In [12] I proved Kneser's conjecture using Borsuk's theorem. Very shortly after this, Bárány [1] found a much shorter way to get Kneser's conjecture from Borsuk's theorem. His proof is given here; then we return to my original proof which in fact yields a more general theorem relating the chromatic number to algebraic topology.

Bárány uses the following theorem of Gale [8].

Theorem *Given n, $k \geq 1$, there exist $2n + k$ points on S^k such that any open hemisphere contains at least n of them.*

With the same value of n and k, let the set S in Kneser's conjecture be these $2n + k$ points on S^k. Assume that the n-subsets of S are $(k + 1)$-colored with colors $0, 1, \ldots, k$. Let U_i denote the set of those points $x \in S^k$ for which the open hemisphere around x contains an n-subset of S with color i. Then by the choice of S, we have $U_0 \cup \cdots \cup U_k = S^k$ and, obviously, $U_0, \ldots, U_k$ are open. By Borsuk's theorem, one of the U's, say U_0, contains antipodal points. But the open hemispheres around these points are disjoint and both contain n-subsets of S with color 0. So there are two disjoint n-subsets of S with color 0. ∎

After this extremely simple proof, let us see how Borsuk's theorem can be employed to deduce a general bound on the chromatic number (one which applies to all graphs, not just $K(n, k)$). To this end, let the *neighborhood complex* of a graph G be defined as the simplicial complex whose vertices are the vertices of G and whose simplices are those subsets of $V(G)$ which have a neighbor in common in G. We recall the following definition from topology: A topological space T is called *n-connected*, if for every j, $0 \le j \le n$, every continuous mapping of S^j into T extends to a continuous mapping of the $(j + 1)$-ball with boundary S^j into T.

Theorem *If the neighborhood complex of a graph G is $(k - 1)$-connected then G is not $(k + 1)$-colorable.*

Note that 0-connected means arcwise connected, so the first case of the theorem asserts that if the neighborhood complex is connected then G is not 2-colorable. For connected graphs this condition is necessary and sufficient, and very easily verified in both directions. The first nontrivial case is the following: If the neighborhood complex is connected and simply connected then G is not 3-colorable.

It is not trivial to see that $K(n, k)$ satisfies the condition of the above theorem. For the proof see [12]. Surely this theorem can be applied to graphs other than $K(n, k)$ (for example, it applies to Borsuk's graph; but this is uninteresting since Borsuk's theorem is used in the proof). However, the graph-theoretical meaning of the condition is quite difficult to understand and further work in this direction is necessary.

To sketch the proof of this last theorem we state (without proof) the lemma which is used.

Lemma *Let G be a graph whose neighborhood complex N is $(k - 1)$-connected. Then there exists a continuous mapping $f: S^k \to N$ such that for every point $x \in S^k$, all common neighbors of the vertices of the least simplex of N containing $f(x)$ are contained among the vertices of the least simplex of N containing $f(-x)$.*

Provided this mapping f exists, it is easy to show that G cannot be $(k + 1)$-colorable. For consider any coloration of the points of G with colors $0, 1, \ldots, k$. Let S_i denote the set of points of S^k for which the vertices of the least simplex of N containing $f(x)$ have a neighbor of color i in common. Then trivially $S_0 \cup \cdots \cup S_k = S^k$ and $S_0, \ldots, S_k$ are closed. So by Borsuk's theorem, one of these sets, S_0, say, contains antipodal points x and $-x$. Now by the definition of S_0, the vertices of the smallest simplex containing $f(x)$ have a neighbor with color 0 in common, and a similar assertion holds for $f(-x)$. But these two vertices are then two adjacent vertices with the same color. ∎

3

Several colleagues have pointed out to me that the use of topology in the previous two examples is probably avoidable. In fact, the proofs of the topological theorems used there are achieved by considering a triangulation and then applying combinatorial (or algebraic) arguments. Clearly, the translation to a continuous problem and then back again could be saved. However, this may result in a complicated proof with no visible underlying idea. So I think it may be useful to draw attention to the topological picture of what is going on.

But it also may happen that getting rid of the unnecessary topological methods will lead to greater insight and, probably, to better results. Let me describe an example of this. We shall again deal with a version of the neighborhood complex; this suggests that a similar analysis of the main theorem of the previous paragraph could be carried through, although I have been unable to do so.

Let G be a critically 4-chromatic graph on n points. Denote by $\tau(G)$ the minimum number of points covering all edges of G. Simonovits [21] proved that

$$\tau(G) \geq c_1 n^{2/5}, \qquad (1)$$

while Brown and Moon [3] constructed critically 4-chromatic graphs G with

$$\tau(G) \leq c_2 n^{1/2}. \qquad (2)$$

In [13] the method of Simonovits was improved to get

$$\tau(G) \geq c_3 n^{1/2}. \qquad (3)$$

More generally, if G is critically k-chromatic then

$$\tau(G) \geq c_4 \cdot k \cdot n^{1/(k-2)},$$

and an appropriate generalization of the construction in [3] shows that this bound is essentially the best possible.

Let us see first the proof of Simonovits. Let A be a set of $\tau(G)$ points covering all edges of G and let $B = V(G) - A$. It is quite easy to show that we may assume that all points in B have degree 3.

Consider now the triangles spanned by the neighborhoods of the points in B. Simonovits observes that no subset of them forms a "double pyramid." More generally, no subset of them forms a triangulation of the sphere. For assume indirectly that $\triangle_1, \ldots, \triangle_p$ form a triangulation of the sphere. Let b_i be the point in B with neighborhood $\triangle_i$. Then $G - b_1$ is 3-colorable; consider a 3-coloration of it. In this 3-coloration, $\triangle_2, \ldots, \triangle_p$ get at most two different colors, since $\triangle_i$ misses the color of b_i. On the other hand, $\triangle_1$ gets three different colors, since otherwise the coloration would extend to a coloration of G. This, however, contradicts Sperner's lemma, which asserts (in one possible form) that in any 3-coloring of the vertices of a triangulation of the sphere, the number of triangles with all vertices differently colored is even.

Now it follows from the theory of extremal hypergraphs that a 3-uniform hypergraph on t points containing no double pyramid has at most $ct^{5/2}$ edges. Hence

$$|B| \geq c \cdot |A|^{5/2}$$

from which (1) follows.

One might hope that knowing that the neighborhoods of points in B form no triangulation of the sphere leads to an improvement, but it follows from the results of Brown et al. [4] that this is not the case. However, if we apply the method of proof, rather than the statement of Sperner's lemma, we get the following argument leading to the improved bound (3).

Let $E_1, \ldots, E_N$ be all pairs of points in A $[N = \binom{|A|}{2}]$ and $\triangle_1, \ldots, \triangle_M$ all neighborhoods of points in B ($M = |B|$). Set

$$a_{ij} = \begin{cases} 1 & \text{if } E_i \subset \triangle_j, \\ 0 & \text{otherwise.} \end{cases}$$

Claim *The columns of the matrix (a_{ij}) are linearly independent over $GF(2)$.*

Note that the columns of (a_{ij}) correspond to the points in B; a triangulated sphere formed by the $\triangle$'s would be just a special linear relation between the columns. If the claim is true then the matrix must have at least as many rows as columns, that is

$$M = |B| = n - \tau(G) \leq N = \binom{|A|}{2} = \binom{\tau(G)}{2},$$

whence (3) follows.

To prove the claim assume indirectly that, for instance,

$$a_{i1} = \sum_{j=2}^{M} \lambda_j a_{ij} \quad (i = 1, \ldots, N),$$

with some $\lambda_j \in GF(2)$. Let b_1 be the point with neighborhood $\triangle_1$ and let us consider a 3-coloration of $G - b_1$. As before, $\triangle_1$ is the only neighborhood which gets three different colors. Let

$$\mu_i = \begin{cases} 1 & \text{if } E_i \text{ is colored } (1, 2), \\ 0 & \text{otherwise.} \end{cases}$$

Then

$$\sum_{i=1}^{N} \mu_i a_{ij} = \begin{cases} 1 & \text{if } \triangle_j \text{ has three different colors (that is if } j = 1), \\ 0 & \text{otherwise} \end{cases}$$

[all calculations are over $GF(2)$]. But then

$$0 \neq \sum_{i=1}^{N} \mu_i a_{i1} = \sum_{i=1}^{N} \sum_{j=2}^{M} \mu_i \lambda_j a_{ij} = \sum_{j=2}^{M} \sum_{i=1}^{N} \lambda_j \mu_i a_{ij} = 0,$$

a contradiction. This proves the claim. ∎

4

The claim in the previous proof is an example of the phenomenon that very often vectors associated with edges or points of critical graphs or hypergraphs in an appropriate way are linearly independent. This provides a very useful tool in the study of graphs critical with respect to various properties. Let us look at further examples of this phenomenon.

The following theorem is due to Seymour [20].

Theorem *Let H be a critically 3-chromatic hypergraph with points $x_1, \ldots, x_n$ and edges $E_1, \ldots, E_m$. Consider the incidence matrix of H defined by*

$$a_{ij} = \begin{cases} 1 & \text{if } x_i \in E_j, \\ 0 & \text{otherwise.} \end{cases}$$

Then the rows of (a_{ij}) are linearly independent over the real field.

It is somewhat surprising that the assertion is not true over fields with characteristic $p \neq 0$, as shown by any p-uniform critically 3-chromatic hypergraph. From the theorem one can conclude immediately that $m \geq n$, which implies the following result [14]: If a hypergraph has the property that the union of any k edges has at least $k + 1$ elements ($k = 1, 2, \ldots$) then it is 2-colorable.

Seymour's proof is very simple. Assume indirectly that

$$\sum_{i=0}^{n} \lambda_i a_{ij} = 0 \quad (j = 1, \ldots, m)$$

i.e.,

$$\sum_{x_i \in E_j} \lambda_i = 0 \quad (j = 1, \ldots, m). \tag{4}$$

Let, say, $\lambda_1 = \cdots = \lambda_t = 0$, $\lambda_{t+1}, \ldots, \lambda_s < 0$, $\lambda_{s+1}, \ldots, \lambda_n > 0$. Since we are considering a nontrivial linear combination, $t < n$. By the critical property of H, the partial hypergraph H' spanned by $\{x_1, \ldots, x_t\}$ is 2-colorable; let us 2-color it with red and blue. Color $x_{t+1}, \ldots, x_s$ red, $x_{s+1}, \ldots, x_n$ blue. Then we have a legitimate 2-coloration of H (which is clearly a contradiction). For consider any edge E_j. If $E_j \subset \{x_1, \ldots, x_t\}$ then it contains both red and blue points, since the coloration of H' is good. If $E_j \not\subset \{x_1, \ldots, x_t\}$ then (4) implies that it must contain points x_i with both negative and positive λ_i, that is, with both red and blue color. ∎

A third version of the same idea comes up in the proof [15] of a conjecture of Toft [22]:

Theorem *Let H be a critically 3-chromatic r-uniform hypergraph with m edges and n points. Then*

$$m \le \binom{n}{r-1}.$$

Furthermore, if $E_1, \ldots, E_m$ are the edges of H and $A_1, \ldots, A_N$ all the $(r-1)$-tuples of points of H ($N = \binom{n}{r-1}$), and we set

$$a_{ij} = \begin{cases} 1 & \text{if } A_i \subset E_j, \\ 0 & \text{otherwise,} \end{cases}$$

then the columns of the matrix (a_{ij}) are linearly independent over the real field.

The proof of this is somewhat more lengthy and is not given here.

5

Let us look at an example of the phenomenon mentioned in the previous paragraph where criticality concerns the covering number instead of the chromatic number.

Recall that $\tau(G)$ is the minimum number of points covering the edges of the graph G. G is called τ-critical if $\tau(G') < \tau(G)$ for every proper subgraph

G' of G. The study of τ-critical graphs has led to many interesting results. Here we discuss only the following theorem of Erdös et al. [6]:

Theorem *If G is τ-critical then $|E(G)| \leq \binom{\tau(G)+1}{2}$.*

We state and prove a result which is a generalization of this one in the same fashion as Seymour's theorem in Section 4 is a generalization of the result of [14].

Let us assume that the vertices of G are column vectors of dimension $t = \tau(G)$ such that any t of them are linearly independent. Let $e = (x, y)$ be any edge of G. Set

$$A_e = xy^T + yx^T. \tag{5}$$

Note that the order of multiplication is such that A_e is a $t \times t$ matrix. Clearly it is symmetric.

Theorem *If G is τ-critical then the matrices A_e [$e \in E(G)$] are linearly independent.*

Proof Assume indirectly that

$$A_f = \sum_{e \in E(G) - \{f\}} \lambda_e A_e \tag{6}$$

for some edge f. By hypothesis, there is a set B of $t - 1$ vertices of G covering all edges except f. Let b be a vector perpendicular to the $t - 1$ vectors in B. For every edge $e = (x, y) \neq f$ we have that either $x \in B$ or $y \in B$, and therefore

$$(b^T x)(b^T y) = 0.$$

Thus

$$b^T A_e b = b^T xy^T b + b^T yx^T b = 2(b^T x)(b^T y) = 0. \tag{7}$$

By the same argument we get that

$$b^T A_f b \neq 0. \tag{8}$$

Multiply now (6) by b^T from the left and by b from the right. By (7) we get zero on the right hand side; by (8) the left hand side is nonzero. This is a contradiction. ∎

The previous proof gives slightly more than just the theorem. For a given t, construct the matrices A_e. Suppose that A_f is linear combination of the other matrices. Then e can be deleted from G in the sense that $\tau(G - e) < t$ if and only if $\tau(G) < t$.

A more important feature of this approach is that it works similarly if the vectors representing the points of G are not in general position. In this case $\tau(G)$ should be interpreted as the minimum *rank* of a set of points covering all edges of G. For details and applications we refer the reader to [16].

6

The matrix A_e in (5) above can be considered as the symmetric tensor product of the vectors x and y. Applying more advanced techniques of multilinear algebra we get further methods for handling combinatorial problems. We hope that these extensions will lead to a better understanding of the common background of all applications of linear algebra shown here.

The following result is another generalization of the Erdös–Hajnal–Moon theorem (and of its extension to hypergraphs by Bollobás [2]). For details see [16].

Theorem *Let a collection H of r-subspaces of a linear space have the property that for any $\binom{t+r}{r}$ of them there exists a t-dimensional subspace meeting each of them in a nonzero subspace. Then there exists a t-dimensional subspace meeting each member of H in a nonzero subspace.*

Proof We may assume that the conclusion is false but that it holds if any member of H is deleted. Also we may assume that the whole space has dimension $k + r$, since otherwise we could project to a subspace of dimension $k + r$ from a "general" direction.

Given any subspace A, let us select a basis $a_1, \ldots, a_k$ of A and put

$$\hat{A} = a_1 \wedge \cdots \wedge a_k$$

(where $\wedge$ denotes exterior multiplication). It follows from a basic fact in multilinear algebra that

$$\hat{A} \wedge \hat{B} = 0 \Leftrightarrow A \cap B \neq 0.$$

Claim *The tensors $\hat{A}$ ($A \in H$) are linearly independent.*

For suppose that there is an $A \in H$ such that

$$\hat{A} = \sum_{B \in H - \{A\}} \lambda_B \hat{B}.$$

Let M be a t-subspace which has a nonzero intersection with all members of $H - \{A\}$. By the indirect assumption, $M \cap A = 0$. Multiplying in the preceding equation by $\hat{M}$, we get zero on the right hand side but not on the left. This contradiction proves the claim.

Now the dimension of the space of r-contravariant antisymmetric tensors (to which all the tensors $\hat{A}$, $A \in H$ belong) is $\binom{r+t}{r}$. This proves that $|H| \leq \binom{r+t}{r}$, which contradicts the assumption. ∎

7

The following problem was mentioned to me by R. L. Graham: A k-uniform hypergraph H is called a k-*forest* if for each edge E of H there exists a partition of the underlying set $V(H)$ into k classes which totally separates E but does not totally separate any other edge. For $k = 2$ this means that every edge is a cut-edge, that is, the graph is a forest.

Theorem *The maximum number of edges in a k-forest on n points is $\binom{n-1}{k-1}$.*

Proof The number $\binom{n-1}{k-1}$ is attained, among others, by the k-forest whose edges are all k-tuples containing a given point. So we want to show that no k-forest has more edges than this one.

Let S be the set of vertices of our k-forest. Suppose that the elements of S are linearly independent vectors over, say, the real field. With each pair $e = \{x, y\} \subseteq S$, associate the vector $\check{e} = x - y$ (the sign is irrelevant). Then it is easily seen that $e_1, \ldots, e_k$ are linearly independent, i.e.,

$$\check{e}_1 \wedge \cdots \wedge \check{e}_k \neq 0,$$

if and only if the pairs $e_1, \ldots, e_k$ form a forest (this is just the usual co-ordinatization of the matroid of the complete graph K_n).

Let $X \subseteq S$. Let $e_1, \ldots, e_t$ ($t = |X| - 1$) be the edges of a tree spanning X. Define

$$\check{X} = \check{e}_1 \wedge \cdots \wedge \check{e}_t.$$

Now let $E_1, \ldots, E_M$ be the edges of our k-tree. We claim that $\check{E}_1, \ldots, \check{E}_M$ are linearly independent. For let, say,

$$\check{E}_1 = \sum_{i=2}^{M} \lambda_i \check{E}_i. \tag{9}$$

Consider the partition $\{V_1, \ldots, V_k\}$ which totally separates the points of E_1 but not the points of any other edge. Multiply (9) by $\check{V}_1 \wedge \cdots \wedge \check{V}_k$. On the left hand side, if we use the definitions of $\check{E}_1, \check{V}_1, \ldots, \check{V}_k$, we obtain a product of tensors $\check{e}$, for which the corresponding pairs e form a tree on S. So the left hand side is not zero. The same reasoning yields, however, that all terms on the right hand side are zero. This contradiction proves the claim.

Now $\check{E}_1, \ldots, \check{E}_M$ are $(k - 1)$-contravariant antisymmetric tensors over the $(n - 1)$-dimensional space generated by the vectors $x - y$ ($x, y \in S$). So their number is at most $\binom{n-1}{k-1}$. ∎

8

Finally, I would like to mention some further results in graph theory and combinatorics which are related to the topic of this paper.

(1) It would be very difficult to survey the numerous results of the type that any two things in a given class can be transformed into each other by a sequence of simple transformations. Some of these may have generalizations in the direction that certain complexes associated with graphs are highly connected. Such generalizations would be interesting primarily if they enabled graph-theoretical results to be derived from topological theorems.

(2) The number of examples where 1-dimensional homotopy is applied is much smaller. There is Tutte's famous matroid homotopy theorem [23], Wilson's paper [25], and, more in the direction of embedding theory, Tutte's paper on nonseparating circuits [24]. For these theorems the generalization to higher dimensions is probably more at hand but I do not see any applications right now.

(3) The theory of maps (graphs embedded in surfaces) is a very strong link between graph theory and topology. Is there any relation between these topological aspects and the topological methods applied in the study of chromatic number in this paper (for example, can they be used to prove the four color theorem)?

(4) A very well-known application of linear algebra to combinatorics is the proof of Fisher's inequality and various extensions (see, e.g., Ray-Chaudhuri and Wilson [19]). I hope that a large class of combinatorial inequalities, where the conditions involve intersection properties of sets, can be handled by using linear algebra.

REFERENCES

1. I. Bárány, A short proof of Kneser's conjecture, *J. Combinatorial Theory*, to appear.
2. B. Bollobás, On generalized graphs, *Acta Math. Acad. Sci. Hungar.* **16** (1965) 447–452.
3. W. G. Brown and J. W. Moon, Sur les ensembles de sommets indépendentes dans les graphes chromatiques minimaux, *Canad. J. Math.* **21** (1969) 274–278.
4. W. G. Brown, P. Erdös, and V. T. Sós, On the existence of triangulated spheres in 3-graphs and related problems, *Period. Math. Hungar.* **3** (1973) 221–228.
5. P. Erdös and A. Hajnal, Kromatikus gráfokról, *Mat. Lapok* **18** (1967) 1–4.
6. P. Erdös, A. Hajnal, and J. W. Moon, A problem in graph theory, *Amer. Math. Monthly* **71** (1964) 1107–1110.
7. A. Frank, Some polynomial algorithms for certain graphs and hypergraphs, *Proc. Brit. Combinatorial Confer., 5th, Aberdeen, 1975, Congr. Numerantium* **XV** (1976) 221–226.
8. D. Gale, Neighborly and cyclic polytopes, *Proc. Symp. Pure Math., 7th, 1963*, pp. 225–232.
9. E. Györi, On the division of graphs to connected subgraphs, *Combinatorics, Colloq. Math. Soc. J. Bolyai* **18** (1978) 485–494.

10. M. Kneser, Aufgabe 300, *Jber. Deutsch. Math. Verein* **58** (1955).
11. L. Lovász, A homology theory for spanning trees of a graph, *Acta Math. Acad. Sci. Hungar.* **30** (1977) 241–251.
12. L. Lovász, Kneser's conjecture, chromatic number, and homotopy, *J. Combinatorial Theory*, to appear.
13. L. Lovász, Independent sets in critical chromatic graphs, *Studia Sci. Math. Hungar.* **8** (1973) 165–168.
14. L. Lovász, Graphs and set-systems, *in* "Beiträge zur Graphentheorie," pp. 99–106, Teubner, Leipzig, 1968.
15. L. Lovász, Chromatic number of hypergraphs and linear algebra, *Period. Math. Hungar.*, to appear.
16. L. Lovász, Flats in matroids and geometric graphs, *in* "Combinatorial Surveys" (P. J. Cameron, ed.), pp. 45–86. Academic Press, New York, 1977.
17. S. Maurer, Problem presented, *Brit. Combinatorial Confer.*, *5th, Aberdeen, 1975*.
18. K. Milliken, private correspondence.
19. D. K. Ray-Chaudhuri and R. J. Wilson, On t-designs, *Osaka J. Math.* **12** (1975) 737–744.
20. P. Seymour, On 2-colorings of hypergraphs, *Quart. J. Math. Oxford Set.* **25** (1974) 303–312.
21. M. Simonovits, On colour-critical graphs, *Studia Sci. Math. Hungar.* **7** (1972) 67–81.
22. B. Toft, On colour-critical hypergraphs, *Infinite Finite Sets, Colloq. Math. Soc. J. Bolyai* **10** (1975) 1445–1457.
23. W. T. Tutte, A homotopy theorem for matroids, *Trans. Amer. Math. Soc.* **88** (1958) 144–174.
24. W. T. Tutte, How to draw a graph, *Proc. London Math. Soc.* **13** (1963) 743–767.
25. R. J. Wilson, Graph puzzles, homotopy, and the alternating group, *J. Combinatorial Theory Ser. B* **16** (1974) 86–96.

AMS 05C99

BOLYAI INTÉZET
JÓZSEF ATTILA UNIVERSITY
SZEGED, HUNGARY

All the King's Horses
(A Guide to Reconstruction)

W. T. TUTTE

Dedicated to the memory of Joseph William Tutte, 1901–1977

1. Introduction

In [5] the author has shown that the rank-polynomial of a graph G can (in principle) be calculated from the rank-polynomials of its vertex-deleted subgraphs. It now seems that this result may be of some importance in reconstruction theory.

It is possible to make many variations on the multiplicative theory of [5]. It can be applied to some other graph-polynomials. Either the vertex-set or the edge-set of G can be taken as fundamental. Either connection or non-separability can be regarded as the basic graph-theoretical property.

In this paper we state the multiplicative theory in a form that covers all these cases, and we then apply it to some problems of reconstruction. The proof achieved is so general that it may well apply to systems other than graphs. Nevertheless it is thought that there has been a gain in simplicity over the argument presented in [5].

Among other applications we establish the reconstructability of the rank-polynomial, the characteristic polynomial, and a generalization of the latter that we call the *idiosyncratic polynomial* of G. Moreover we show that if the characteristic or idiosyncratic polynomial of G is reconstructed and found to be prime, then the reconstruction of G can be carried through to completion.

2. Musters and Clusters

Let Q be a finite set of q elements. We use the special term *muster* for a subset of Q. Some non-null musters may be distinguished as *clusters*. The resulting system we call a *clustering on Q*.

Consider any muster M of Q. Let S be a subset of M. We say that S is a *separator* of M if each cluster that is contained in M is contained either in S or in $M - S$. Thus M itself and its null subset are separators of M.

It is clear that the complement in M of a separator of M is a separator of M. Moreover any union or intersection of separators of M is a separator of M.

A separator of M is called *elementary* if it is non-null and has no other non-null separator of M as a subset. It follows from the preceding observations that the elementary separators of M are disjoint and have M as their union. We write $F(M)$ for the set of elementary separators of M. This set is null only for the null muster.

A clustering $C(Q)$ on Q is said to be *regular* if every elementary separator of a muster of Q is a cluster.

In a regular clustering $C(Q)$ any muster consisting of a single element must be a cluster. The null muster is of course not a cluster.

Our multiplicative theory applies to all regular clusterings. If we wish to specialize it to graph theory we must first take note of some regular clusterings associated with a general graph G.

We may, for example, take Q to be the vertex-set $V(G)$ of G. We may then specify the clusters as those non-null subsets of $V(G)$ that induce connected subgraphs of G. Any muster M then induces a subgraph $G(M)$ of G, and its elementary separators induce the components of $G(M)$. The condition for regularity is thus satisfied. We denote the resulting regular clustering on $V(G)$ by $C(G; 0)$.

In another example we take Q to be the edge-set $E(G)$ of G. If S is any subset of $E(G)$ we define the *reduction* $G \cdot S$ of G to S as the subgraph of G made up of the edges of S and their incident vertices. We can define the clusters as those non-null subsets S of $E(G)$ such that $G \cdot S$ is connected. The elementary separators of a muster M then correspond to the components of $G \cdot M$, and the condition for regularity is satisfied. We denote the resulting regular clustering on $E(G)$ by $C(G; 1a)$.

In a variation on the last example we define the clusters as those non-null subsets S of $E(G)$ for which $G \cdot S$ is nonseparable. The elementary separators of a muster M then correspond to the blocks of $G \cdot M$. The resulting clustering on $E(G)$ is again regular. We denote it by $C(G; 1b)$.

The definition of $C(G; 1b)$ is easily generalized from graphs to matroids.

3. Upper and Lower Polynomials

Let $C(Q)$ be a regular clustering on a set Q.

We agree to operate with a given finite set of indeterminates over the ring of integers. Let P denote the set of all polynomials, with integral coefficients, in these indeterminates.

To each cluster K of $C(Q)$ we assign a member $j(K)$ of P. We impose the condition that $j(K)$ must have a zero constant term, but otherwise its choice is arbitrary. We extend the definition of j to a general muster M by postulating that

$$j(M) = \prod_{K \in F(M)} j(K). \tag{1}$$

This, of course, yields no contradiction when M is a cluster. We call $j(M)$ the *lower polynomial* of M.

For each muster M we now define a polynomial $J(M)$ as follows.

$$J(M) = \sum_{X \subseteq M} j(X). \tag{2}$$

We call this the *upper polynomial* of M.

In interpreting these formulae we use the rule that an empty product has the value 1. So if Ω is the null muster we have

$$J(\Omega) = j(\Omega) = 1. \tag{3}$$

For any non-null muster M the lower polynomial $j(M)$ has a zero constant term. It follows from (2) that the upper polynomial $J(M)$ of a muster M is always *monic*, that is with constant term 1.

3.1 *Let S and T be complementary separators of a muster M of $C(Q)$. Then*

$$J(M) = J(S) \cdot J(T).$$

Proof By (1) and (2) we have

$$J(M) = \sum_{X \subseteq M} \left\{ \prod_{K \in F(X)} j(K) \right\}.$$

Instead of summing over the subsets X of M we could sum equivalently over the ordered pairs (X_S, X_T), where X_S and X_T are subsets of S and T, respectively. We can interpret X as the union of X_S and X_T. Then, by the definition of a separator, we have

$$J(M) = \sum_{(X_S, X_T)} \left\{ \prod_{K \in F(X_S)} j(K) \cdot \prod_{K \in F(X_T)} j(K) \right\}$$

$$= J(S) \cdot J(T) \qquad \text{by (1) and (2).} \qquad \blacksquare$$

By repeated application of 3.1 we obtain the following theorem.

3.2 *If M is any muster of $C(Q)$, then*

$$J(M) = \prod_{K \in F(M)} J(K).$$

4. Multiplicative Theorems

For each muster M of the regular clustering $C(Q)$ we define $Y(M)$ as the class of all sets

$$\{K_1, K_2, \ldots, K_r\} \qquad (4)$$

of distinct clusters K_i of $C(Q)$ such that the union of the K_i is M.

We define $Z(M)$ in the same way, except that each set (4) is required to have at least two members.

For each cluster K we define a power series $\lambda(K)$ in our indeterminates. It is to be given by the following recursive formula.

$$\lambda(K) = -j(K) + \sum_{X \in Z(K)} \left\{ \prod_{L \in X} \{-\lambda(L)\} \right\}. \qquad (5)$$

We can discuss this equation in terms of a sequence of "approximations" to $\lambda(K)$. The first approximation is

$$\lambda_1(K) = -j(K). \qquad (6)$$

The second approximation $\lambda_2(K)$ is obtained by replacing $\lambda(L)$ on the right of (5) by its first approximation $-j(L)$, and so on. Thus

$$\lambda_2(K) = -j(K) + \sum_{X \in Z(K)} \left\{ \prod_{L \in X} j(L) \right\}. \qquad (7)$$

It is clear that in passing from $\lambda_n(K)$ to $\lambda_{n+1}(K)$ we add only terms whose degrees in the indeterminates exceed n. This is because each term in the lower polynomial $j(K)$ is of at least the first degree, for each cluster K, and because each X has at least two members L. Hence a given product of powers of the indeterminates, of degree n, has the same coefficient in all the approximations $\lambda_m(K)$ from the nth onwards. As part of the definition of $\lambda(K)$ we take this to be its coefficient in $\lambda(K)$. We thus obtain $\lambda(K)$ as a well-defined power series in the indeterminates, satisfying (5).

We go on to find some other equations relating λ to j and J. First however we extend the definition of λ as follows.

4.1 *If M is any muster of $C(Q)$ that is not a cluster, then, by definition, $\lambda(M) = 0$.*

4.2 *If K is any cluster of $C(Q)$, then*

$$j(K) = \sum_{X \in Y(K)} \left\{ \prod_{L \in X} \{-\lambda(L)\} \right\}.$$

Proof Denote the expression on the right by $\hat{j}$. Then, by the definitions of $Y(K)$ and $Z(K)$,

$$\hat{j} = -\lambda(K) + \sum_{X \in Z(K)} \left\{ \prod_{L \in X} \{-\lambda(L)\} \right\}$$
$$= -\lambda(K) + \{\lambda(K) + j(K)\} \quad \text{by (5)}$$
$$= j(K). \qquad \blacksquare$$

4.3 *If M is any muster of $C(Q)$, then*

$$j(M) = \sum_{X \in Y(M)} \left\{ \prod_{L \in X} \{-\lambda(L)\} \right\}.$$

Proof Denote the expression on the right by $\hat{j}$.

If M is null the product on the right is empty. Hence $\hat{j} = 1 = j(M)$, by (3).

In the remaining case let us enumerate the elementary separators of M as $K_1, K_2, \ldots, K_r$.

Consider any particular X. Each member L of X is contained in one of the elementary separators K_i. Let X_i denote the set of those members L of X that are contained in K_i. Then we can write

$$\hat{j} = \sum \left\{ \prod_{i=1}^{r} \prod_{L \in X_i} \{-\lambda(L)\} \right\},$$

where the sum is over all sequences $(X_1, X_2, \ldots, X_r)$ such that X_i is a member of $Y(K_i)$. Thus

$$\hat{j} = \prod_{i=1}^{r} \left\{ \sum_{X_i} \left\{ \prod_{L \in X_i} \{-\lambda(L)\} \right\} \right\}$$
$$= \prod_{i=1}^{r} j(K_i) \quad \text{by 4.2}$$
$$= j(M),$$

by the definition of the lower polynomial. $\blacksquare$

If M is any muster of $C(Q)$ let us write $A(M)$ for the class of all clusters L of $C(Q)$ such that $L \in M$. Let us also write $|S|$ for the cardinality of a set S.

4.4 *If K is any cluster of $C(Q)$, then*
$$J(K) = \prod_{L \in A(K)} \{1 - \lambda(L)\}.$$

Proof Denote the expression on the right by J. Then
$$J = \sum_U \left\{ (-1)^{|U|} \prod_{L \in U} \lambda(L) \right\},$$
where U can be any subclass of $A(K)$. Thus
$$J = \sum_{M \subseteq K} \left\{ \sum_{X \in Y(M)} \left\{ \prod_{L \in X} \{-\lambda(L)\} \right\} \right\}$$
$$= \sum_{M \subseteq K} j(M) \quad \text{by 4.3}$$
$$= J(K). \qquad \blacksquare$$

It is convenient to make the following slight extension of this result.

4.5 *If M is any muster of $C(Q)$, then*
$$J(M) = \prod_{L \in A(M)} \{1 - \lambda(L)\}.$$

This is a consequence of 3.2 and 4.4.

We can if we wish make a formal change in the expression on the right of 4.5, no longer restricting L to be a cluster but allowing it to range over all subsets of M. This is justified by the definition 4.1.

Exercise for the Reader Letting M be any muster of $C(Q)$ show that
$$1 - \lambda(M) = \prod_{X \subseteq M} \{J(X)\}^{a(X)}, \quad \text{where} \quad a(X) = (-1)^{|M-X|}.$$

5. Reconstruction

Let G be any graph having at least three vertices. Loops and multiple joins are allowed. Let the n vertices of G be enumerated as $v_1, v_2, \ldots, v_n$. For each vertex v_i let G_i be the graph obtained from G by deleting v_i and all its incident edges.

In the reconstruction problem we suppose given the isomorphism classes of the n graphs G_i. Thus we may assume that we are given a diagram of each G_i, but there is no labeling to indicate which vertices in one drawing correspond to which vertices in another.

ALL THE KING'S HORSES (A GUIDE TO RECONSTRUCTION)

The reconstruction conjecture asserts that the isomorphism class of G is uniquely determined by the list of isomorphism classes of the graphs G_i. It is not yet proved or disproved. One way of attacking it is to discover properties of G that can be inferred from the list of graphs G_i. Such properties of G are called *reconstructible*. Three trivial examples are dealt with in the following theorem.

5.1 *The numbers of vertices, loops, and links of G are reconstructible.*

Proof The number n of vertices of G is simply the number of the graphs G_i. The number of loops of G is evidently obtained by summing the numbers of loops of the graphs G_i and dividing the result by $n - 1$. Similarly we obtain the number of links of G by summing the numbers of links of the graphs G_i and dividing by $n - 2$. ∎

Let S be any subset of $V(G)$. We write G_S for the graph derived from G by deleting the members of S and their incident edges. If S is null then $G_S = G$. If S has only the one member v_i, then $G_S = G_i$. In any case we describe G_S as a *vertex-deleted* subgraph of G of order $|S|$.

5.2 *Let s be any positive integer. Then, given the isomorphism classes of the graphs G_i, we can determine the isomorphism classes of the vertex-deleted subgraphs of G of order s, in the sense that we can say how many such subgraphs there are in each isomorphism class.*

Proof We may suppose $s > 1$. We make a list of the vertex-deleted subgraphs of G_j of order $s - 1$, for each suffix j. Combining the n lists we obtain a list of vertex-deleted subgraphs of G of order s. Moreover each isomorphism class has exactly s times as many members in this list as it does in the list of distinct vertex-deleted subgraphs of G of order s. ∎

Let C be one of the regular clusterings $C(G; 1a)$ and $C(G; 1b)$. It may seem at first that C, being based on edges and not vertices, cannot be relevant to the reconstruction problem. However we can relate it to the sets of vertices of G as follows.

If S is any subset of $V(G)$ we write $B(S)$ for the set of all clusters K of C such that

$$V(G \cdot K) = S.$$

Having defined j, J, and λ in C we write

$$h(S) = \prod_{K \in B(S)} \{1 - \lambda(K)\}. \tag{8}$$

It will be convenient to write $G(S)$ for the subgraph of G induced by S, that is, for the subgraph G_T, where $T = V(G) - S$.

5.3 *If $C = C(G; 1a)$ and $G(S)$ is disconnected, or if $C = C(G; 1b)$ and $G(S)$ is separable, then*

$$h(S) = 1.$$

This is because the product on the right of (8) is then empty.

5.4 *In the regular clustering C we have*

$$J(E(G)) = \prod_{S \subseteq V(G)} h(S).$$

This follows from (8), with the help of 4.5.

In a reconstruction problem we would know all the factors $h(S)$ on the right except $h(V(G))$. We shall find that in some favourable but important cases enough partial information about $h(V(G))$ is available to make possible the calculation of $J(E(G))$ from the above equation.

In using 5.4 it may be necessary to distinguish between the graphs G and $G \cdot E(G)$. They are identical if and only if G is without isolated vertices. We therefore conclude this section with another trivial theorem on reconstruction.

5.5 *The number of isolated vertices of G is reconstructible. Moreover if this number is not zero then the isomorphism class of G can be determined.*

Proof The number $|E(G)|$ is reconstructible by 5.1. The number of isolated vertices of G is the number of graphs G_i such that $|E(G_i)| = |E(G)|$. If such a G_i exists then G is determined as the disjoint union of G_i and a vertex-graph. ∎

6. Indicators

Let $C(Q)$ be a regular clustering on a set Q. We define an *indicator* of $C(Q)$ as a mapping α of the set of clusters of $C(Q)$ onto the set of integers, satisfying the following conditions.

(i) $\alpha(K)$ *is non-negative.*
(ii) *If a cluster K is expressed as a union of two or more distinct clusters $K_1, K_2, \ldots, K_r$, then*

$$\alpha(K) < \sum_{i=1}^{r} \alpha(K_i).$$

We now take note of some examples of indicators for the regular clusterings $C(G; 1a)$ and $C(G; 1b)$ associated with a graph G. If K is a cluster of either of these we write $n(K)$ for the number of vertices of $G \cdot K$.

6.1 *In each of the clusterings $C(G; 1a)$ and $C(G; 1b)$ the function $n(K)$ is an indicator.*

Proof Condition (i) is trivial. Condition (ii) follows from the fact that $G \cdot K$ is connected and therefore each of the graphs $G \cdot K_i$ must have a vertex in common with one of the others. ■

6.2 *In $C(G; 1b)$ the function $\rho(K) = n(K) - 1$ is an indicator.*

Proof Condition (i) is trivial. We verify condition (ii) as follows.

Assume that for some positive integer $s < r$ the union of the subgraphs $G \cdot K_1$ to $G \cdot K_s$ is a connected graph H_s. Since $G \cdot K$ is connected we can adjust the notation so that $G \cdot K_{s+1}$ has at least one vertex in common with H_s. Then the union of the subgraphs $G \cdot K_1$ to $G \cdot K_{s+1}$ is a connected graph H_{s+1}, the union of H_s and $G \cdot K_{s+1}$.

Suppose first that H_s and $G \cdot K_{s+1}$ have only one vertex v in common. If they have also a common edge E, then E must be a loop on v. But then the set $\{E\}$ must be identical with K_{s+1}, and with one of the sets K_1 to K_s, by nonseparability. This is contrary to the requirement in condition (ii) that the r sets K_i must be distinct. It is now evident that

$$|V(H_{s+1})| - 1 = |V(H_s)| - 1 + \rho(K_{s+1})$$

and that $G \cdot K_{s+1}$ is a block of H_{s+1}.

Since $G \cdot K$ is nonseparable the above case cannot arise when $s = r - 1$. In the remaining case we must have

$$|V(H_{s+1})| - 1 < |V(H_s)| - 1 + \rho(K_{s+1}).$$

Starting with $s = 1$ we repeat the above argument until we arrive at $H_r = G \cdot K$. Combining the results of each step we complete the required verification. ■

We return to the general indicator $\alpha(K)$ of $C(Q)$.

We assume in the remainder of this section that

$$j(K) = t^{\alpha(K)} f(K) \tag{9}$$

for each cluster K, where t is one indeterminate, and $f(K)$ is a polynomial in the others or is a constant. Let us call t the *indicative variable*.

In what follows we use the symbol T to denote any power series in the indeterminates. It need not represent the same power series in each of its occurrences, even within one formula.

Let us now study the power series λ associated with the lower polynomial j and the upper polynomial J in $C(Q)$.

6.3 *If K is any cluster of $C(Q)$, then*

$$\lambda(K) = -t^{\alpha(K)} f(K) + t^{\alpha(K)+1} T.$$

Proof We first establish the formula

$$\lambda_n(K) = -t^{\alpha(K)}f(K) + t^{\alpha(K)+1}T \qquad (10)$$

for each approximation $\lambda_n(K)$ of $\lambda(K)$.

Formula (10) holds when $n = 1$, by (6) and (9). Assuming it true for a particular n we obtain $\lambda_{n+1}(K)$ by substituting the nth approximation of $\lambda(L)$ on the right of (5). This gives

$$\lambda_{n+1}(K) = -t^{\alpha(K)}f(K) + \sum_{X \in Z(Y)} t^{d(X)}T,$$

where

$$d(X) = \sum_{L \in X} \alpha(L).$$

But, by the definition of $Z(M)$ in Section 4, each member X of $Z(K)$ has at least two clusters L as elements. Moreover the clusters in X are distinct and have K as their union. So, by condition (ii) for an indicator, we have

$$d(X) > \alpha(K).$$

Accordingly formula (10) remains valid when n is replaced by $n + 1$. It follows in general by induction.

The theorem now follows from the relation between $\lambda(K)$ and its approximations. ∎

From now on we restrict $C(Q)$ to be $C(G; 1a)$ or $C(G; 1b)$, and we proceed to the power series h defined in Section 5. Applying 6.3 to formula (8) we obtain the following theorem.

6.4 *For each subset S of $V(G)$ we have*

$$h(S) = 1 + \sum_{K \in B(S)} t^\alpha f(K) + t^{\alpha+1}T,$$

where α is $|S|$ or $|S| - 1$ according as $C(Q)$ is $C(G; 1a)$ or $C(G; 1b)$.

We now apply this to the reconstruction problem. Suppose we wish to reconstruct $J(E(G))$ to as many terms as possible.

We restrict ourselves to the indicators $n(K)$ for $C(G; 1a)$ and $n(K) - 1$ for $C(G; 1b)$, as permitted by 6.1 and 6.2. We take note of one elementary theorem.

6.5 *Let W be a subset of $E(G)$ such that $G \cdot W$ is connected. Then the sum of the numbers $n(K) - 1$, taken over all the blocks $G \cdot K$ of $G \cdot W$, is $|V(G \cdot W)| - 1$.*

This result can be established by building $G \cdot W$ block by block, as in the construction of 6.2.

For maximum convenient generality we take $f(K)$ to be an indeterminate, called the *class-variable*, associated with the isomorphism class of $G \cdot K$. The class-variables associated with the different isomorphism classes corresponding to the clusters are taken to be independent of one another, and of the indicative variable t.

Any term in the formal infinite sum $h(S)$ of 6.4, or in $J(E(G))$, is of the form

$$mt^a \prod_K \{f(K)\}^{a(K)}, \tag{11}$$

where the product is restricted to a complete set of clusters K corresponding to nonisomorphic reductions $G \cdot K$. Here m is an integer, called the *coefficient* of the term, and the indices a, $a(K)$ are non-negative integers. We say that the term has *indicative degree* a and *class-degree* $\sum a(K)$. We note that every term of $h(S)$ or $J(E(G))$, save only for the constant term 1, has class-degree at least 1. Here "every term" means, as usual, "every term with a nonzero coefficient."

Let us write $p = n$ if $C(Q) = C(G; 1a)$ and $p = n - 1$ if $C(Q) = C(G; 1b)$. Using 6.5 in the latter case we see that each term of $J(E(G))$ has indicative degree not exceeding p.

6.6 *The coefficients of all the terms of $J(E(G))$ are reconstructible, save possibly for those terms, all of indicative degree p, which involve a cluster K satisfying $\alpha(K) = p$.*

Proof By 6.4 all the terms of $h(V(G))$ of indicative degree not exceeding p are trivially reconstructible, except for those involving a cluster K which is a spanning subgraph of G. Referring to 5.4 we see that the product on the right can be reconstructed, except for the single factor $h(V(G))$, by 5.2. Hence 5.4 implies 6.6, there being no term in $J(E(G))$ with indicative degree greater than p. ∎

Consider $C(G; 1a)$. By (1) and (2) we can interpret the coefficient of a term of $J(E(G))$ as follows. It is the number of reductions $G \cdot W$ of G such that the number of components of $G \cdot W$ in any isomorphism class is the index of the corresponding indeterminate. By 6.6 this number can be reconstructed if $G \cdot W$ has fewer than n vertices or if, having n vertices, it is disconnected. In the former case $G \cdot W$ can be made into a spanning subgraph of G by adjoining one or more vertex-graphs as new components. We can therefore interpret 6.6 as follows.

6.7 *The number of disconnected spanning subgraphs of G having a specified number of components in each isomorphism class is reconstructible.*

If the spanning subgraphs are to have one or more vertex-graphs as components their number can be found by Kelly's lemma [2], that is by the kind of counting argument used in 5.2. So the more interesting case of 6.7 is that concerned with spanning reductions.

There is a similar interpretation for the case of $C(G; 1b)$. We state it only for spanning connected subgraphs, referring the disconnected ones to 6.7. The spanning connected subgraphs are the reductions corresponding to the terms of $J(E(G))$ with indicative degree $n - 1$, by 6.5. Our second interpretation of 6.6 follows.

6.8 *The number of connected spanning subgraphs of G having a specified number of blocks in each isomorphism class is reconstructible, provided that each such block has fewer vertices than G.*

In particular we may take all the components in 6.7, and all the blocks in 6.8, to be link-graphs. Then, since n is at least 3, we have the following specializations.

6.9 *The number of 1-factors and the number of spanning trees of G are reconstructible.*

Exercise for the Reader Deduce from 6.7 that the number of components of G is reconstructible, and that if this number exceeds 1 the isomorphism class of G is reconstructible.

7. The Rank Polynomial

The *coboundary-rank* $\rho(H)$ of a graph H is the number of vertices of H minus the number of components of H. Hence it takes the value 0 for a null graph, a vertex-graph, or a loop-graph. For a cluster K of $C(G; 1b)$, $\rho(G \cdot K)$ is the indicator $n(K) - 1$.

The *cycle-rank* $r(H)$ of H is the number of edges of H minus the coboundary-rank. Using 6.5 in each component of H that is not a vertex-graph we can deduce the following rule.

7.1 *The coboundary-rank of a graph H is the sum of the coboundary-ranks of its blocks, and therefore the cycle-rank of H is the sum of the cycle-ranks of its blocks.*

Let us now consider a particular graph G, of at least three vertices, and its regular clustering $C(G; 1b)$.

Taking indeterminates t and z we deduce from 7.1 that there is an acceptable lower polynomial j of $C(G; 1b)$ such that

$$j(M) = t^{\rho(G \cdot M)} z^{r(G \cdot M)} \tag{12}$$

for each muster M.

For the corresponding upper polynomial J we note that

$$J(E(G)) = \sum_{M \subseteq E(G)} t^{\rho(G \cdot M)} z^{r(G \cdot M)}. \tag{13}$$

The expression on the right is called the *rank polynomial* of G. It is sometimes denoted by $R(G; t, z)$. Clearly its degree in t is $\rho(G \cdot E(G))$, which is identical with $\rho(G)$ even if G has isolated vertices. The degree in z is at most $|E(G)|$.

We now establish a simple property of the rank polynomial that is of crucial importance in reconstruction theory. We write e for $|E(G)|$.

7.2 $R(G; z, z) = (1 + z)^e$.

Proof By the definitions of the two ranks we have

$$R(G; z, z) = \sum_{M \subseteq E(G)} z^{|M|}$$

$$= \sum_{i=0}^{e} \binom{e}{i} z^i = (1 + z)^e. \blacksquare$$

7.3 *If the terms of $R(G; t, z)$ of degree less than $\rho(G)$ in t are given, then $R(G; t, z)$ is uniquely determined.*

Proof

$$R(G; t, z) = R_1 + \sum_{i=0}^{e} b_i t^{\rho(G)} z^i,$$

where R_1 is a known polynomial in t and z, and the b_i are unknown integers. Substituting z for t and applying 7.2 we find that

$$(1 + z)^e = R_2 + \sum_{i=0}^{e} b_i z^{i + \rho(G)},$$

where R_2 is a known polynomial in z. Accordingly b_i can be determined, for each i, as the coefficient of $z^{i + \rho(G)}$ in the polynomial $(1 + z)^e - R_2$. $\blacksquare$

7.4 *The rank polynomial of G is reconstructible.*

Proof We can take t as the indicative variable. For any cluster K the indicator $\alpha(K)$ is $\rho(G \cdot K) = n(K) - 1$. We deduce from 6.4 that the terms of $h(V(G))$ of degree less than $\rho(G) = n - 1$ in t are trivially reconstructible. Hence, by 5.2 and 5.4, the terms of $J(E(G))$ of degree less than $\rho(G)$ in t are reconstructible. The reconstructibility of $J(E(G))$ now follows from 7.3. $\blacksquare$

We must take note here of two important specializations. One is the *chromatic polynomial* $P(G, u)$ of G, giving the number of ways of colouring G in u colours. It is pointed out in [1] that $P(G, u) = u^n R(G; -u^{-1}, -1)$, and that the dually related *flow polynomial* is $(-1)^e R(G; -1, -u)$. We may therefore deduce the following theorem from 7.4.

7.5 *The chromatic polynomial and flow polynomial of G are reconstructible.*

The reconstructibility of the chromatic polynomial implies that of the chromatic number of G.

Having proved 7.4 we observe that 5.4 can be used to reconstruct the entire corresponding power series $h(V(G))$. By 6.4 the sum of the terms of $h(V(G))$ of degree $\rho(G)$ in t is

$$\sum_K t^{\rho(G)} z^{r(G \cdot K)},$$

where the sum is over all subsets K of $E(G)$ such that $G \cdot K$ is a spanning non-separable subgraph of G. But then

$$r(G \cdot K) = |K| - n + 1$$

by the definition of the two ranks. We deduce the following theorem.

7.6 *The number of nonseparable spanning subgraphs of G with a specified number of edges is reconstructible.*

The case in which the number of edges is n is of special importance. It can be stated as follows.

7.7 *The number of Hamiltonian circuits of G is reconstructible.*

Research Problem Is it possible to get further significant information about the structure of G from terms of $h(V(G))$ of degree exceeding $\rho(G)$ in t?

8. The Characteristic Polynomial

As usual let G be a graph of $n > 2$ vertices. Let the vertices be enumerated as $v_1, v_2, \ldots, v_n$.

Let $A(G)$ be the $n \times n$ matrix for which the entry in the rth row and sth column is the number a_{rs} of edges joining v_r and v_s. We interpret a_{rr} as the number of loops on v_r. Then $A(G)$ is called the *adjacency matrix* of G.

A graph Λ is called *sesquivalent* if each of its components is a link-graph or a circuit. We write $s(\Lambda)$ for the number of circuits of Λ having two or more vertices, $g(\Lambda)$ for the number of circuits of Λ of even length, and $s_2(\Lambda)$ for the number of 2-circuits. We also write $l(\Lambda)$ for the number of components of Λ that are link-graphs.

8.1
$$\det A(G) = \sum_\Lambda (-1)^{g(\Lambda) + l(\Lambda)} 2^{s(\Lambda)},$$

where the sum is over all spanning sesquivalent subgraphs Λ of G.

ALL THE KING'S HORSES (A GUIDE TO RECONSTRUCTION)   29

Proof   Unfortunately the proof of this theorem in [1] applies only to the special case of "simple graphs." The following modification covers all cases.

We have
$$\det A(G) = \sum_\pi \operatorname{sgn}(\pi) \cdot A_\pi, \tag{14}$$

$$A_\pi = a_{1\pi 1} a_{2\pi 2} \cdots a_{n\pi n}, \tag{15}$$

where the sum is over all permutations π of the sequence of integers from 1 to n.

The permutation π can be expressed uniquely as a composition of disjoint cycles. A cycle (i) of length 1 is said to *conform* to any 1-circuit, that is to any loop-graph, whose vertex is v_i. A cycle (i, j) of length 2 *conforms* to any link-graph or 2-circuit with vertices v_i and v_j. A cycle $(b, c, \ldots, z)$ of greater length *conforms* to any circuit with vertices $b, c, \ldots, z$ in which two vertices are adjacent when and only when one is the immediate successor of the other in the cycle of suffixes. Finally π *conforms* to any spanning sesquivalent subgraph Λ of G such that each cycle of π conforms to a component of Λ.

In all the above cases when X conforms to Y we say also that Y conforms to X. All the sesquivalent subgraphs Λ conforming to a given π have the same circuits of each length other than 2, and have the same value of $s_2(\Lambda) + l(\Lambda)$. Two permutations conform to the same Λ if and only if one is obtained from the other by reversing one or more cycles of length exceeding 2.

Consider any permutation π. For a cycle (i) the number a_{ii} is the number of 1-circuits conforming to the cycle. For a cycle (i, j) the number
$$a_{ij}a_{ji} = a_{ij}^2 = a_{ij} + a_{ij}(a_{ij} - 1)$$
is the number of link-graphs conforming to (i, j) plus twice the number of 2-circuits conforming to it. For any other cycle $(b, c, d, \ldots, z)$ the product $a_{bc}a_{cd}\cdots a_{zb}$ is the number of conforming circuits. If therefore $U(\pi)$ is the set of all sesquivalent spanning subgraphs of G conforming to π we have
$$\det A(G) = \sum_\pi \left\{ \sum_{\Lambda \in U(\pi)} (-1)^{g(\Lambda) + l(\Lambda)} 2^{s_2(\Lambda)} \right\}.$$

But it is clear that the number of permutations π conforming to a given Λ is 2^k, where $k = s(\Lambda) - s_2(\Lambda)$. The theorem follows. ∎

8.2   *The determinant of the adjacency matrix of G is reconstructible.*

Proof   The number of disconnected sesquivalent spanning subgraphs Λ of G with a specified number of components of each kind is reconstructible, by 6.7. The remaining sesquivalent spanning subgraphs are the Hamiltonian circuits of G, and their number is reconstructible by 7.7. Hence $\det A(G)$ can be reconstructed from the formula of 8.1. ∎

The *characteristic polynomial* $\chi(G, \lambda)$ of G is defined as follows.

$$\chi(G, \lambda) = \det(A(G) - \lambda I_n), \tag{16}$$

where I_n is the unit matrix of order n. Here λ is an indeterminate, not to be confused with the λ of Section 4.

Let S be any subset of $V(G)$. Then the submatrix of $A(G)$ obtained by deleting the rows and columns corresponding to the members of S is identical with $A(G_S)$. So we can write the expansion of $\chi(G, \lambda)$ in powers of λ as follows.

$$\chi(G, \lambda) = \sum_{S \subseteq V(G)} (-\lambda)^{n-|S|} \det A(G_S). \tag{17}$$

8.3 *The characteristic polynomial of G is reconstructible.*

Proof The part of the sum on the right of (17) corresponding to non-null subsets S is reconstructible by 5.2. The remaining term, $(-\lambda)^n \det A(G)$, is reconstructible by 8.2. ∎

The reconstructive relationship between $\chi(G, \lambda)$ and the number of Hamiltonian circuits was pointed out by Pouzet [4].

9. A Method of Generalization

Let $G^{(m)}$ be the graph derived from G by adding m new links joining each pair of vertices of G that are not adjacent in G. The same operation applied to G_i gives a graph $G_i^{(m)}$, and this is identical with the graph obtained from $G^{(m)}$ by deleting v_i and its incident edges. Accordingly any reconstructible property of $G^{(m)}$ can be determined when the G_i are known. When expressed as a property of G it can be added to the list of reconstructible characteristics of G.

As an example let us consider the number $\eta(G)$ of Hamiltonian circuits of G. We write η_k for the number of Hamiltonian circuits of $G^{(1)}$ having exactly k edges outside G. Then evidently

$$\eta(G^{(m)}) = \sum_{i=0}^{\infty} \eta_i m^i. \tag{18}$$

The value of this polynomial in m can be reconstructed for every positive integer m, by 7.7. The polynomial can therefore be added to the list of reconstructible parameters of G. In this list we can of course replace the variable integer m by an ordinary indeterminate.

9.1 *The number of Hamiltonian paths of G with nonadjacent ends is reconstructible.*

The required number is the coefficient of η_1 in (18).

Likewise we can reconstruct $\chi(G^{(m)}, \lambda)$ for each positive integer m. After replacing m by an indeterminate α we refer to the new reconstructible parameter of G, a polynomial in λ and α as the *idiosyncratic polynomial* of G. It is the determinant of the matrix obtained from $A(G) - \lambda I_n$ by replacing each nondiagonal zero entry by α.

When G is simple there seems to be no further significant generalization of $\chi(G, \lambda)$ along these lines. In the general case we can, by an analogous procedure, replace each integer occurring in one or more nondiagonal positions in $A(G) - \lambda I_n$ by its own associated indeterminate. We can then do the same, with different indeterminates, for the diagonal integers of $A(G)$.

10. On Prime Characteristic Polynomials

Let us write χ for $\chi(G, \lambda)$, and χ_i for $\chi(G_i, \lambda)$.

We make the abbreviation $X(G)$ for $A(G) - \lambda I_n$, and we write C_{rs} for the cofactor of the (r, s) position in $X(G)$. Evidently $C_{rs} = C_{sr}$ and $C_{rr} = \chi_r$. We write $Y(G)$ for the adjugate matrix of $X(G)$, whose entry in the (r, s) position is C_{rs}.

We note that χ and χ_r are nonzero polynomials in λ of degrees n and $n - 1$, respectively. If r and s are unequal the degree of C_{rs} cannot exceed $n - 2$.

Let S and T be complementary non-null subsets of $V(G)$. We write $X_S(G)$ and $Y_S(G)$ for the matrices derived from $X(G)$ and $Y(G)$, respectively, by striking out the rows and columns corresponding to the members of S. By a well-known identity in the theory of determinants we have

$$\chi^{|S|-1} \det X_S(G) = \det Y_T(G). \tag{19}$$

It is a curious fact that if $\chi(G, \lambda)$ is reconstructed and found to be prime, that is irreducible over the integers, then the reconstruction of the isomorphism class of G can be carried through to completion. This result is contained in the following theorem. The statement of the theorem is somewhat stronger than is actually needed for reconstructive purposes, in that the graphs G_i are not supposed given. Only their characteristic polynomials are assumed to be known.

10.1 *If χ and the n polynomials χ_i are known, and if χ is prime, then the isomorphism class of G is uniquely determined.*

Proof If some parameter P is uniquely determined by χ and the χ_i let us call it *accessible*. Parameters proved to be accessible will be described as "known," in spite of whatever practical difficulties may attend their actual evaluation. We proceed by a sequence of lemmas. ∎

Lemma I *If r and s are distinct, then C_{rs} is nonzero.*

Proof By (19) we have
$$\chi \cdot \det X_{\{r,s\}}(G) = \chi_r \chi_s - C_{rs}^2. \tag{20}$$
Hence if $C_{rs} = 0$, then χ must divide either χ_r or χ_s. This is impossible since these polynomials are nonzero and lower in degree than the prime polynomial χ. This establishes Lemma I. ∎

Lemma II *If r and s are distinct, then C_{rs}^2 is accessible.*

Proof The unknowns in (20) are C_{rs}^2 and the determinant on the left. If C_{rs}^2 is not uniquely determined by the equation there must be a polynomial P in λ with the following properties.

(i) P^2 differs from C_{rs}^2.
(ii) The degree of P does not exceed $n - 2$.
(iii) There is a polynomial Q in λ such that
$$\chi Q = \chi_r \chi_s - P^2.$$
But then we can subtract this equation from (20), finding that χ divides $(P + C_{rs})(P - C_{rs})$. This however is impossible by the limitations on degree and the fact that P^2 is not equal to C_{rs}^2. The lemma follows. ∎

Lemma III *If r, s, and t are distinct, then $C_{rs} C_{st} C_{tr}$ is accessible.*

Proof As a special case of (19) we have
$$\chi^2 \det X_{\{r,s,t\}}(G) = \chi_r \chi_s \chi_t + 2 C_{rs} C_{st} C_{tr}$$
$$- \chi_r C_{st}^2 - \chi_s C_{rt}^2 - \chi_t C_{rs}^2. \tag{21}$$
By Lemma II the only unknown on the right of this equation is $C_{rs} C_{st} C_{tr}$. Even here the absolute value is known: only the sign is in doubt. So if the lemma is false there must be a polynomial P in λ such that the equation continues to hold when the left side is replaced by $\chi^2 P$ and the term $C_{rs} C_{st} C_{tr}$ on the right is replaced by its negative. But then we can subtract the new equation from (21), finding that χ^2 divides $4 C_{rs} C_{st} C_{tr}$. This is impossible by Lemma I and the limitations on degree. ∎

Given a cyclic sequence $(b, c, d, \ldots, z)$ of m distinct suffixes we define the corresponding *cyclic product* as $C_{bc} C_{cd} \cdots C_{zb}$ and say that the *length* of this product is m. Thus C_{rs}^2 and $C_{rs} C_{st} C_{tr}$ are cyclic products of lengths 2 and 3, respectively.

Lemma IV *Every cyclic product is accessible.*

Proof We prove this by induction on the length of the product. The lemma holds for lengths 2 and 3, by Lemmas II and III. Assume it to hold for some length $m > 2$. Consider a cyclic product
$$B = C_{bc} C_{cd} C_{de} \cdots C_{zb}$$

of length $m + 1$. Then

$$BC_{bd}^2 = (C_{bc}C_{cd}C_{db})(C_{bd}C_{de} \cdots C_{zb})$$

and the product on the right is accessible by the inductive hypothesis. But then B is accessible, by Lemma I. So the induction succeeds and Lemma IV is established. ∎

Given any two complementary non-null subsets S and T of $V(G)$ we can expand det $Y_T(G)$ as a sum of terms each of which is a product of factors taken from the powers of -1, the polynomials χ_i, and the cyclic products. Accordingly det $Y_T(G)$ is accessible by Lemma IV. Hence det $X_S(G)$ is accessible, by (19).

Evaluating det $X_S(G)$ for the cases in which $|T| = 1$ we determine the diagonal elements of $A(G)$. Then we evaluate det $X_S(G)$ for the cases in which $|T| = 2$ and so determine the nondiagonal elements a_{rs} of $A(G)$. (Actually we determine a_{rs}^2 and have to use our knowledge that a_{rs} is non-negative.) We thus reconstruct $A(G)$ and so determine the isomorphism class of G.

Exercise for the Reader Assuming only that χ and χ_r are coprime show that $\chi_s = \chi_t$ if $C_{rs}^2 = C_{rt}^2$.

The proof of Theorem 10.1 remains valid when the characteristic polynomials are replaced by the corresponding idiosyncratic polynomials, or by the generalizations noticed at the end of Section 9. Perhaps sometimes the idiosyncratic polynomial is prime even though the characteristic polynomial is composite. It should be noted however that the characteristic polynomial of G is composite whenever G has an automorphism not leaving all the vertices fixed (see [3]), and that compositeness of this kind must extend to the idiosyncratic polynomial.

REFERENCES

1. N. Biggs, "Algebraic Graph Theory," Cambridge Tracts No. 67, Cambridge Univ. Press, London and New York, 1974.
2. P. J. Kelly, A congruence theorem for trees, *Pacific J. Math.* 7 (1957) 961–968.
3. A. Mowshowitz, The adjacency matrix and the group of a graph, in "New Directions in the Theory of Graphs" (F. Harary, ed.), pp. 129–148. Academic Press, New York, 1973.
4. M. Pouzet, Note sur le problème de Ulam, *J. Combinatorial Theory Ser. B*, to appear.
5. W. T. Tutte, On dichromatic polynomials, *J. Combinatorial Theory* 2 (1967) 301–230.

AMS 05C15, 05C99

DEPARTMENT OF MATHEMATICS
UNIVERSITY OF WATERLOO
ONTARIO, CANADA

GRAPH THEORY AND RELATED TOPICS

Hadwiger's Conjecture and Six-Chromatic Toroidal Graphs

MICHAEL O. ALBERTSON

and

JOAN P. HUTCHINSON

Suppose G is a connected graph with V vertices which is embedded on the torus. It is well known that the vertices of G can be properly colored in seven colors and that seven colors are necessary if and only if G actually contains K_7 as a subgraph [5]. This is a best possible characterization of seven-chromatic toroidal graphs. Hadwiger's conjecture [8] and the four color theorem [3] provide a necessary condition for a toroidal graph to be five-chromatic. Specifically Hadwiger conjectured that if a graph is r-chromatic then it contracts to K_r. Wagner showed that Hadwiger's conjecture in the case $r = 5$ is equivalent to the four color theorem [9, 10]. Thus a five-chromatic graph must contract to K_5. The purpose of this paper is to discuss six-chromatic toroidal graphs. The theorems will be followed by comments indicating the nature of the omitted proofs. Twice we sketch one case of a many-case proof. We will close by proving Hadwiger's conjecture for six-chromatic toroidal graphs.

To place this material in context we begin by recalling three theorems of Dirac. First, every six-chromatic graph which embeds on the torus, the Klein bottle, or the Möbius strip contracts to K_5 [7]. Second, let S_n denote the surface of genus n and $H(n)$ its Heawood number. If G is an $(H(n) - 1)$-chromatic graph which embeds on S_n then G contracts to $K_{H(n)-1}$ if n is at least three [4]. Third, if G is six-chromatic then either G contracts to K_6 or else a suitably chosen set of at least eleven edges can be deleted from G leaving a graph which contracts to K_6 minus one edge [6].

Thus it is reasonable to begin a study of six-chromatic toroidal graphs by considering Hadwiger's conjecture. It is natural to attempt a proof of this result by induction on the number of vertices. One finds that a toroidal graph which contains a vertex of degree less than six either contains K_6 as a subgraph or a configuration which is reducible with respect to five colors. (See case 1 of the proof of Theorem 5.) Thus among the six-chromatic toroidal graphs those which are regular of degree six are of particular interest. It is not immediately evident that there exist any such graphs. Figure 1 exhibits one such graph J which arose in our investigation of independence ratios [1]. J does not contain K_5 as a subgraph though it easily contracts to K_6.

Theorem 1 *If $G \neq J$, K_7 is a toroidal graph which is regular of degree six and has $V \leq 12$, then G can be five-colored.*

Comments The proof is by modified exhaustion. We sketch what happens if $V = 11$. In [1] we showed that a toroidal graph which is regular of degree six and not equal to J must have at least three independent vertices. Removing three independent vertices from G leaves a graph G' with eight vertices and

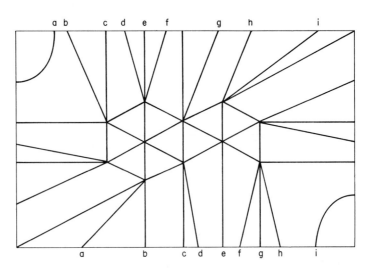

Figure 1

fifteen edges. Examination of the four possible degree sequences for G' leads to the conclusion that G' must be four-colorable. Thus the original graph was five-colorable.

We now describe a method for characterizing all six-chromatic toroidal graphs which are regular of degree six. Suppose G is a graph which is embedded on the torus and C is a cycle in G. C is said to be noncontractible or "nc" if C is not homotopic to a point. If C is an nc cycle in G we form from G an embedded planar graph $G(C)$ which is useful in generalizing the width of G as described below. We cut the torus along C and add two disks keeping a copy of C on each boundary: call these two copies C_1 and C_2. For $i = 1, 2$ let $N_i(C)$ denote the set of vertices in G and $G(C)$ which are not on C but which are adjacent to at least one vertex of C_i in $G(C)$. In G each $N_i(C)$ induces a cycle homotopic to C.

The width of G, $W(G)$, is defined to be the length of the shortest nc cycle in the embedded graph G [1]. If C is a minimum length nc cycle in G whose vertices are labeled $(X_1, \ldots, X_m)$, let $(X_1', \ldots, X_m')$ and $(X_1'', \ldots, X_m'')$ denote the corresponding vertices of C_1 and C_2, respectively. For $k = 1, \ldots, m$ let $d(X_k', X_k'')$ be the length of the shortest path in $G(C)$ joining X_k' to X_k''. This path will be an nc cycle in G which goes around the torus in a direction other than that of C. Set

$$W_C^*(G) = \underset{k}{\operatorname{Min}} \{d(X_k', X_k'')\} \tag{1}$$

and

$$W^*(G) = \underset{C}{\operatorname{Min}} \{W_C^*(G)\}. \tag{2}$$

Equation (2) defines the second width of G: the minimum is with respect to all nc cycles of minimum length. Thus we can assign to each toroidally embedded graph G an ordered pair of natural numbers $(W(G), W^*(G))$ called the widths of G. One can easily check that $(W(J), W^*(J)) = (3, 4)$. The widths are crucial to our discussion of six-chromatic graphs which are regular of degree six as shown by the following three theorems.

Theorem 2 *If G is regular of degree six and embedded on the torus with widths $(W(G), W^*(G))$, then $V \leq W(G) \cdot W^*(G)$.*

Comments The proof proceeds by induction on the second width. Small cases are treated by exhaustion. The inductive step is accomplished by identifying the vertices of C, a minimum length nc cycle, with those of $N_1(C)$ to obtain a new graph G'. Menger's theorem, topological facts about triangulations of the torus, and case arguments are used to show that we can arrange such a contraction so that G' will be regular of degree six, and at least one width of G' will be smaller than the corresponding width of G.

Theorem 3 *If G is a graph which is regular of degree six and embedded on the torus with $W(G)$ even, then G can be five-colored.*

Comments We sketch the proof of the simplest case. Suppose C is a minimum length nc cycle. Let $G-C$ denote the graph obtained from G by deleting all vertices of C and their incident edges and adding two vertices a_i with a_i adjacent to all vertices of $N_i(C)$ ($i = 1, 2$). Note that $G-C$ is a planar graph and hence can be four-colored. If a four-coloring of $G-C$ assigns the same color to a_1 and a_2 then at most three colors are used on the vertices of $N_1(C)$ and $N_2(C)$. There are then two colors left for C which is an even cycle.

If every four-coloring of $G-C$ assigns a_1 and a_2 different colors we alter the coloring of $G-C$. Suppose $G-C$ is four-colored using the colors 1, 2, 3, and 4 with a_1 colored 1 and a_2 colored 4. It must be the case that there is a Kempe chain joining vertices a_1 and a_2 using colors 1 and 4. If not, we could interchange colors in the component of the 1–4 subgraph of $G-C$ containing a_2. Since such chains exist we will break them as follows. In the four-coloring of $G-C$ look at every vertex which is colored 1 and is adjacent to some vertex of $N_1(C)$. Color each of these vertices with a 5. This must break every Kempe chain joining a_1 and a_2 with colors 1 and 4. We interchange colors in the component of the 1–4 subgraph of $G-C$ containing a_2. If no vertex on $N_2(C)$ is colored 5 we can extend the five-coloring to C using colors 1 and 5. If some vertex on $N_2(C)$ is colored 5 then a_1 and a_2 must be joined by a path of length three. In this case suppose $W(G) = M = 2K$. Then $W^*(G) \le 3 + K$ since the path from a_1 to a_2 together with no more than one-half of the vertices of C is a cycle in G whose length must be at least $W^*(G)$. Since $W^*(G) \ge W(G)$ we have $3 + K \ge W^*(G) \ge W(G) = 2K$. Hence $M \le 6$. Thus we have shown that if $W(G)$ is even and at least eight then G can be five-colored. The cases where $W(G)$ is four or six are similar but require more care.

Theorem 4 *If G is a graph which is embedded on the torus and regular of degree six with widths other than (3, 3), (3, 4), and (5, 5) then G can be five-colored.*

Comments The proof is similar to the proof of Theorem 3 but more intricate.

Thus we can combine Theorems 2 and 4 to show that if G is regular of degree six with more than 25 vertices then G can be five-colored.

Conjecture J is the only six-chromatic toroidal graph which is regular of degree six.

We turn now to the theorem which together with previously known results gives that Hadwiger's conjecture holds for all toroidal graphs.

Theorem 5 *If G is a six-chromatic toroidal graph then G contracts to K_6.*

Proof Suppose G is a counterexample to the theorem with the fewest number of vertices. G must be vertex critical and hence contains no vertex whose degree is less than five.

Case 1 G is not regular of degree six.

Suppose X is a vertex of degree five. If each pair of neighbors of X is adjacent then G contains K_6. Suppose a and b are neighbors of X which are not adjacent. Contract along the edges joining X to a and X to b to form a new vertex X' in a smaller graph G'. If G' contracts to K_6 so does G. If G' does not contract to K_6 then G' can be five-colored. A five-coloring of G' can be extended to a five-coloring of G by assigning to both a and b the color assigned X' in G'. At this point every vertex in G except X is colored using only five colors but the five neighbors of X use no more than four colors. Thus G can be five-colored.

Case 2 G is regular of degree six.

Suppose the neighbors of a vertex V when listed clockwise in the embedding are (a, b, c, d, e, f). Suppose $[a, c, e]$ is an independent set of vertices. Contract along the edges joining V to a, V to c, and V to e forming a new vertex V' in a smaller graph G'. If G' contracts to K_6 so does G. If G' does not contract to K_6 then G' can be five-colored. A five-coloring of G' can be extended to a five-coloring of G by assigning to a, c, and e the color assigned V' in G'. At this point every vertex in G except V is colored using five colors but the six neighbors of V use no more than four colors. Thus G can be five-colored.

If $[a, c, e]$ is not an independent set of vertices then some pair of these vertices, say a and c, are joined by an edge. Thus (a, c, V) form a 3-cycle. This 3-cycle is not a face boundary by our assumption about the embedding. This cycle must be nc since a contractible 3-cycle which is not a face boundary must contain in its interior a vertex of degree less than six. If (a, c, V) form an nc 3-cycle and G is six-chromatic then Theorem 4 tells us that G has widths (3, 3) or (3, 4). Theorem 2 tells us that G has no more than 12 vertices. Theorem 1 tells us that such a graph must be J, but J contracts to K_6. ∎

Proofs and related results are contained in [2].

Postscript

Theorem 4 has been strengthened to show that a six-chromatic graph on the torus which is regular of degree six must have widths (3, 3) or (3, 4). This result together with Theorem 1 proves the conjecture that J is the only six-chromatic toroidal graph which is regular of degree six. Hadwiger's conjecture has also been proved for graphs which embed on Klein's bottle and the projective plane.

REFERENCES

1. M. O. Albertson and J. P. Hutchinson, The independence ratio and genus of a graph, *Trans. Amer. Math. Soc.* **226** (1977) 161–173.
2. M. O. Albertson and J. P. Hutchinson, On six chromatic toroidal graphs, preprint.
3. K. Appel and W. Haken, Every planar map is four colorable, *Bull. Amer. Math. Soc.* **82** (1976) 711–712.
4. G. A. Dirac, Map color theorems related to the Heawood colour formula, *J. London Math. Soc.* **31** (1956) 460–471.
5. G. A. Dirac, Short proof of a map color theorem, *Canad. J. Math.* **9** (1957) 225–226.
6. G. A. Dirac, On the structure of 5- and 6-chromatic abstract graphs, *J. Reine Angew. Math.* **214** (1963) 43–52.
7. G. A. Dirac, Theorems related to the four colour conjecture, *J. London Math. Soc.* **29** (1954) 143–149.
8. H. Hadwiger, Über eine Klassifikation der Streckenkomplexe, *Vierteljschr. Naturforsch. Ges. Zürich* **88** (1943) 133–142.
9. K. Wagner, Bemerkungen zum Vierfarbenproblem, *Jber. Deutsch. Math.-Verein* **46** (1936) 26–32.
10. K. Wagner, Bemerkungen zu Hadwigers Vermutung, *Math. Ann.* **141** (1960) 433–451.

AMS 05C15, 05C10, 55A15

DEPARTMENT OF MATHEMATICS
SMITH COLLEGE
NORTHAMPTON, MASSACHUSETTS

Chromatic Polynomials and the Internal and External Activities of Tutte

RUTH A. BARI

1. Introduction

Let G be a connected graph, $P(G, \lambda)$ the chromatic polynomial of G, and $T(G; x, y)$ its Tutte polynomial, or dichromate.

It is known that $T(G; x, y)$ may be used to find the chromatic polynomial of G, and that, if G is planar, $T(G; x, y)$ will also yield the chromatic polynomial of the planar dual of G. Thus $T(G; x, y)$ generalizes $P(G; x, y)$.

In 1932, Whitney [4] characterized the coefficients of $P(G; x, y)$ in terms of certain spanning subgraphs of G. To simplify the counting of these subgraphs, he introduced the concept of broken circuits.

In this paper, we show how broken circuits and broken cocircuits may be used to define the coefficients of $T(G; x, y)$, and how to compute $P(G, \lambda)$ from $T(G; x, y)$ using Whitney's theorem on broken circuits.

2. Basic Definitions and Theorems

Let G be a connected (p, q)-graph, that is, a graph with p points and q lines, and label the lines of G with integers $1, 2, \ldots, q$. Then G is a *line-labeled* graph. In this paper, every graph G is line-labeled.

Definition 2.1 (a) Let Z be a cycle of G, and let m be the largest integer assigned to the lines of Z.

The path $P = Z - m$ is the *broken circuit* associated with Z. The line m is the *break* in the broken circuit P.

(b) Let Z^* be a cocycle of G, and let n be the largest label attached to the lines of Z^*.

The subgraph $P^* = Z^* - n$ is the broken cocircuit associated with Z^*. The line n is the *break* in the broken cocircuit.

Note A one-line circuit is a loop, and a one-line cocircuit is an isthmus or bridge.

It will be convenient to consider the loop a circuit break and an isthmus a cocircuit break.

In 1932, Whitney [4] proved the following theorem that characterized the coefficients of the chromial of a graph G.

Theorem 2.1 (**Whitney—Broken Circuits**) *Let G be a line-labeled (p, q)-graph. Then*

$$P(G, \lambda) = \sum_{i=0}^{p-1} (-1)^i m_i \lambda^{p-1},$$

where m_i is the number of i-line spanning subgraphs of G that contain no broken circuits.

Theorem 2.2 (**Planar Dual of Whitney's Theorem**) *Let G be a planar (p, q)-graph, and let G^* be the planar dual of G.*

If r is the number of regions of G, then

$$P(G^*, \lambda) = \sum_{i=0}^{r-1} (-1)^i n_i \lambda^{r-i},$$

where n_i is the number of i-line spanning subgraphs of G that contain no broken cocircuits.

Proof Since the regions of G correspond to the points of G^*, and the broken cocircuits of G correspond to the broken circuits of G^*, the theorem follows. ∎

Definition 2.2 Let G be a connected (p, q)-graph, and let t be a spanning tree of G.

(a) The deletion of a line i of t decomposes t into two components, X_i and Y_i. The line i is *internally active in t* if $i \geq r$ for all lines r that have one end in X_i, and the other in Y_i.

The *internal activity of t* is the number of internally active lines in t.

(b) Let $t^* = G - t$ be the cotree of t, and let j be a line of t^*. Then there is a unique cycle Z in $t + j$. The line j is *externally active in t* if $j \geq s$ for all lines s in Z.

The *external activity of t* is the number of externally active lines of t.

Definition 2.3 The *dichromate*, or *Tutte polynomial* of G is the function

$$T(G; x, y) = \sum_{i,j} a_{i,j} x^i y^j,$$

where $a_{i,j}$ is the number of spanning trees of G with internal activity i and external activity j.

The next two theorems express the internal and external activities of the spanning trees of G in terms of broken cocircuits and broken circuits, respectively.

Theorem 2.3 *Let G be a connected (p, q)-graph, t a spanning tree of G, $t^* = G - t$, the cotree of t. If i is a line of t, there is a unique cocycle Z^* in $t^* + i$.*

Then line i is internally active in t if and only if i is the break in the cocycle Z^*, or equivalently every line in the broken cocircuit $P^* = Z^* - i$ is in t^*.

Proof Since $i \in t$, $t - i$ has two components, X_i and Y_i. The set of lines with one end in X_i, the other in Y_i, is the unique cocycle Z^* in $t^* + i$.

But then i is internally active in t if and only if $i \geq r$ for all lines r in Z^*, that is, if i is the break in the cocycle Z^*.

Since i is the only line of $Z^* \cap t$, every line of the broken circuit $P^* = Z^* - i$ is in t^*. ∎

Theorem 2.4 *Let G be a connected (p, q)-graph, t a spanning tree of G, $t^* = G - t$ the cotree of t. If j is a line of t^*, then there is a unique cycle Z in $t + j$.*

The line j is externally active in t if and only if j is the break in the broken circuit $P = Z - j$, or equivalently if every line in the broken circuit P is in t.

Proof If $j \in t^*$, j is externally active in t if and only if $j \geq s$ for all lines s in Z, that is, if j is the break in the cycle Z of $t + j$.

Since j is the only line of $Z \cap t^*$, $P = Z - j$ is in t. ∎

Example 2.1 *The Internal and External Activities of a Tree* Let $G = K_4 - e$, labeled as shown in Fig. 2.1.

We compute the internal and external activities of the tree $t = 1$–3–4, where the symbol $H = e_1$–e_2–$\cdots$–e_r means that the spanning subgraph H has precisely the lines $e_1, e_2, \ldots, e_r$.

If $t = 1$–3–4, then $t^* = G - t = 2$–5. The lines of t are shown in boldface in Fig. 2.1.

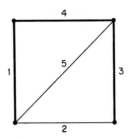

Figure 2.1 $G = K_4 - e$

A basis for the cocycles of G can be found by adding one line of t at a time to the lines of t^*. The cocycle Z_i^* is the unique cocycle in $t^* + i$, $i \in t$. $Z_{i,j}^*$ is the symmetric difference of Z_i^* and Z_j^*.

In the following table, if a broken cocircuit P_i^* is in t^*, it contributes 1 to the internal activity of t, 0 otherwise.

Cocycles	Broken cocircuits	Internal activity
$Z_1^* = 1\text{–}2\text{–}5$	$P_1^* = 1\text{–}2$	0
$Z_3^* = 2\text{–}3$	$P_2^* = 2$	1
$Z_4^* = 2\text{–}4\text{–}5$	$P_3^* = 2\text{–}4$	0
$Z_{1,3}^* = 1\text{–}3\text{–}5$	$P_4^* = 1\text{–}3$	0
$Z_{1,4}^* = 1\text{–}4$	$P_5^* = 1$	0
$Z_{3,4}^* = 3\text{–}4\text{–}5$	$P_6^* = 3\text{–}4$	0
$Z_{1,3,4}^* = 1\text{–}2\text{–}3\text{–}4$	$P_7^* = 1\text{–}2\text{–}3$	0
		$i = 1$

A basis for the cycles of G can be found by adding one line of t^* at a time to t. The cycle Z_i is the unique cycle of G in $t + i$. $Z_{i,j}$ is the symmetric difference of Z_i and Z_j.

In the following table, if a broken circuit P is in t, it contributes 1 to the external activity of t, otherwise 0.

Cycles	Broken circuits	External activity
$Z_2 = 1\text{–}2\text{–}3\text{–}4$	$P_1 = 1\text{–}2\text{–}3$	0
$Z_5 = 1\text{–}4\text{–}5$	$P_2 = 1\text{–}4$	1
$Z_{2,5} = 2\text{–}3\text{–}5$	$P_3 = 2\text{–}3$	0
		$j = 1$

Thus $t = 1\text{–}3\text{–}4$ contributes 1 to $a_{1,1}$ in $T(G; x, y)$.

3. A Recursive Definition of the Tutte Polynomial

In his paper, Codichromatic Graphs [3], Tutte introduces the dichromate with the following recursive definition.

Definition 3.1 With any graph G, there is associated a polynomial $T(G; x, y)$ in two variables x and y having the following properties I and II.

I. If G has l loops, i isthmuses, and no other edges, then
$$T(G; x, y) = x^i y^l.$$

II. If A is an edge of G that is neither a loop not an isthmus, then
$$T(G; x, y) = T(G_A'; x, y) + T(G_A''; x, y).$$

Here G_A' is the graph obtained from G by deleting edge A, and G_A'' the graph obtained from G by contracting A, with its two ends, to a single vertex.

Theorem 3.1 *Let G be connected. Then $T(G; 1, 1)$ counts the number of spanning trees of G.*

Proof Let $\tau(G)$ denote the number of spanning trees of G.

1. If every line of G is a loop, G has exactly one point, and the only spanning tree of G is the trivial tree.
Hence $\tau(G) = T(G; 1, 1) = 1$.
2. If every line of G is an isthmus, G is a tree, so $\tau(G) = T(G; 1, 1) = 1$.
3. If G has a line e that is neither a loop nor an isthmus, and t is a spanning tree of G, then either T does or does not contain line e.

If t contains line e, there is a spanning tree t_e'' in G_e'' corresponding to t, where t can be derived from t_e'' by restoring line e. Thus $\tau(G_e'') = (T(G_e''; 1, 1)$.

If t does not contain line e, then t is a tree of $G_e' = G - e$. Thus $\tau(G_e') = T(G_e'; 1, 1)$.

It follows immediately that
$$\tau(G) = \tau(G_e') + \tau(G_e'') = T(G_e'; 1, 1) + T(G_e''; 1, 1)$$
$$= T(G; 1, 1). \qquad \blacksquare$$

If G is a connected, line-labeled graph, the recursion formula for finding $\tau(G) = T(G; 1, 1)$ can be used to obtain a list of all spanning trees of G, and of the broken circuits and cocircuits in each spanning tree.

Starting with the graph G, we construct a binary tree of graphs, as follows:

1. The first entry is the graph G. Let $L(G)$ be the set of lines of G that are neither isthmuses nor loops.
If $L(G) = \emptyset$, stop.
If $L(G) \neq \emptyset$, let e be the line of $L(G)$ with the smallest label.

2. The second line of the binary tree of graphs consists of two graphs, G_e' and G_e''.

To indicate that the spanning trees of G_e' represent the spanning trees of G that *do not contain* line e, place the symbol $\widehat{e}$ near G_e'. Thus if t' is a spanning tree of G_e', the symbol $\widehat{e}$ tells us that the line e is not in the corresponding spanning tree t of G, but is in $t^* = G - t$.

To indicate that the spanning trees of G_e'' represent the spanning trees of G that *contain* line e, place the symbol $\boxed{e}$ near G_e''. Thus, if t'' is a spanning tree of G_e'', the symbol $\boxed{e}$ tells us that e is a line of the corresponding spanning tree t of G.

3. If $L(G_e') = \emptyset$, delete and contract the line of $L(G_e')$ with the smallest label.

If $L(G_e'') \neq \emptyset$, delete and contract the line of $L(G_e'')$ with the smallest label.

4. Whenever a graph H is reached such that $L(H) = \emptyset$, stop.

Denote by K the collection of graphs H_i, such that $L(H_i) = \emptyset$, derived from G by the procedure described above. The collection K has the following properties:

1. If $H_i \in K$, every line of H_i is either an isthmus or a loop, since $L(H_i) = \emptyset$.

2. There is a one-to-one correspondence between the graphs in K and the spanning trees of G, by Theorem 3.1.

3. If $H_i \in K$ corresponds to a tree t_i of G, then the lines of t_i are represented in H_i by the isthmuses of H_i and the lines denoted by a symbol $\boxed{e_j}$ near H_i.

The cotree $t_i^* = G - t_i$ is represented by the loops of H_i and the lines denoted by a symbol $\widehat{e_k}$ near H_i.

The last line is the list of spanning trees of G. The trees with a single underline have no loops. Those with a double underline have no isthmuses.

Example 3.1 Spanning trees of the labeled graph $G = K_{4-e}$ are shown in Fig. 3.1.

If a line e of G is an isthmus or a loop, it will obviously be represented by an isthmus or a loop in G_e' and in G_e'', as well as in every graph derived from G in the collection K.

If e is an isthmus of G, e is a one-line cocycle of G. If e is a loop of G, e is a one-line cycle of G. In either case, $e \geq r$ for all lines r in the cocycle or cycle, so e is a cocircuit break or a circuit break.

In the next theorem, we prove that if e is represented by an isthmus or a loop in any graph H of K, e is a cocircuit break or a circuit break.

CHROMIALS AND INTERNAL AND EXTERNAL ACTIVITIES 47

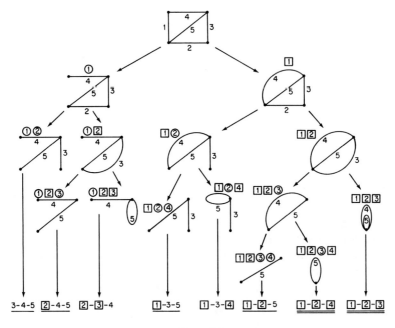

Figure 3.1

Theorem 3.2 *Let G be a connected graph, and let $H \in K$ represent the tree t of G. Then*

1. *a line i is an isthmus of H if and only if i is the break in a broken co-circuit of G, and*
2. *a line e is a loop of G if and only if e is the break in a broken circuit of G.*

Proof 1. If i is an isthmus of H, $i \in t$. Hence $t - i$ has two components X_i and Y_i, and there is a unique cocycle Z^* in $t^* + i$ whose lines have one end in X_i, the other in Y_i.

A line i is an isthmus of H if it is the only line of Z^* in H. Since the lines of Z^* are deleted in increasing numerical order, this can occur if and only if $i \geq r$ for all $r \in Z^*$.

Thus i is an isthmus of H if and only if i is the break in the cocircuit Z^*.

2. If a line e is a loop of H, $e \in t^* = G - t$, and there is a unique cycle Z in $t + e$.

A line e of Z is a loop of H if and only if it is the only line of Z in H. Since the lines of Z are deleted and contracted in increasing numerical order, this can occur if and only if $e \geq r$ for all lines r in Z, so e is a loop of H if and only if e is the break in the cocircuit Z. ∎

Corollary *The line i is internally active in a tree t of G if and only if it is represented by an isthmus in H.*

The line e is externally active in a tree t of G if and only if it is represented by a loop in H.

Proof Apply Theorems 2.4 and 2.5. ∎

Definition 3.2 Let t be a tree of G. If i is a line of t, and i is not internally active in t, then i is *internally passive* in t.

If e is a line of $t^* = G - t$, and e is not externally active in t, then e is *externally passive* in t.

Example 3.2 In Fig. 3.1, the last two lines give the spanning trees and cotrees of G.

The lines that are internally passive are those whose labels are enclosed in squares in the list of trees of G.

The lines that are externally passive are those whose labels are encircled in the list of cotrees of G.

4. Chromials and Spanning Trees

Whitney's theorem on broken circuits, Theorem 2.1, expresses the chromial of G in terms of the spanning subgraphs of G that contain no broken circuits.

The fact that a spanning tree of G is a maximal acyclic subgraph of G suggests that we may express $P(G, \lambda)$ in terms of the spanning trees of G that contain no broken circuits, that is, trees with $j = 0$.

We find the spanning trees of G by applying the recursive definition of $T(G; 1, 1)$. At each step, we partition the spanning trees of a graph into those that do not contain a specified line i, and those that contain line i. Line i is then internally passive in every tree in the latter set.

To count the spanning subgraphs of G that contain no broken circuits, we partition these subgraphs in a similar manner.

If line i is passive in a tree t, every spanning subgraph S of G chosen from t *must* contain line i. On the other hand, if i is active in t, S may or may not contain line i.

Example 4.1 In Example 3.1, the spanning trees 3–4–5 and $\boxed{2}$–4–5 have external activity 0.

The spanning subgraph 4–5 *may* be chosen from 3–4–5, but *may not* be chosen from $\boxed{2}$–4–5, since every subgraph chosen from $\boxed{2}$–4–5 *must* contain line 2.

Example 4.2 In Example 3.1, the spanning tree $t = \boxed{2}\text{-}4\text{-}5$ has internal activity 2 and external activity zero.

Since t has one passive line, every spanning subgraph of G chosen from t contains that line.

Thus the number of 2-line spanning subgraphs of G we may choose from t is $\binom{2}{1}$, since we must include line $\boxed{2}$ and may include either of the lines 4 or 5.

Theorem 4.1 *Let G be a connected (p, q)-graph, and let $P(G, \lambda)$ be the chromial of G. Then*

$$P(G, \lambda) = \sum_{s=1}^{p-1} (-1)^r a_{s,0} \lambda(\lambda - 1)^s,$$

where $a_{s,0}$ is the number of spanning trees of G with internal activity s and external activity zero, and $r = p - 1 - s$ is the number of internally passive lines in a tree with s internally active lines.

Proof By Whitney's theorem on broken circuits,

$$P(G, \lambda) = \sum_{i=1}^{p-1} (-1)^i m_i \lambda^{p-i},$$

where m_i is the number of spanning subgraphs of G that have i lines and no broken circuits.

If t is a spanning tree of G with external activity zero, then t contains no broken circuits.

Let t have r internally passive lines and s internally active lines. Then we choose from t every spanning subgraph of G with $r + l$ lines, $0 \le l \le s$, containing all of the r passive lines of t and any selection of l lines of t chosen from the s active lines.

Thus the contribution of t to $P(G, \lambda)$ is

$$(-1)^r \left[\binom{s}{0} \lambda^{p-r} - \binom{s}{1} \lambda^{p-r-1} + \cdots + (-1)^s \binom{s}{s} \lambda^{p-r-s} \right].$$

Since t has $p - 1$ lines, $r + s = p - 1$, so $p - r = s + 1$, and $p - r - s = 1$. Therefore t contributes

$$(-1)^r \left[\binom{s}{0} \lambda^{s+1} - \binom{s}{1} \lambda^s + \cdots + (-1)^s \binom{s}{s} \lambda \right] = (-1)^r \lambda(\lambda - 1)^s.$$

Every spanning subgraph of G with no broken circuits can be chosen from the spanning trees of G with external activity zero. If a tree t has external activity zero and internal activity s, $0 \le s \le p - 1$. Hence

$$P(G, \lambda) = \sum_{s=1}^{p-1} (-1)^r a_{s,0} \lambda(\lambda - 1)^s. \qquad \blacksquare$$

Example 4.3 From Example 3.1, we list the spanning trees of $G = K_4 - e$ with external activity zero in the following tabulation.

tree	s = internal activity	$r = 3 - s$
$t_1 = $ 3–4–5	3	0
$t_2 = $ ⊡–3–5	2	1
$t_3 = $ ⊡–4–5	2	1
$t_4 = $ ⊡–⊡–5	1	2

These trees partition the spanning subgraphs of G with external activity zero into four classes, $C_1, C_2, C_3,$ and C_4.

Let S be a spanning subgraph of G with external activity zero. Then

$S \in C_1$ if S does not contain lines 1 or 2,
$S \in C_2$ if S contains line 1 but not line 2,
$S \in C_3$ if S contains line 2 but not line 1,
$S \in C_4$ if S contains both line 1 and line 2.

In each case, $S \in C_i$ if and only if S is a subgraph of t_i and S contains every internally passive line of t_i, $i = 1, 2, 3,$ or 4.

We use Theorem 4.1 to compute $P(G, \lambda)$. From the table of trees above, we see that $a_{3,0} = 1, a_{2,0} = 2,$ and $a_{1,0} = 1$. Hence

$$P(G, \lambda) = \lambda(\lambda - 1)^3 - 2\lambda(\lambda - 1)^2 + \lambda(\lambda - 1)$$

$$= \binom{3}{0}\lambda^4 - \binom{3}{1}\lambda^3 + \binom{3}{2}\lambda^2 - \binom{3}{3}\lambda$$

$$- 2\binom{2}{0}\lambda^3 + 2\binom{2}{1}\lambda^2 - 2\binom{2}{2}\lambda$$

$$+ \binom{1}{0}\lambda^2 - \binom{1}{1}\lambda$$

$$= \lambda^4 - 5\lambda^3 + 8\lambda^2 - 4\lambda.$$

A similar computation yields $P(G^*, \lambda)$, where G^* is the planar dual of G. From Example 3.1, we list the spanning cotrees of $G = K_4 - e$ with internal activity zero in the following tabulation.

CHROMIALS AND INTERNAL AND EXTERNAL ACTIVITIES

Cotree	j = external activity	$k = 2 - j$
4–5	2	0
③–5	1	1

Hence $a_{0,2} = 1$, $a_{0,1} = 1$, and

$$P(G^*, \lambda) = \lambda(\lambda - 1)^2 + \lambda(\lambda - 1)$$

$$= \binom{2}{0}\lambda^3 - \binom{2}{1}\lambda^2 + \binom{2}{2}\lambda$$

$$\quad - \binom{1}{0}\lambda^2 + \binom{1}{1}\lambda$$

$$= \lambda^3 - 3\lambda^2 + 2\lambda.$$

Postscript

We have seen that the Tutte polynomial $T(G; x, y)$ counts the number of circuit breaks and cocircuit breaks in the spanning trees of G. This suggested to me that Tutte may have introduced the dichromate to generalize chromatic polynomials and Whitney's theorem on broken circuits.

Since I am interested in the intuitions that lead to mathematical discoveries or inventions, I asked Tutte whether or not this was the case. It was interesting to learn that it was *not* his intent.

After working with the deletion–contraction algorithm, Tutte decided to use this algorithm to associate a polynomial with a graph G. This led to the recursive definition 3.1 for $T(G; x, y)$.

When he applied this method to several small graphs, he found that if $G = K_3$, a triangle,

$$T(G; x, y) = x^2 + x + y,$$

that is, the contribution of each spanning tree of G is different, although the trees of K_3 are isomorphic to one another.

To explain the different contributions of these trees, Tutte labeled the lines of G and defined the internal and external activities of a tree. He then proved that $T(G; x, y)$ is independent of the labeling.

Not only does $T(G; x, y)$ yield the chromial of G and of the planar dual G^* of G if G is planar, but it also gives such additional information as the number of subgraphs of G and the number of spanning forests of G.

REFERENCES

1. W. T. Tutte, A contribution to the theory of chromatic polynomials, *Canad. J. Math.* **6** (1953) 80–91.
2. W. T. Tutte, On dichromatic polynomials, *J. Combinatorial Theory* **2** (1967) 301–320.
3. W. T. Tutte, Codichromatic graphs, *J. Combinatorial Theory Ser. B* **16** (1974) 168–174.
4. H. Whitney, A logical expansion in mathematics, *Bull. Amer. Math. Soc.* **38** (1932) 572–579.

AMS 05C15

DEPARTMENT OF MATHEMATICS
GEORGE WASHINGTON UNIVERSITY
WASHINGTON, D.C.

Chromatic Equations

ARTHUR BERNHART

1. Basic Counting Theorems

Cubic Graphs A map drawn on any surface, of Euler characteristic G, with R regions, A arcs, and P points satisfies Euler's formula $R - A + P = G$. This equation may be presented as the 2×2 array

$$\boxed{\begin{array}{cc} R & A \\ G & P \end{array}},$$

where the diagonal sum $R + P$ counting even dimensional objects equals the sum $A + G$.

In the case of a polyhedron $G = 2$, and central projection converts it into a combinatorially equivalent global map, and both can be transformed into a plane graph by stereographic projection. The common difference $R - G = H = A - P$ may be called the *height* of the graph. For example, in the case of the Egyptian pyramid these parameters have the values $G^2 R^5 A^8 P^5 H^3$.

An algebraic equivalent is the face adjacency matrix M_{ik} with R rows and R columns in which $M_{ik} = 1$ whenever regions R_i and R_k are adjacent. Otherwise $M_{ik} = 0$. Duals like the *cube* $G^2 R^6 A^{12} P^8 H^4$ and the *octahedron* $G^2 R^8 A^{12} P^6 H^6$ interchange R and P values.

Let R_n be the number of n-gons; and let P_v be the number of vertices with valence v. Then the census $\sum nR_n = N$ over all $R = \sum R_n$ regions counts each edge twice and each vertex v times. A survey $\sum vP_v$ over all $P = \sum P_v$ vertices also counts each edge twice and each polygon n times. Hence

$$2A = \sum nR_n = \sum vP_v = N.$$

The average values are $\bar{n} = N/R$ and $\bar{v} = N/P$. For the pyramid we have $N = 4 + 3 + 3 + 3 + 3 = 16$ so that $\bar{n} = \frac{16}{5} = \bar{v}$. If all vertices have the same valence then $N = vP$; if all polygons have the same order $N = nR$. The cube has both these properties. In a *cubic* map each vertex has valence $v = 3$ so that

$$2A = N = 3P = 6H.$$

Since N is divisible by both 2 and 3, therefore $N/6$ is an integer n, the height of the graph. The cubic parameters $G^2 R^{n+2} A^{3n} P^{2n} H^n$ are readily verified for a regular prism, which has two n-gon bases and n rectangular walls.

Any cubic can be reduced to a prism by a succession of *switch* transformations [9] which replace

In the dual representation we replace

which can be described as switching diagonals in the diamond. Two cubics G_1 and G_2 may be called switch adjacent if G_1 can be converted to G_2 by a single switch transformation.

The average value $\bar{n}$ over all n-gons is given by

$$\bar{n} = 6 - 12/R.$$

Thus $\bar{n}$ is always less than six but the difference is small if R is large. Considering $n = 6$ as *par* for polygons, a pentagon is one below par, while an octagon is two above par. Such deviations from par are proportional to what Heesch [11] calls the *charge* on the polygon. Heesch, Stromquist, Allaire, Máyer, and others choose the proportionality constant so as to avoid fractions, and usually change the sign so that each pentagon has a positive charge. In our notation the total charge for the whole map is -12.

CHROMATIC EQUATIONS 55

Kempe [12] showed that a cubic map which avoids all minor polygons with $n < 5$ must contain at least 12 pentagons. Haken and Appel [1] argue that if it could avoid each configuration in a specified list U, then $\bar{n}$ would exceed six.

Edge Partitions The formula $A = 3H$ challenges us to partition the edges into three classes, with H edges in each class. Such a partition is easy to obtain for the cube. Its 12 edges fall quite naturally into three classes according as they are parallel to the x, y, or z axes. Indeed at each vertex there is one representative from each class. This is called a 3-coloring.

Can such edge partitions be guaranteed for arbitrary cubics? The answer seems to be yes (Appel et al. [1]) but a simple rule for doing it is hard to find.

The plane dual of a cubic map is a triangulated graph in which any tour through every region of the map becomes a spanning cycle. Whitney [18] proposed a *normal* form for such Hamiltonian graphs. Starting with a regular polygon of $n + 2$ vertices, one can triangulate it by drawing $n - 1$ inside diagonals and also $n - 1$ outside arcs. If we mark the inside, onside, and outside edges with *scores* 5, 6, 7, respectively, and regard them as lengths, then every Whitney triangle is isosceles. To get a 3-coloring, however, every triangle must be scalene.

Consider the special case when the map is a prism. The lateral faces determine another cycle having only n nodes. By inserting nodes B and B' representing the two base polygons, and by changing only six scores (incidentally eliminating equilateral triangles), we can get a triangulation which is *strictly* isosceles. Finally, by alternating the scores 5, 7 in the Chvatal fans issuing from B and B' a genuine 3-coloring is achieved.

Can this prism algorithm be generalized to fit other cubics? There is a switch path connection which does not break the Whitney cycle, but unfortunately even one diagonal switch may introduce isosceles triangles. So, although each arbitrary cubic may be derived from a prism by switch transformations, these canonical derivations do not provide an edge 3-coloring.

Vertex Partitions The formula $p = 2n$ challenges us to divide the vertices into two groups of n each. The eight vertices of a cube are readily divided into two classes, marked $+1$ and -1, so that any two vertices incident on the same edge have opposite signs. But such a bipartitite division of the vertices is possible iff each n-gon is even. Accordingly the 2-coloring of vertices is either impossible or trivial.

Kempe had observed that a pyramid can be reduced to a cube by cutting off the corner whose valence is four. For any polyhedron cutting a corner with valence v adds one region, v arcs, and $v - 1$ points, while the Euler parameter $R - A + P$ is unchanged. If one cuts a corner where the valence is already three, a triangular region is added; if one cuts k corners in a cubic graph of height H we obtain another cubic graph of height $H + k$. If four alternate corners of a cube are cut, the new polyhedron has four triangles and six hexagons. Its edges can be 3-colored by scoring them 5, 6, 7 around each polygon. Shrinking each triangle to a point produces a 3-coloring on the original edges. Heawood's problem for an arbitrary cubic is: Which corners should be cut so that the order of each polygon becomes divisible by 3?

If the edges have already been 3-colored, Heawood [10] assigns the index -1 to any vertex whenever the scores 5, 6, 7 of its incident edges have a clockwise orientation, otherwise the index is $+1$. This index is therefore the *mod* 3 change in edge scoring, as one proceeds clockwise around the perimeter of any polygon. After all edges of a polygon are traversed the cumulative mod 3 increment must be 0.

Thus for a quadrilateral the indices must be $(1, 1, -1, -1)$ in some permutation. For a pentagon one index must be different from the other four. For a hexagon there are three options: (a) all six indices $+1$, (b) each -1, (c) three $+1$, three -1. As already noted, if each face of a polyhedron is an n-gon where n is divisible by 3, it may be trivially colored by using the same orientation at each vertex. In that case it is immaterial whether we cut every corner, or none of them.

Each 3-coloring of edges corresponds to a 4-coloring of the regions. Pick any region in the map and assign it one of the four colors 1, 2, 3, 4. Let the colors x and y assigned to adjacent regions satisfy the *compatibility* equation

$$x + y = c,$$

where either c or its ten's complement $10 - c$ is the score s of the separating edge. (s may be called the *chromatic sum* of x and y.)

Put $s = 5$ when the sum $x + y$ is either $4 + 1$ or $2 + 3$;
let $s = 6$ when the sum is either $4 + 2 = 6$ or $1 + 3 = 10 - 6$; and
set $s = 7$ when the sum is either $4 + 3 = 7$ or $1 + 2 = 10 - 7$.

Conversely, any 4-coloring of the regions immediately determines a 3-coloring of the edges.

For each selected value of s any two colors whose chromatic sum is s are *s-conjugates*. Thus (1, 3) are 6-conjugates, forming the compatible triple $(x, s, y) = (1, 6, 3)$.

The foregoing remarks indicate that three problems are equivalent: (1) to assign one of the two indices $(+1; -1)$ to each of the $2n$ vertices (subject to the mod 3 condition); (2) to assign one of three scores (5, 6, 7) to each of $3n$ edges; or (3) to assign one of four colors (1, 2, 3, 4) to each of $n + 2$ regions. Our analysis uses the third approach, but employs insights gained from the other formulations.

CHROMATIC EQUATIONS

2. The Derivation of Chromatic Equations

Scope of Our Problem In the quantitative analysis of n-ring configurations we introduce two kinds of variables: *elementary* frequencies $\{x_k\}$ and *isotopic* parameters $\{u_i\}$. Each of these variables is a non-negative integer, since it counts the number of ways certain regions may be colored subject to specified restrictions. The number of elements L is a function of the ring order n.

n:	4	5	6	7	8	9	10
$L(n)$:	4	10	31	91	274	820	2461

Each element is the sum of certain associated isotopes

$$x_k = \sum_{i=1}^{J} u_i, \qquad \$(x, u)$$

$$u_i \geq 0.$$

The entire system $\$(x, u)$ breaks up into a series of subsystems of various orders of complexity p. Let $m = 2^{p-1}$. Each subsystem $\$$ of order p involves m^2 elements. The number J of u-terms on the right side of $\$$ depends on the order.

p:	0	1	2	3	4	5
$J(p)$:	1	1	2	5	14	42
m^2:			4	16	64	256
$I = mJ$:			4	20	112	672

The accompanying table has been extended back to $J = 0$ in order to show how the J values can be generated recursively. The scalar product of the vector $[J_0 J_1 \cdots J_{p-1}]$ with its palindrome $[J_{p-1} \cdots J_1 J_0]$ yields J_p. Thus

$$J_2 = [1, 1] \cdot [1, 1] = 2, \qquad J_3 = [1, 1, 2] \cdot [2, 1, 1] = 5,$$

and $\quad J_4 = [1, 1, 2, 5] \cdot [5, 2, 1, 1] = 14.$

The m^2 elements can be arranged in an $m \times m$ chromatic matrix (or box) together with rules for identifying the I isotopes. The total number of $m \times m$ boxes is the binomial coefficient $C(n, 2p)$. Thus, for the study of 7-rings there are $C(7, 4) = 35$ 2×2 boxes; and there are $C(7, 6) = 7$ 4×4 boxes. In a 10-ring there are 210 2×2 boxes, 210 4×4 boxes, 45 8×8 boxes, and one 16×16 box. The table indicates, for example, that each 4×4 box involves 16 elements and 20 isotopes, where every x_k is the sum of five isotopes, and each isotope u_i belongs to four different elements.

On eliminating the parameters u_i from the subsystem $\$(x, u)$ we obtain implicit equations

$$\sum_{i=1}^{m} x_i = \sum_{j=1}^{m} x_j \qquad \$(x)$$

which are linear homogeneous in the elementary variables. Thus the x_i are not independent.

The chromatic equations $\$(x)$ may be compared with the boundary value problems of physics, particularly when the differential equations are approximated by linear difference equations. We are invited to prepare a collection of eigensolutions ϕ_i and then to satisfy additional constraints by an appropriate linear combination $\sum c_i \phi_i$. The open sets of Tutte exploit this analogy.

However, a set of natural numbers $x_i \geq 0$ which satisfy $\$(x)$ will not necessarily lead to a solution of the parametric system $\$(x, u)$. The inequalities $u_i \geq 0$ have yet to be reckoned with. One needs to verify that the linear program $\$(x, u)$ is *feasible*. Such algorithms are well known. We postpone further comments on the solution of system $\$$ to the last section. Our current task is to show how the chromatic equations are derived.

Coloring n-Rings Kempe had considered a polygon G_0 surrounded by its ring of cyclically adjacent neighbors. Birkhoff [6] noticed that the analysis is much the same if G_0 is a *cluster* of k regions, such as 5–555, for any value of k. Even $k = 0$ can be included. Let the cluster be bounded by a cycle $R_1 R_2 \cdots R_n$ of n regions where each R_i adj R_{i+1} and R_n adj R_1. If these are the only adjacencies between the n regions the ring is called *proper*.

We first assign colors to each of the n ring regions and label the particular coloring scheme in some recognizable way. An actual listing of the string of n colors is unambiguous; but it is more convenient to tabulate all relevant schemes and tell them apart by their place in the table. In 10-ring analysis there are 2461 different schemes. Let (k) indicate one of these. Regarding the colors assigned by (k) to the ring regions as fixed boundary conditions, let x_k count the number of different ways that the interior cluster can be colored. Clearly x_k is a non-negative integer which vanishes if there is no way to color the interior regions subject to the given boundary conditions.

As a start let us focus our attention on four 10-ring schemes:

$$\#76 = 12 \quad 1213 \quad 4134, \qquad \#83 = 12 \quad 1213 \quad 4234,$$
$$\#78 = 12 \quad 1213 \quad 4143, \qquad \#85 = 12 \quad 1213 \quad 4243.$$

Scheme $\#85$ identifies the boundary condition $R_1 = R_3 = R_5 = 1$; $R_2 = R_4 = R_8 = 2$; $R_6 = R_{10} = 3$; $R_7 = R_9 = 4$. The elementary variable x_{85} is the number of different ways in which the interior regions can be colored.

CHROMATIC EQUATIONS

We propose to show that both the inside and the outside of any 10-ring configuration satisfies the chromatic equation

$$x_{76} + x_{85} = x_{78} + x_{83}.$$

In particular consider the inside of the *oriole* cluster 7–5665. (The name "oriole" was coined by my son Frank who had discovered that the "redwing" cluster 7–5565 was reducible. Starting with Winn's 7–5555 cluster, these form a ramp of 8-ring, 9-ring, and 10-ring configurations.) Figure 1a shows a

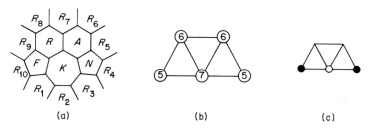

Figure 1 Oriole, (a) Map configuration with five interior polygons. (b) Dual representation with five nodes. (c) Heesch notation.

7-gon K, two adjacent hexagons R and A, flanked by two pentagons F and N. Figure 1b represents each polygon by a point (graphically indicated as a dot, cross, or small circle). The numbers 7–5665 indicate the number of incident edges at each node.

Since the adjacencies in the oriole cluster are known the value of x_{85} can be found by actual counting in Fig. 2. The choice $R = 3$ forces $K = 4$,

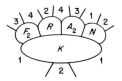

Figure 2

$N = 3$ but the choice $R = 1$ allows $K = 4$, $N = 3$ and also $K = 3$, $N = 4$. We have found three valid colorings for the interior regions FRANK, namely 21234, 21243, and 23234. Therefore $x_{85} = 3$. In like manner we find $x_{76} = 0$, $x_{78} = 1$, and $x_{83} = 2$. These integers clearly satisfy the chromatic equation

$$\begin{aligned} x_{76} - x_{78} &= x_{83} - x_{85} \\ 0 - 1 &= 2 - 3 \end{aligned},$$

an identity which may be displayed as the 2 × 2 matrix

$$\$078 = \begin{array}{|cc|} \hline 76 & 83 \\ 78 & 85 \\ \hline \end{array}$$

whose diagonal sums $x_{76} + x_{85}$ and $x_{78} + x_{83}$ are equal. (Curiously the subscripts often satisfy the same chromatic equation!)

It is worth noting that from the knowledge that oriole is *colorable* subject to each of the boundary conditions #78, #83, and #85 we cannot predict the result for #76. But the chromatic equation predicts that x_{76} must vanish.

The notation $078 indicates that this is equation number 78 among the 210 chromatic equations of 2 × 2 type. Another example is

$$\$080 = \begin{array}{|cc|} \hline 87 & 89 \\ 88 & 90 \\ \hline \end{array}$$

corresponding to the schemes

#87 = 12 1213 4313,   #89 = 12 1213 4323,
#88 = 12 1213 4314,   #90 = 12 1213 4324.

Since oriole is not colorable using schemes #87 or #88 we write $x_{87} = x_{88} = 0$ and conclude that $x_{89} = x_{90} = c$. By actual count we find that $c = 1$.

Chromatic Matrices When the interior structure is unknown the elementary frequencies x_k are likewise unknown, but they must satisfy certain linear homogeneous equations. In particular we shall show that

$$x_{87} + x_{90} = x_{88} + x_{89}.$$

This equation will usually be abbreviated as a 2 × 2 chromatic box

$$\$080 = \begin{array}{|cc|} \hline \#87 & \#89 \\ \#88 & \#90 \\ \hline \end{array}.$$

In a running text we occasionally employ the Thomas notation 87 + 90 = 88 + 89, or the multiplicative notation $\frac{87}{88} = \frac{89}{90}$.

Each color scheme generates three chromatic matrices, one for each edge score s, unless the corresponding path order is less than two. The color

CHROMATIC EQUATIONS

scheme #87 = 12 1213 4313 determines ten edge scores. For each pair of consecutive colors the simple sums are indicated by subscripts:

$$1_3 2_3 \quad 1_3 2_3 1_4 3_7 \quad 4_7 3_4 1_4 3_4.$$

But edge scores are the chromatic sums:

$$1_7 2_7 \quad 1_7 2_7 1_6 3_7 \quad 4_7 3_6 1_6 3_6.$$

Designate one score, say, $s = 7$, as *neutral* and consider the other two scores as *active*. Drop neutral subscripts and replace active scores by a dot. We find that the cyclic coloring is separated into four periods 12121.343.1.3. which Kempe thinks of as belonging to chains using the color pair (1, 2) or (3, 4). Let us rather look at edges. Active edges may be *directed* by keeping the color 4 (and its *s*-conjugate) on the right. The p directed paths alternately enter and exit the cluster, with their $2p$ terminals located where we have placed dots. It is precisely the cubic property of the map which guarantees that these paths will never fork nor dead-end inside the ring. Each vertex has three edges: one enters, another exits, the third is neutral.

Knowing the ring colors determines the entrances and exits but not which belong to the same path. This isotopic information depends on how the internal cluster has been colored. The set of all cluster colorings compatible with a particular boundary scheme A are partitioned into isotopes A^1, A^2, etc., according to how the internal regions are colored. We put two colorings in the same isotope whenever the entering and exiting edges are paired off in the same way. Each elementary frequency x_k is the sum of its isotopic frequencies x_{ki}.

The resolution of scheme #87 into four periods can be abbreviated 00001.003.1.3. saving only the last color in each period as significant. The occurrence of these four significant digits in the 10-ring is just one of $C(n, 4) = 210$ ways to split a 10-ring into 4 periods. Each of these will lead to a 2×2 chromatic equation, here labeled \$001–\$210. A more scientific label is provided by the pattern 0000100110 which actually identifies which ring edges are active. This binary number can be written more compactly as an octal \$(46). There are $C(n, 6) = 210$ six period combinations, each leading to a 4×4 box with equations \$301–\$510. There are 45 8×8 boxes \$601–\$645; and one 16×16 box \$800. Only the binomial coefficients $C(n, r)$ where r is even are needed. The number $p = r/2$ is the path order. When there is only one path no chromatic equations are determined. If the score s produces $p(s)$ paths we have

$$p(5) + p(6) + p(7) = n.$$

For element #87 we have $2 + 3 + 5 = 10$.

2 × 2 Boxes By retaining only the significant digits the #87 pattern of four periods can be further abbreviated to 1313. Place this residual pattern in any slot of a 2 × 2 array. By interchanging conjugate colors (1, 2) or (3, 4) we fill the other slots: change the third period, 1313 → 1323, and put the result in another column. Change the fourth period, 1313 → 1314, and put the result in another row. Making both changes completes the array pattern

1313	1323		A	B		#87	#89
1314	1324		C	D		#88	#90

Such 2 × 2 chromatic matrices are produced whenever there are precisely two paths and four periods.

We have yet to show that $A + D = B + C$. The proof is due to Kempe [12]. Let the four periods be P.Q.R.S. Either the first period P is connected to the third period R by a chain of neutral edges (odd isotope) or the second period Q is connected to the fourth period S (even isotope). Indicating the frequency of isotope number i by (i) we have

$$A = (1) + (2) \qquad B = (1) + (4)$$
$$C = (3) + (2) \qquad D = (3) + (4)$$

Clearly $A + D = (1) + (2) + (3) + (4) = B + C$.

Actually Kempe was satisfied with a logical corollary: "If either (1) or (2) is positive, so is A; and conversely." Lewis [13] sent me such equations in 1947.

The 2 × 2 equations defining four elements A, B, C, D in terms of four isotopes $u_i = (i)$ is the simplest example of a parametric system $\$(x, u)$. The derived system $\$(x)$ here consists of the single equation $A + D = B + C$. Whenever this elementary equation is satisfied, a feasible solution for the parametric system can always be found.

4 × 4 Boxes We can illustrate the construction of a 4 × 4 chromatic array with #87 choosing $s = 6$. The canonical rules are these:

(1) The 16 entries must be schemes which have the same "periodicity" with six significant digits, and $n - 6$ ciphers in the same ring positions.

(2) The columns are marked 111, 113, 131, 133; the rows 222, 244, 224, 242. The odd digits determine the column and the even digits the row. Since #87 has the periodicity pattern 1.2.1.2.03.4.000 it is placed in column 113 and row 224.

CHROMATIC EQUATIONS

(3) There are 20 isotopes in the array; each isotope contains four schemes; each scheme belongs to five isotopes.

```
┌─────────┐   ┌─────────┐   ┌──────────────┐
│ 1 1 1 1 │   │ 5 6 7 8 │   │ 9  9  10 10  │
│ 2 2 2 2 │   │ 5 6 7 8 │   │ 9  9  10 10  │
│ 3 3 3 3 │   │ 5 6 7 8 │   │ 11 11 12 12  │
│ 4 4 4 4 │   │ 5 6 7 8 │   │ 11 11 12 12  │
└─────────┘   └─────────┘   └──────────────┘
   Row          Column         Quarters
 isotopes       isotopes
```

```
┌──────────────┐   ┌──────────────┐
│ 13 14 13 14  │   │ 17 18 18 17  │
│ 15 16 15 16  │   │ 19 20 20 19  │
│ 13 14 13 14  │   │ 19 20 20 19  │
│ 15 16 15 16  │   │ 17 18 18 17  │
└──────────────┘   └──────────────┘
    Thirds              Fifths
```

(4) Each row and each column is an isotope. The remaining isotopes are easily recognized. Pick any two rows say i and j. Then either pick columns i and j or the other two columns. The 2×2 submatrix is an isotope.

```
            ─────────────────────────
                  111  113  131  133
            222
            244
            224          #87
            242
            ─────────────────────────
```

The accompanying arrangement is not compulsory. Other displays with modified rules would provide the same information. Our choice which we call *palindrome* is motivated by the observation that the transpose of the matrix should correspond to a clockwise path around the ring.

Regarded as a mnemonic for the 20 parametric equations $\$(x, u)$ the palindrome display admits several automorphisms. Row i and row j may be permuted provided another two rows or two columns with the same chromatic sum $s = i + j$ are also permuted. Also the matrix may be transposed, thereby interchanging rows and columns. Thus any single element M_{ij} from row i column j can be moved to the (1, 1) slot, and vice versa.

1	2	3	4
5	6	7	8
9	10	11	12
13	14	15	16

Palindrome

1	12	14	7
15	6	4	9
8	13	11	2
10	3	5	16

Magic

The arrangement on the right is obtained by interchanging six pairs but keeping the same principal diagonal. Each of the 20 isotopes now occurs once in every row and column, which accounts for the magic square property. The six chromatic equations are easily written down by equating magic row sums (spoors) and magic column sums.

Frank Bernhart [5] found a simpler rule. Using the palindrome display subtract row 1 from the other three: then subtract column 1 from each of the others. The residual 3 × 3 is a symmetric matrix and each element on the principal diagonal is the sum of the off-diagonal elements in the same row. This *pivot rule* shows that the eight variables in any two rows (or columns) are related by a simple equation. Unlike the spoor identities read from the magic square, these do not involve all 20 isotopes. The pivot equations become especially simple if one row vanishes. (See the following boxes.) Thus if $x_{13} = x_{14} = x_{15} = x_{16} = 0$, then $x_1 + x_4 = x_2 + x_3$ and $x_5 + x_7 = x_6 + x_8$ and $x_9 + x_{10} = x_{11} + x_{12}$. These have the form of 2 × 2 chromatic equations. The chromatic equations also become simple if each element vanishes in some isotope other than a row or column.

Suppose the elements in the southeast quarter of the palindrome display all vanish. Using the Thomas [17] notation we get $3 = 4, 7 = 8, 9 = 13$, and $10 = 14$. We also have $1 - 2 = 5 - 6 = 9 - 10 = 13 - 14$. In words, the difference between the first column vector and the second column vector is a constant. The same is true for the first two rows: $1 - 5 = 2 - 6 = 3 - 7 = 4 - 8$.

1	2	3	4
5	6	7	8
9	10	11	12
x	x	x	x

1	2	3	4
5	6	7	8
9	10	x	x
13	14	x	x

CHROMATIC EQUATIONS

8 × 8 Boxes The palindrome algorithm has a natural extension to higher path orders. The 8 × 8 rules are typical: Label rows and columns from 0 to 7 writing them as binary numerals. (Thus the third row and seventh column are identified as row 010, column 110). Place an initial scheme having path order 3 in any position (i, j), usually (000, 000). If an odd period is altered place the result in the same row at (i, j^*); if an even period is altered place it at (i^*, j), where i^* differs from i, and j^* from j, according to the following rules.

Period ordinal	1	2	3	4	5	6	7	8
Change in j	x		100		010		001	
Change in i		x		110		011		001

The change from i to i^* affects only one binary digit. We write $i^* = i$ nim 100 to indicate that the leftmost digit (place value 4) is altered. In *nim* addition binary digits in the same column are added (mod 2) without carry. The digits are treated as coefficients of a polynomial in a Galois field of 2^n elements.

These rules permit placing a given element in any (i, j) slot of the matrix. A canonical form for the complete display labels columns in the natural order 111, 113, 131, 133; 311, 313, *331*, 333 where an initial 1 for period ordinal 1 is assumed. The palindrome row order is 222, 224, *244*, 242; 444, 442, 422, 424 with prefix 2 understood. Thus the element in the third row and seventh column ($i = 010, j = 110$) has the pattern 1232 3414.

To identify 8 × 8 isotopes build the display

$$J = \begin{vmatrix} A & A \\ A & B \end{vmatrix}, \quad \text{where} \quad A = \begin{vmatrix} 0 & 0 & 0 & 0 \\ 0 & 1 & 0 & 1 \\ 0 & 0 & 1 & 1 \\ 0 & 1 & 1 & 0 \end{vmatrix} \quad \text{and} \quad B = \begin{vmatrix} 1 & 1 & 1 & 1 \\ 1 & 0 & 1 & 0 \\ 1 & 1 & 0 & 0 \\ 1 & 0 & 0 & 1 \end{vmatrix}.$$

Both rows and columns are given binary addresses in the natural order 000, 001, 010, 011; 100, 101, 110, 111. The complete 8 × 8 is symmetric, so that any rule stated for rows-and-columns holds also for columns-and-rows.

The *nim sum* for rows i and j are the same as for rows $i + k$ and $j + k$. Thus row 011 is 0110 0110 and row 110 is 0011 1100. Their nim sum gives row 101 which is 0101 1010.

Each row is an isotope. There are two 2 × 4 isotopes for every pair of rows i and j except i nim j = 7. Write the nim sum of row i and row j twice.

Thus, for the fourth row $i = 011$ and seventh row $j = 110$ we obtain (replacing 1 by X)

```
0  X  0  X    X  0  X  0
0  X  0  X    X  0  X  0
```

The slots marked 0 in the chosen rows constitute one isotope; those marked X constitute another. These 2×4 isotopes may be regarded as two half-rows. Together with the columns and half-columns we have 14 types of isotopes. There are eight mutually disjoint isotopes of each type.

In row 001 two elements (i, j) and (i, j^*) are in different half-rows whenever j nim $j^* = 001$ but remain in the same half-row if the nim sum is 010 or 110. Those elements of rows 000 and 001 belong to the same half-row isotope whose column binaries $j = [abc]$ have the same c. Likewise for row 000 and 010, elements belong to the same isotope if b is the same; while for rows 000 and 100 the value of digits a is constant.

16 × 16 Boxes Each 16×16 array contains those associated schemes whose colors alternate odd–even, odd–even, etc., in ten periods forming five paths. The columns correspond to odd period colors in lexicographic order. The palindrome row ordering is less natural: ignore the odd prefix 1 and even prefix 2 the accompanying table indicates how ring colors are assigned to column j or row i whose binary address is $[abcd]$.

cd \ ab	00	01	10	11
00	1111 2222	1311 4422	3111 4222	3311 2422
01	1113 4444	1313 2244	3113 2444	3313 4244
10	1131 4442	1331 2242	3131 2442	3331 4242
11	1133 2224	1333 4424	3133 4224	3333 2424

$$J = \begin{matrix} A & A & A & A \\ A & B & A & B \\ A & A & B & B \\ A & B & B & A \end{matrix}$$

The isotopes can be identified from the pattern J, where A and B represent the same 4×4 blocks used for 8×8 boxes. The composite 16×16 matrix J is symmetric, with the disposition of isotopes invariant when rows and columns are interchanged. F. Bernhart has found elegant rules [5] for identifying all the isotopes. In this paper a few comments must suffice. Again each row is an isotope. Any two rows i and j contain two half-row isotopes except when i nim j has one of the taboo values 7, 11, 13, 14, 15. With interchange of rows and columns this accounts for 22 of the 42 isotope types. The remaining isotopes are appropriate 4×4 submatrices of the 16×16 box.

Each particular scheme (k) assigns colors to the n regions of its ring, and these determine the compatible scores for the ring edges. Let $\sigma(s)$ be the number of times score s occurs around the ring. Let $\pi(s)$ be the number of paths when s is neutral. Then

$$\sigma(5) + \sigma(6) + \sigma(7) = n = \pi(5) + \pi(6) + \pi(7)$$

so that $\sigma(5)$, $\sigma(6)$, $\sigma(7)$, and n have the same parity. If each π is at least 2, the variable x_k appears in 3 chromatic matrices and their implied equations.

3. Irreducible Ring Configurations

5-Rings It is easy to verify for known clusters that the chromatic equations are satisfied, by an actual count of all possible colorings. But these equations must also be satisfied by ringed configurations whose geometric structure is not known. Thus if a pentagon is deleted from a cubic map the complementary regions also constitute a 5-ring cluster. If the map were minimal 5-chromatic its complementary ring is called an anti-pentagon. Any boundary scheme for which the pentagon is colorable would only use 3 colors on the ring, and for such schemes the anti-pentagons cannot be 4-colorable. The chromatic equations show quickly that the frequency y_k for coloring the anti-pentagon must be zero whenever the corresponding pentagon scheme x_k is positive, but that the remaining y_k must be equal. There are geometrical constraints which exclude $y_k = 0$ for all k, so that the colorability of the anti-pentagon is essentially known. If all the chromatic equations $\$(x, u)$ are satisfied, and if all the constraints are satisfied, we call an n-ring configuration *irreducible*. Some authors use the term reducible in a narrower sense (disregarding the constraints), suggesting the term *feasible* when the inequalities are also satisfied. This corresponds to the linear programming nomenclature.

6-Rings Since 1945 this author has been engaged in a systematic search for irreducible configurations [4]. The only irreducible 5-ring is the pentagon (Birkhoff [6]). The ten 2×2 boxes in ten unknowns can be set up and solved by hand in a half hour. The six irreducible 6-rings were found [2] in February, 1944, but the entire chromatic analysis can be completed in two hours. Cohen [7] repeats these computations in his thesis.

7-Rings Six irreducible 7-rings with their complements are well known. If configurations for each irreducible 6-ring exist (geometrically) then a score of irreducible 7-rings can be identified.

Numerical values for both sides of 157 different 7-rings were completed in 1950 (after some 3000 man-hours). Each of these has been verified to be *irreducible* and *feasible* subject to all known constraints. All 91 elementary variables appear in the seven 4 × 4 boxes. In most cases each x_k and y_k is either 0 or 1, but in the known geometrical instances this is not the case. Later Rector [14] studied the linear system (x). A computer verification that this list is indeed complete is desired (A. Bernhart [4]).

8-Rings By solving the chromatic equations $ for the 274 variables in an 8-ring this author discovered that the configuration 6–565 was reducible [3], but that others such as 6–556 were irreducible. Later Goldbeck [8] recomputed the colorings for some 30 8-rings, and in addition verified that all irreducible solutions of $ were also feasible.

9-Rings More recently Rill [15], F. Bernhart [5], and others have studied some 150 9-rings, determining which are reducible, and proving that the others are both irreducible and feasible.

Oriole During the past five years the 10-ring configuration oriole 7–5665 has been investigated both by hand computation and by computer. The project is not yet completed. We conjecture that oriole will be irreducible with a dozen or more independent parameters. Swart [16] is studying the same configuration hoping to show that it is reducible.

For small n-rings with fewer variables almost any procedure gets the job done. We have considered both linear programming and integer programming algorithms but it seems better to choose an algorithm tailor made for the investigation of chromatic systems.

First one considers all 2461 variables and sets each $x_k = 0$ for which the inside is colorable. Then certain others must vanish in the outside system $, due to the isotope requirements. For oriole we can reduce the remaining variables by a factor of 2 due to the symmetry. Taking one chromatic equation at a time we solve for that x_k which has the largest subscript and eliminate it from the other equations. The method should ultimately lead to a residual core of x_i which may be taken as independent parameters in terms of which all other frequencies can be expressed as linear combinations. Our incomplete solution has already determined many equivalence classes of variables. The chromatic equations prove that several of these classes must vanish, results which are not obtainable by crude chaining.

There is a disjoint pair of open sets for the 6-ring which does not satisfy the stronger chromatic equations. But since the excluded case had already been listed as a variant solution, all six irreducible 6-rings survive.

What is desired is a configuration which is not reducible by the Boolean logic of open sets, but which can be reduced by chromatic equations. Perhaps oriole will be such an example.

REFERENCES

1. K. Appel, W. Haken, and J. Koch, Every Planar Map is Four Colorable, Preprint, University of Illinois, Urbana, 1976.
2. A. Bernhart, Six-rings in minimal five-color maps, *Amer. J. Math.* **69** (1947) 391–412.
3. A. Bernhart, Another reducible edge configuration, *Amer. J. Math.* **70** (1948) 144–146.
4. A. Bernhart, Irreducible rings in minimal five-color maps, *Proc. Internat. Congr. Mathematicians, Cambridge, 1950*, **2** (1952) 521.
5. F. Bernhart, The five color conjecture and related topics in graph theory, Ph.D. Thesis, Kansas State University, Manhattan, 1974; also private communications.
6. G. D. Birkhoff, The reducibility of maps, *Amer. J. Math.* **35** (1913) 114–128.
7. D. I. A. Cohen, Small rings in critical maps, Ph.D. Thesis, Harvard University, Cambridge, Massachusetts, 1975.
8. B. T. Goldbeck, 8-rings in minimal maps, Ph.D. Thesis, University of Oklahoma, Norman, 1957.
9. W. J. Growney, Edge conjugation and coloration in cubic maps, Ph.D. Thesis, University of Oklahoma, Norman, 1970.
10. P. J. Heawood, On extended congruences connected with the four colour map theorem, *Proc. London Math. Soc.* **33** (1932) 253–286.
11. H. Heesch, "Untersuchungen zum Vierfarbenproblem," Bibliographisches Institut, Mannheim, 1969, based on Habilitationsschrift, Technische Hochschule, Hanover, 1958; also Chromatic Reductions, *J. Combinatorial Theory* **13** (1972) 46–55.
12. A. B. Kempe, On the geographical problem of the four colours, *Amer. J. Math.* **2** (1879) 193–204.
13. D. C. Lewis, private communication; also G. D. Birkhoff and D. C. Lewis, Chromatic polynomials, *Trans. Amer. Math. Soc.* **60** (1946) 355–451.
14. R. W. Rector, Fundamental linear relationships for the seven ring, Ph.D. Thesis, University of Maryland, 1956.
15. M. Rill, private correspondence; also Sigma Xi award, University of Oklahoma, Norman, 1972.
16. E. R. Swart, private communication, University of Rhodesia, Salisbury (1977).
17. J. M. Thomas, The Four Color Theorem, Preprint, Philadelphia, 1969.
18. H. Whitney, A theorem on graphs, *Ann. of Math.* **32** (1931) 378–390.

AMS 05C15

DEPARTMENT OF MATHEMATICS
UNIVERSITY OF OKLAHOMA
NORMAN, OKLAHOMA

Planar Colorings: A Theory

FRANK R. BERNHART

1. The Central Question

Symbol $Q(n)$ or Q_n is used below to denote a standardized circuit of length n, carrying a standard cyclic labeling of the vertices: $1, 2, 3, \ldots, n$. A *planar configuration* of order $n \geq 2$ (for short, *n-configuration*) is a connected graph in the plane whose outer boundary is identified with Q_n for some n. If the configuration is not degenerate, the subgraph induced by the boundary vertices is indeed Q_n. Two kinds of degeneracies can occur: (i) the configuration includes a diagonal of the circuit Q_n, and (ii) the configuration coalesces into one vertex a pair of Q_n vertices that would otherwise be distinct. It is convenient to treat both types of degeneracy as features not of the boundary, but of the interior; hence in all cases Q_n will be shown as a circuit. Where degeneracies of type (i) or (ii) occur, they will be shown as in the following diagram, with a long equal sign or doubled diagonal indicating a merger. So to "represent" introduces a species of fiction both innocuous and convenient, and which could be removed at the cost of a lengthy circumlocution.

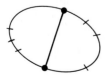 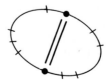

If a configuration, say A, is colored using any suitable number of colors, then automatically Q_n is also colored. We will suppose it unnecessary for many purposes to distinguish between a color pattern for Q_n and any variant of this pattern obtained by permuting colors. Thus define *color scheme* as a standardized color pattern for Q_n: taking the vertices in the sequence of labeling, and writing down their colors, using the least unused positive integer for each new color. For $n = 4$ exactly four schemes occur,

$$1212, \quad 1213, \quad 1232, \quad 1234.$$

Suppose that the number of colors has an upper limit r (for "rainbow"). Let symbol J_n^r denote the set of possible n-circuit colors schemes in r or fewer colors. Also write $A^{(n)}$ to indicate that configuration A has order n. The function $j(A^{(n)}, r, s)$ is defined as the number of different colorings of $A^{(n)}$, including differences due to permutation of colors, such that the coloring restricted to the boundary Q_n is identical to the color scheme $s \in J_n^r$ in its canonical form. Write $J^r(A)$ for the set of color schemes s such that $j(A^{(n)}, r, s) > 0$. Any subset $U \subseteq J_n^r$ such that $U = J^r(A)$ for some $A = A^{(n)}$ will be called *r-realizable*. The case $r = 4$ is of special importance; in this case r will be omitted from the notation, thus: $J(A)$, "realizable," and so on.

Which Are the Realizable Subjects of J_n for Each n? That is the question which we propose to establish as the central "axis" of the theory of planar colorings. Function $j(A^{(n)}, r, s)$ can also be studied as a means to an end: that of finding a simple and satisfying characterization. Moreover, the four color conjecture may be phrased (in several different ways, as will be seen) as a particular instance of the general question. It might be thought that a solution of the conjecture would furnish a grip on the characterization problem, but unfortunately it does not. Not only does the Appel–Haken proof not provide many hints for a characterization, but it seems that a thorough investigation of realizability might help with the toughest parts of the proof.

One should not get the idea however that the scope of the theory is limited to the four color theorem. It actively groups the theorem together with related results. There are undoubtedly many ways to combine the four color conjecture with related studies. But it does not appear that many of them are likely to preserve the importance of the Kempe chain conditions, or bring attention to the strengths and weaknesses of the theory of reduction. Our survey will attempt to clear up these main points. Some familiarity with the idea of Kempe chains will be helpful if the reader is to fill in the gaps in our discussion. To attempt to remove them ourselves would have resulted in a long technical digression, and we prefer to refer the interested investigator to Whitney and Tutte [13]. Our exposition is aimed primarily at the general character and usefulness of the theory.

Nearly all the facts currently available about Kempe chain entanglements are included in this theory. The open sets defined by Whitney and Tutte lead to a theory of union and splicing [4, 11]. Furthermore, one can give a more algebraic and concise description of classical reduction, and the degrees of reduction of Heesch [8]. Some progress has also been made in the study of irreducibility [3].

Several papers have recently studied the question of the minimal size of realizable and open sets [1, 10] and found interesting lower bounds.

W. T. Tutte and the present author have studied a group of identities called "flattening equations" which hold among the constrained chromatic polynomials of planar graphs. Their existence helps to solve problems of "ring" analysis posed in the monograph of Birkhoff and Lewis [7]. These equations can also be viewed as a "quantified" study of Kempe chains. The realizable sets of colorings happen to be the support sets for functions that arise naturally in this approach. If one takes the set of functions satisfying suitable conditions (in fact a linear program) then the condition for a set of colorings to be a support of such a function is sometimes stronger than the condition of being an open set. Several workers have been attempting to show that with these improved conditions certain configurations that were "borderline" irreducible under the classical approach become reducible. Hopefully there will soon be reports or papers available describing this "breakthrough."

2. Kempe Chains and Realizability

Let us begin with a simple question. Why cannot every subset of J_n be a realizable subset? We expect to find limitations, but it is startling to discover that almost the only source of any restrictions is the theory of Kempe chains.

This theory is so called because it first appears in the 1879 paper of Kempe [9], a paper whose claim to have solved the four color problem was accepted completely for ten years. Ironically, the 1976 proof of Appel and Haken [2] may be described as closing the gap in Kempe's proof, using subtle and involved ramifications of his methods! We mention Birkhoff [6] and Heesch [8] as responsible for important extensions of the method.

First suppose that configuration $A = A^{(n)}$ is colored in some fashion with four colors. Call the four colors a, b, c, d. Pick any two of the four colors and amalgamate them as *black*. Do the same with the remaining two, and call the result *white*. We now have a pseudo-coloring of vertices which is not a proper coloring. Rather, it is a division of the vertices into two groups. Each group defines a subgraph containing one or more components. Each component is a *Kempe chain*. The set of black and white components impose a partition σ on the vertices of the boundary Q_n of A. By varying the way in which colors a, b, c, d are associated with black and white, three partitions $\sigma_1, \sigma_2, \sigma_3$ are obtained. We will call these partitions *K-chain dispositions*.

The set of all possible K-chain dispositions is finite, and for relatively small values of n not difficult to tabulate. It is even simpler to start with a scheme for Q_n, and tabulate the possibilities. This amounts to segmenting the ring Q_n in three ways (take A as Q_n temporarily, with the given scheme as coloring), then joining the segments in different ways through the interior. It is essentially the same problem as joining $2m$ points on a circle with m noncrossing chords; it follows that the number of relevant K-chain dispositions for a given scheme s is the sum of three Catalan numbers. The value of m differs in the three segmenting operations, but the three values add up to n. The single scheme $1212 \cdots 12$ is recognized as an anomaly, but not a serious one. Our description here is suggestive but incomplete. The interested reader may check Whitney and Tutte [13], or Swart [12], or the author's thesis [5, Chaps. 5 and 6].

We write $j(A^{(n)}, \sigma, s)$ for the number of ways that A may be colored in at most four colors, satisfying two conditions: (i) The coloring of the boundary is exactly s in canonical form, and (ii) σ is one of the three partitions $\sigma_1, \sigma_2, \sigma_3$, of the vertices of Q_n which derive from the coloring. Also write $J_\sigma(A)$ for the set of schemes s such that $j(A, \sigma, s) > 0$. The members of $J_\sigma(A)$ are distinguished in the manner of an *isotopic* variety: the outer "electron shell" represented by Q_n behaves identically, but also certain aspects of "nuclear" arrangment are the same as well. What is especially important is the fact that $J_\sigma(A)$ if not empty can be reconstructed from one member! It is only necessary to substitute for A, which may be of unknown structure, a suitable configuration with no interior vertices (a *stand-in* for A), say, A', such that $j(A', \sigma, s) > 0$. Alternately, one can work directly with the partition σ. In any case, we have indicated the reason why the K-chain analysis leads to an effective algorithm.

Suppose that set U is realizable and has a member s. It must be possible to pick three partitions σ_i, one for each segmenting, such that the three isotopes one gets are all contained in U. This is why U cannot have just one member, and in fact must contain at least a minimum number. The restriction we have just stated will be called the K-chain condition for member s of U.

Now take any U, realizable or not, satisfying the K-chain condition for each of its members. We call U *open*. Immediately from the definition, the union of open sets (as subsets of J_n) is again open, and both $\emptyset$ and J_n are open. Call the complements of open sets *closed*. The system of open sets can be associated with a Kuratowski interior operator, and the system of closed sets with the dual closure operator. Many interesting facts flow from this central one. Swart [12] has studied the consequences for $n = 4$ and $n = 5$ with some thoroughness. The term "open" is due to Whitney and Tutte [13].

PLANAR COLORINGS: A THEORY

Remark 1 *The union of open subsets of J_n is open.*

Conjecture 1 *The union of two realizable subsets of J_n is realizable.*

The remark is true in part because of the weakness of the definition of open, or rather the unrelatedness of K-chain structure from one member of a set of schemes U to another. There seems no convincing reason to believe the conjecture, except for its conformity with evidence. There do exist a few examples of natural union. Consider a vertex of degree four with the surrounding Q_4, called a *four-wheel*. It has a realizable set $U \subseteq J_4$ with three members that is the union of two smaller sets (see the accompanying diagram). This fact forms part of the argument commonly used to remove vertices of small degree from planar graphs, in elementary four color reduction.

There is another important property of open sets which is an immediate generalization from the realizable case. Consider the following diagram, which is to be taken as schematic.

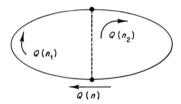

We may suppose that a left circuit $Q(n_1)$ and a right circuit $Q(n_2)$ are joined along a path of k edges to form a graph with boundary $Q(n)$, $n = n_1 + n_2 - 2k$. If left and right circuits are supplied or "filled in" with appropriate configurations A_1 and A_2 then a configuration A or order n is formed. We say that A is formed as a *splice* of A_1 and A_2. Of course when splicing is referred to in the following, it must be assumed that an appropriate diagram is given. We also allow extreme cases like the two shown in the following diagram. For more precision, we can refer to a k-splice. Note that for $k = 0$ and $k = 1$ the splice result is a degenerate configuration.

 =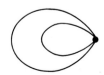

Remark 2 *It follows from the Jordan curve theorem that the class of planar configurations can be generated using splicing and starting with the set of circuits $Q(n)$.*

There seems to be something natural and satisfying about splicing as a basic operation. Topologically, two disks with exactly a boundary arc in common combine to form a new disk. Using the splicing concept, we can have it all in pure combinatorial form.

Implicit splicing seems to have been first discovered by Shimamoto and has since been explicitly developed by several investigators. Stromquist [11] gives a complete development, and [4] can also be consulted.

Splicing is immediately applicable to the study of color schemes. It takes only a minute to realize that if A is formed by splicing A_1 with A_2, and if we have the diagram that details exactly how the boundaries of the A_i overlap, then the set $J(A)$ realized by A can be directly defined from $J(A_1)$ and $J(A_2)$ without regard to the inner structure of A_1 and A_2! The reader is invited to form the definition of the splice of two sets of color schemes. Immediately we have

Remark 3 *The splice of two realizable sets is realizable.*

Remark 4 *Splicing distributes over union.*

The next question is, what happens if two open sets are spliced? The fortunate answer is given by the following.

Theorem 1 *The splice of two open sets is open.*

Proof can be found in Stromquist [11] or in [4]. It is implicitly present in the analysis of penta-triangulations in Whitney and Tutte [13] and may therefore be credited indirectly to Shimamoto. The idea behind the proof is as follows. Each way of setting up a possible disposition of Kempe chains for an open set can be exhibited using a suitable configuration. In the splicing diagram the combination of two dispositions in the left and right circuits can therefore yield an appropriate disposition in the larger circuit.

The same operation can be approached more generally: we can completely define the function $j(A, n, r, s)$ of scheme s in terms of functions $j(A_i, n_i, r, s_i)$ and show that there are underlying bilinear properties. Some aspects have been studied by Rolle (unpublished work at the University of Waterloo). Basic to the general theory is the fact that the general splice of realizable sets is again realizable.

Recall now the meaning of a degenerate configuration A: two vertices which are nonconsecutive on the boundary Q_n are merged, or made adjacent by a diagonal, within A. Differently put, A is a 0-splice or a 1-splice of two subconfigurations. In these cases say that A is 0-*degenerate* or 1-*degenerate*.

We can frame a corresponding definition for open sets. Say that open set U is 0-*degenerate* (or 1-*degenerate*) if there are two nonconsecutive vertices of Q_n (we assume $U \subseteq J_n$) such that in every member of U the two vertices have the same (have different) colors.

Clearly if configuration A is degenerate, then $U = J(A)$ is degenerate in a corresponding way. However, the converse fails in the worse way. There is an open set of order $n = 7$ which is 1-degenerate, but which is not even a 1-splice of smaller open sets. One can show by contrast that a 0-degenerate open set is a 0-splice of two open sets.

Theorem 2 *Let $k = 0$ or $k = 1$. There exists a nondegenerate configuration A such that $U = J(A)$ is k-degenerate if and only if the four color conjecture is false.*

Proof We consider $k = 0$ (case $k = 1$ is similar). In vertices u, v on the boundary of A are such that in $U = J(A)$ the colors are always the same, but u, v are not merged or adjacent in A, then we can add a uv-edge to A on the outside, and get a non-four-colorable planar graph G (without loops). Conversely, if G is given, we obtain A by removing any edge. ∎

If G were a given non-four-colorable planar graph, it would also be possible to "cleave" G along a minimal separating circuit and get two configurations. Hence

Theorem 3 *The four color conjecture is equivalent to the nonexistence of two nonempty disjoint realizable sets, both nondegenerate.*

3. Classical Reduction

Using the language of open sets, the gist of classical reduction can be phrased in compact algebraic form.

Recall that the union of open sets is open, and (vacuously) the empty set of schemes is open. For a given subset S of J_n, there is therefore a unique largest open subset S^o. Moreover, since the complement of an open set is a closed set, there is also a unique smallest closed set $\bar{S}$ which contains S. Call S^o and $\bar{S}$ the *interior* and the *closure* of S.

Suppose that T is a triangulation of the sphere or plane. If Q is any circuit of T, then T is decomposed into two configurations with Q as the common boundary, say A and B. We then have realizable sets $U = J(A)$ and $V = J(B)$ on Q. Moreover, T can be four-colored just in case U, V are *not* disjoint. We will assume that U, V are nonempty, which will certainly be true on the assumption that all triangulations (and hence all planar graphs) with fewer vertices than T are colorable. We make no assumption about T. If for some choice of configuration A it follows that T is four-colorable, regardless of configuration B, then A is said to be *reducible*. The four color theorem, in short, states that every finite triangulation T contains at least one of a list of known reducible configurations.

Let A be a given nondegenerate configuration of modest size. Then $U = J(A)$ can be computed by a simple-minded computer program (first phase). Then with a straightforward but more complex program (second phase) one can compute the unique largest open set disjoint from U, which we call the *conjugate*, and denote $\tilde{U}$. The conjugate is the complement in J_n of the closure $\overline{U}$, so that it is essentially the same thing to compute $\overline{U}$ as to compute $\tilde{U}$.

Traditionally $\overline{U}$ is computed in stages, beginning with $W = U$. Members s of the difference $J_n \setminus W$ are examined individually. When the Kempe chain dispositions do not allow for the possibility that s can be in an open set disjoint from W, then s is added to W. Set W grows dynamically and eventually becomes $\overline{U}$ at the point that the process fails to add new schemes.

For many configurations A, one finds that $\tilde{U} = \emptyset$ (i.e., $\overline{U} = J_n$). Thus if $V = J(B)$ is nonempty, it must meet U, and T is 4-colorable. This kind of reducibility is highlighted by Heesch [8], and termed by him "*D*-reducibility," and by others "colorable." Except for the labor of computing $\overline{U}$ it is the simplest kind of reducibility.

For the more general reduction technique, it is necessary to supply a configuration A' called the *reducer*. The total number of vertices of A' must be less than A, and if $U' = J(A')$, we must have $U' \subseteq \overline{U}$. In other words, any open set V disjoint from U must be also disjoint from U'. But replacing A with A' gives a smaller triangulation than T, say T'. The assumed colorability of T' means that U' and V have nonempty intersection, and therefore V meets U also (since V does not satisfy $V \subseteq \tilde{U}$). The most useful reducers are generally extremely degenerate; it is even unusual for any other kind to be necessary. Some care is required with a degenerate reducer. It is best if B can be assumed to be nondegenerate. Fortunately, for small enough A, a degeneracy in B would combine with the structure of A to produce in T a small circuit of forbidden type (forbidden because it is reducible). For larger sizes, there exist various technical results, not to be discussed here, which patch the difficulty.

The search for a suitable reducer (if $\tilde{U} \neq \emptyset$) is phase three for the computer approach. The number of candidates grows so rapidly with n, that a thorough search is almost impossible for $n > 12$. A theoretical shortcut would be very helpful.

Sometimes it happens that A' exists with $U' \subseteq U$ ("direct reduction" and "A-reduction"). If this can be discovered before computing $\overline{U}$, some gain in efficiency is obtained. But these configurations are few.

REFERENCES

1. M. O. Albertson and H. S. Wilf, Boundary values in the four color problem, *Trans. Amer. Math. Soc.* **181** (1973) 471–482.
2. K. Appel and W. Haken, Every planar map is four colorable, *Bull. Amer. Math. Soc.* **82** (1976) 711–712.
3. F. R. Bernhart, Irreducibility and the four color conjecture, "Theory and Applications of Graphs (Proceedings of the International Graph Theory Conference, Kalamazoo, 1976)," *in* Lecture Notes Mathematics, No. 642, pp. 37–52. Springer-Verlag, Berlin and New York, 1978.
4. F. R. Bernhart, Splicing and the Four Color Conjecture, Res. Rep. CORR 75-18, Department of Comb. & Opt., University of Waterloo, Ontario, Canada, 1975.
5. F. R. Bernhart, Topics in graph theory related to the five color conjecture, Ph.D. Thesis, Kansas State University, Manhattan, 1974.
6. G. D. Birkhoff, The reducibility of maps, *Amer. J. Math.* **35** (1913) 114–128.
7. G. D. Birkhoff and D. C. Lewis, Chromatic polynomials, *Trans. Amer. Math. Soc.* **60** (1946) 355–451.
8. H. Heesch, "Untersuchungen zum Vierfarbenproblem," Bibl. Inst. AG, Mannheim, 1969.
9. A. B. Kempe, On the geographical problem of the four colours, *Amer. J. Math.* **2** (1879) 193–204.
10. J. W. Schlesinger, Sets of colorings of circuits, *Bull. Amer. Math. Soc.* **81** (1974) 721–723.
11. W. Stromquist, Some aspects of the four color problem, Ph.D. Thesis, Harvard University, Cambridge, Massachusetts, 1975.
12. E. R. Swart, Some basic theorems on the abstract theory of Kempe–Chain interchanges, *Utilitas Math.* **10** (1976) 209–228.
13. H. Whitney and W. T. Tutte, Kempe chains and the four color problem, *Utilitas Math.* **2** (1972) 241–281. (Reprinted in *MAA Studies in Math.* **11**, Part II (1976) 378–413.)

AMS 05C15, 55A15

CAUSE PROJECT
KINGS COLLEGE
WILKES-BARRE, PENNSYLVANIA

On the Algebra of Graph Types

NORMAN BIGGS

1. Introduction

For several years I have been trying to understand the paper "On Dichromatic Polynomials" which Professor Tutte published in 1967 [5]. One reason for my interest was the belief that the results he obtained in that paper are related to a technique used in theoretical physics and known as "cluster expansions." Unfortunately, I found it no easier to follow the arguments of the physicists [1].

In this paper I shall survey a theory developed for the purpose of linking the two approaches. The physical background is explained in my monograph on "Interaction Models" [2], and more details of the theory (including proofs of the statements in Sections 2 and 3) may be found in a forthcoming paper [3]. The fourth section of the present paper is intended to make explicit the relationship between the theory and Tutte's new formulation, appearing as the second paper in this volume under the title "All the King's Horses." (It may be worth remarking that the theory is developed here for the concrete "$C(G; 1b)$" interpretation of Tutte's muster–cluster formalism.)

2. Star Types, Graph Types, Whitney's Theorem

We define a *star type* to be an isomorphism class of nonseparable simple graphs, and use pictographs $|$, $\triangle$, $\square$, $\boxslash$, ..., to denote the first few star types. The set of all star types will be denoted by St.

Let $\mathbb{N}$ be the set of natural numbers, including zero. A *graph type* is a function g: St $\to \mathbb{N}$ which takes the value zero except on a finite subset of St. We shall say that a finite simple graph is *of type g* if it has $g(\sigma)$ blocks of type σ, for each σ in St. We remark that graphs of the same type need not be isomorphic, although there is a one-to-one correspondence between their blocks such that corresponding blocks are isomorphic. In Fig. 1 there is an

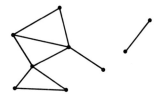

Figure 1

example of a graph of type g, where $g(|) = 2$, $g(\triangle) = 1$, $g(\boxslash) = 1$, and $g(\sigma) = 0$ otherwise. We use the abbreviation $g = \|\triangle\boxslash$ for this type, and other similar abbreviations. It should be noted that there is a null graph type o, defined by $o(\sigma) = 0$ for all σ in St. Also, it is convenient to regard St as a subset of Gr, identifying the star type σ with the graph type whose value is 1 at σ and zero elsewhere. (Gr denotes the set of graph types.)

Let G be a finite simple graph. Each subset S of the edge-set of G defines a *subgraph* $\langle S \rangle$ of G, consisting of the edges in S and those vertices of G incident with them. Let $n_t(G)$ denote the number of subgraphs of G which are of type t. It is clear that the numbers n_t satisfy certain identities which do not depend on G. For example, the number $n_{|\triangle}$ is equal to $n_\triangle(n_| - 3)$, and

$$n_{\|\triangle} = \tfrac{1}{2} n_\triangle (n_| - 3)(n_| - 4) - 2 n_{\boxslash}$$
$$= 6 n_\triangle - \tfrac{7}{2} n_| n_\triangle - 2 n_{\boxslash} + \tfrac{1}{2} n_|^2 n_\triangle. \tag{1}$$

As long ago as 1932, Whitney [6] proved that, for each graph type t, the number n_t is a polynomial function of the numbers n_σ ($\sigma \in$ St), with no constant term. His proof was rather obscure, and the significance of the coefficients of the polynomials was not clear. Our first task is to translate Whitney's theorem into an algebraic setting, which makes it clear how the coefficients can be calculated.

Let **X** denote the vector space of real-valued functions defined on the set St, and let **Y** denote the analogous space of functions defined on the set Gr.

When St is regarded as a subset of Gr we have an induced projection $J: \mathbf{Y} \to \mathbf{X}$, given by $(J\mathbf{y})(\sigma) = \mathbf{y}(\sigma)$ for each σ in St.

In order to express the theorem of Whitney in this context, we define (for a given graph G, of type g) a vector $\mathbf{c}_g$ in $\mathbf{Y}$ by the rule that $\mathbf{c}_g(t) = n_t(G)$. It is easy to see that this vector depends only on g, not on G. The projection $J\mathbf{c}_g$ of $\mathbf{c}_g$ on $\mathbf{X}$ gives the numbers of nonseparable subgraphs of G. Whitney's theorem asserts that there is an operator $W: \mathbf{X} \to \mathbf{Y}$ with the property that

$$W(J\mathbf{c}_g) = \mathbf{c}_g \quad \text{for all} \quad g \text{ in Gr.} \tag{2}$$

In other words, the numbers of separable and disconnected subgraphs may be recovered if the numbers of nonseparable ones are known. What is more, the operator W is universal and does not depend upon g.

It will be noted that W cannot be a linear mapping of $\mathbf{X}$ into $\mathbf{Y}$, since this would imply that the projection J is invertible. However, inspection of the examples given above leads us to expect that the nonlinearity of W is of the polynomial kind. Specifically, if $\mathbf{x}$ is in $\mathbf{X}$, then $W\mathbf{x}(t)$ should be a polynomial function ϕ_t of the values $x_|, x_\triangle, x_\square, \ldots$, of $\mathbf{x}$:

$$W\mathbf{x}(t) = \phi_t(\mathbf{x}) = \phi_t(x_|, x_\triangle, x_\square, \ldots). \tag{3}$$

For example,

$$\phi_{\|\triangle}(\mathbf{x}) = 6 x_\triangle - \tfrac{7}{2} x_| x_\triangle - 2 x_\boxtimes + \tfrac{1}{2} x_|^2 x_\triangle,$$

in accordance with the formula (1).

Now a polynomial is simply a linear combination of monomials, and so we introduce the monomial operator $U: \mathbf{X} \to \mathbf{Y}$ given by

$$(U\mathbf{x})(t) = \prod \mathbf{x}(\sigma)^{t(\sigma)}. \tag{4}$$

The product is taken over all star types σ for which $t(\sigma) \neq 0$, and so, by the definition of t, it is a finite product. For example, $(U\mathbf{x})(\|\triangle) = x_|^2 x_\triangle$. The operator W is a composite AU, and it turns out that it is simpler to begin by defining A^{-1}, rather than A itself.

We define an infinite matrix $B = (b_{st})$ (where s and t are graph types) by the rule that b_{st} is the number of ways of covering a graph of type t with the blocks of a graph of type s. To be precise, let T denote a given graph of type t. Then b_{st} is the number of ordered families

$$\mathscr{S} = \{S_1, S_2, \ldots, S_l\},$$

where each S_i is an nonempty set of edges of T, and the following three conditions are satisfied:

(i) the union of the sets S_i covers the edge-set of T;
(ii) each subgraph $\langle S_i \rangle$ is nonseparable;
(iii) the first $s(|)$ subgraphs $\langle S_1 \rangle, \ldots,$ are of star type $|$, the next $s(\triangle)$ are of star type $\triangle$, and so on.

We remark that the number l appearing above must equal the number of blocks of s. To complete the definition for the null graph type o, we put $b_{oo} = 1$ and $b_{so} = b_{ot} = 0$ for all $s \neq o$ and $t \neq o$. A portion of the matrix B is displayed in Fig. 2a, with its inverse $A = B^{-1}$ in Fig. 2b. Both matrices define linear mappings $Y \to Y$ in the usual way, and we shall use the same letters to represent these mappings.

	o	$\mid$	$\mid\mid$	$\triangle$	$\mid\mid\mid$	$\square$	$\mid\triangle$	$\mid\mid\mid\mid$
o	1							
$\mid$	0	1						
$\mid\mid$	0	1	2					
$\triangle$	0	0	0	1				
$\mid\mid\mid$	0	1	6	6	6			
$\square$	0	0	0	0	0	1		
$\mid\triangle$	0	0	0	3	0	0	1	
$\mid\mid\mid\mid$	0	1	14	36	36	24	24	24

(a)

	o	$\mid$	$\mid\mid$	$\triangle$	$\mid\mid\mid$	$\square$	$\mid\triangle$	$\mid\mid\mid\mid$
o	1							
$\mid$	0	1						
$\mid\mid$	0	$-\frac{1}{2}$	$\frac{1}{2}$					
$\triangle$	0	0	0	1				
$\mid\mid\mid$	0	$\frac{1}{3}$	$-\frac{1}{2}$	-1	$\frac{1}{6}$			
$\square$	0	0	0	0	0	1		
$\mid\triangle$	0	0	0	-3	0	0	1	
$\mid\mid\mid\mid$	0	$-\frac{1}{4}$	$\frac{11}{24}$	3	$-\frac{1}{4}$	-1	-1	$\frac{1}{24}$

(b)

Figure 2

In order to prove Whitney's theorem, as expressed in our formula (2), it is sufficient to show that $Bc_g = UJc_g$, which is a fairly straightforward piece of work. Then applying the mapping $A = B^{-1}$ to both sides, and putting $W = AU$, we have the desired result.

In concrete terms, the theorem tells us that the coefficients of the polynomial ϕ_t are the entries in row t of the matrix A:

$$\phi_t(\mathbf{x}) = \sum_s a_{ts}(U\mathbf{x})(s).$$

For example, $\phi_{\mid\triangle}(\mathbf{x}) = -3x_\triangle + x_\mid x_\triangle$, in agreement with the formula $n_{\mid\triangle} = n_\triangle(n_\mid - 3)$, and our formula (1) may be checked similarly.

3. Expansions and Functionals

If μ is a real-valued graph function which is type-invariant [that is, $\mu(G_1) = \mu(G_2)$ when G_1 and G_2 are of the same type] then there is a corresponding vector $\boldsymbol{\mu}$ in $\mathbf{Y}$. Each such vector has an *ordinary expansion*

$$\mu(g) = \sum_t c_g(t)\mathbf{m}(t), \qquad (5)$$

for some $\mathbf{m}$ in $\mathbf{Y}$. [To prove this, we just define $\mathbf{m} = C^{-1}\boldsymbol{\mu}$, where $C = (c_{st})$ is the lower triangular matrix given by $c_{st} = c_s(t)$.] We may interpret the ordinary expansion (5) by saying that each subgraph of type t contributes an amount $\mathbf{m}(t)$ to the sum.

For example, suppose that $\boldsymbol{\mu}(g)$ is the probability that an arbitrary "colouring" of a graph of type g, with z colours available, is a "proper" colouring; that is,

$$\boldsymbol{\mu}(g) = C(G; z)/z^{|VG|},$$

where $C(G; z)$ is the chromatic polynomial. Then the Birkhoff–Whitney expansion tells us that the contributions $\mathbf{m}(t)$ are given by

$$\mathbf{m}(t) = (-1)^{\text{no. edges of } t}(1/z)^{\text{rank of } t}.$$

In theoretical physics, the vector $\boldsymbol{\mu}$ usually represents the partition function of some interaction model. The expansion (5) is regarded as a means of approximating to $\boldsymbol{\mu}$, since the smallest subgraphs make the most significant contributions, and they are the simplest to count.

Returning to the general case of (5), we define a linear functional $\mathcal{M}$ by the formula

$$\mathcal{M}(\mathbf{y}) = \sum_t \mathbf{y}(t)\mathbf{m}(t). \tag{6}$$

The domain of definition of $\mathcal{M}$ will be a subspace of $\mathbf{Y}$ containing, in particular, the vectors with finite support. Since $\mathcal{M}(\mathbf{c}_g) = \boldsymbol{\mu}(g)$, we see that $\mathcal{M}$ extends the graph function $\boldsymbol{\mu}$. In this interpretation, it is helpful to think of the elements of $\mathbf{Y}$ as "generalized graphs," while each vector $\mathbf{c}_g$ represents a "real graph" of type g.

Suppose that $\boldsymbol{\mu}$ is both *positive* [$\boldsymbol{\mu}(g) > 0$, all g] and *multiplicative*:

$$\boldsymbol{\mu}(g \oplus h) = \boldsymbol{\mu}(g)\boldsymbol{\mu}(h),$$

where $g \oplus h$ is the graph type defined by $(g \oplus h)(\sigma) = g(\sigma) + h(\sigma)$. Then we have the following result.

Theorem [3] *Suppose that $\boldsymbol{\mu}$ is positive and multiplicative, $\mathcal{M}$ is the corresponding linear functional defined by (6), and W is the Whitney operator. Then there is a linear functional $\mathcal{L}$ (defined on a nontrivial subspace of $\mathbf{X}$) such that*

$$\mathcal{M}(W\mathbf{x}) = \exp \mathcal{L}(\mathbf{x}). \tag{7}$$

The point of the theorem is apparent when we substitute $J\mathbf{c}_g$ for $\mathbf{x}$ in Eq. (7). We obtain, since $WJ\mathbf{c}_g = \mathbf{c}_g$,

$$\boldsymbol{\mu}(g) = \mathcal{M}(\mathbf{c}_g) = \mathcal{M}(WJ\mathbf{c}_g) = \exp \mathcal{L}(J\mathbf{c}_g).$$

In other words, the composite $\exp \circ \mathcal{L}$ also extends the graph function $\boldsymbol{\mu}$, and now the real graphs are represented by the vectors $J\mathbf{c}_g$. Thus it is only necessary to count the numbers of nonseparable subgraphs in order to evaluate $\boldsymbol{\mu}$. Of course, it is the existence of the operator W which makes the theorem possible.

There is a natural interpretation of $\mathscr{L}$ in terms of an expansion. We have

$$\mathscr{L}(\mathbf{x}) = \sum_\sigma \mathbf{x}(\sigma)\mathbf{l}(\sigma);$$

in fact, $\mathbf{l}(\sigma)$ is just the value of $\mathscr{L}$ on the characteristic vector $\mathbf{e}_\sigma$ of the type σ. This leads to the *cluster expansion* of $\boldsymbol{\mu}$:

$$\mu(g) = \exp\mathscr{L}(J\mathbf{c}_g) = \exp\left\{\sum_\sigma (J\mathbf{c}_g)(\sigma)\mathbf{l}(\sigma)\right\}. \tag{8}$$

In this expansion, each nonseparable subgraph of type σ contributes an amount $\mathbf{l}(\sigma)$ to the sum, and the separable and disconnected subgraphs do not contribute. Thus the contributing subgraphs are far fewer in number than in the ordinary expansion (5); however, the contributions $\mathbf{l}(\sigma)$ are more complicated than the ordinary contributions $\mathbf{m}(t)$. It can be shown [3] that the relation between $\mathbf{m}$ and $\mathbf{l}$ is

$$\mathbf{m} = B'V\mathbf{l}. \tag{9}$$

In Eq. (9) B' is the transpose of the matrix B defined in Section 2, and $V: \mathbf{X} \to \mathbf{Y}$ is defined by

$$(V\boldsymbol{\mu})(t) = (1/b_{tt})(U\mathbf{x})(t).$$

Since $A = B^{-1}$, the Eq. (9) may be inverted, giving

$$\mathbf{l} = JA'\mathbf{m}. \tag{10}$$

Explicitly,

$$\mathbf{l}(\sigma) = \sum_\sigma a_{t\sigma}\mathbf{m}(t). \tag{11}$$

We remark that the coefficients occurring here are just the numbers a_{pq} which appeared in the polynomials $\phi_t(\mathbf{x})$ [Eq. (3)]. But now the sum involves a column of the infinite matrix A, and so the Eq. (11) is an infinite series for $\mathbf{l}(\sigma)$. Also, it is important to notice that the first nonzero coefficient in (11) is $a_{\sigma\sigma}$, so that the graph types which are "smaller" than σ do not appear in the series. This fact corresponds to Tutte's theorem of vanishing coefficients.

4. Together Again

In this section we shall try to explain how the foregoing theory links up with the recent work of Tutte, in his paper "All the King's Horses."

We begin by factorizing the matrix B. Let $P = (p_{st})$ be defined as follows. The entry p_{st} is the number of ways of covering a given graph of type t with the blocks of a graph of type s, in the same way as in the definition of B, except that now each block of s must imbed onto a different subgraph. In other words, the family $\mathscr{S}$ does not contain any repetitions.

Let D be the diagonal matrix whose entries are the diagonal terms of B:

$$d_{tt} = b_{tt} = \prod_\sigma t(\sigma)!,$$

and let $\Phi = (\phi_{st})$ be defined by

$$\phi_{st} = \prod_\sigma S(s(\sigma), t(\sigma)),$$

where S denotes the Stirling number of the second kind.

Proposition 1 *With the notation introduced above,*

$$B = \Phi D P. \tag{12}$$

Proof We note first that $(\Phi D)_{sq}$ is equal to a product of terms $S(s(\sigma), q(\sigma)) \cdot q(\sigma)!$, each of which gives the number of ways of mapping a set of cardinality $s(\sigma)$ onto a set of cardinality $q(\sigma)$ [4, p. 205]. Thus $(\Phi D)_{sq}$ is the number of ways of collecting up like blocks of s so that a graph of type q results.

Suppose that a graph type s and a graph T of type t are given. Let $\mathscr{S}$ be a family of subsets of the edge-set ET, satisfying the three conditions stated in Section 2, so that $\mathscr{S}$ is one of the families contributing to b_{st}. For each star type σ, there are $s(\sigma)$ members of $\mathscr{S}$ which give subgraphs of T having star type σ, but they need not all be distinct. Suppose that $q_\mathscr{S}(\sigma)$ of them are distinct: this defines a graph type $q_\mathscr{S}$ associated with $\mathscr{S}$. The result of the previous paragraph shows that, for a given type q, there are $(\Phi D)_{sq}$ families $\mathscr{S}$ such that $q_\mathscr{S} = q$. Thus, listing the families contributing to b_{st} according to the associated type q, we obtain

$$b_{st} = \sum_q (\Phi D)_{sq} P_{qt} = (\Phi D P)_{st},$$

as required. ∎

Suppose we have a vector $\boldsymbol{\mu}$ in $\mathbf{Y}$, and its associated vector $\mathbf{m}$, as defined in Section 3. Our $\mathbf{m}$ corresponds to an expression j, in the notation of "All the King's Horses," while $\boldsymbol{\mu}$ corresponds to J. Tutte's definition (2) of J in terms of j is just our expansion (5). Starting from the $\mathbf{l}$ vector associated with $\boldsymbol{\mu}$ and $\mathbf{m}$, we shall now define a new vector $\boldsymbol{\lambda}$ in $\mathbf{X}$, which will turn out to coincide with Tutte's $-\lambda$. The definition is:

$$1 + \lambda(\sigma) = \exp \mathbf{l}(\sigma). \tag{13}$$

In the proof of the following proposition we shall need a useful identity involving the Stirling numbers of the second kind [4, p. 206]:

$$\sum_{n=0}^{\infty} S(n, k) \frac{x^n}{n!} = \frac{(e^x - 1)^k}{k!}. \tag{14}$$

Proposition 2 *Let $V: \mathbf{X} \to \mathbf{Y}$ be the operator defined in Section 3, and let $\Phi': \mathbf{Y} \to \mathbf{Y}$ be the operator induced by the transpose of the matrix Φ. If λ and $\mathbf{l}$ are related as in Eq. (13), we have*

$$\Phi' V \mathbf{l} = V \lambda. \tag{15}$$

Proof From the definitions, we have

$$(\Phi' V \mathbf{l})(t) = \sum_s \phi_{st}(V\mathbf{l})(s)$$

$$= \sum_s \prod_\sigma S(s(\sigma), t(\sigma)) \frac{\mathbf{l}(\sigma)^{s(\sigma)}}{s(\sigma)!}.$$

Using a well-known distributive identity [4, p. 127], we may rewrite this as

$$\prod_\sigma \sum_n S(n, t(\sigma)) \frac{\mathbf{l}(\sigma)^n}{n!},$$

and by (14) we obtain

$$\prod_\sigma \frac{\{e^{\mathbf{l}(\sigma)} - 1\}^{t(\sigma)}}{t(\sigma)!} = \prod_\sigma \frac{\lambda(\sigma)^{t(\sigma)}}{t(\sigma)!} = (V\lambda)(t).$$

Hence $\Phi' V \mathbf{l} = V \lambda$. ∎

According to Eq. (9), the relation between $\mathbf{m}$ and $\mathbf{l}$ is $\mathbf{m} = B'V\mathbf{l}$. The results (12) and (15) yield a consequent relation between $\mathbf{m}$ and λ:

$$\mathbf{m} = B'V\mathbf{l} = P'D\Phi'V\mathbf{l} = P'DV\lambda.$$

That is, since $U = DV$,

$$\mathbf{m} = P'U\lambda. \tag{16}$$

At this point we refer to the equations in Section 4 of Tutte's paper, remembering that our $\mathbf{m}$ is his $\mathbf{j}$. It is best to look at the statement of his Theorem 4.3, when we find an exact correspondence with our (16), given that λ replaces $-\lambda$. Thus the relationship between the two theories is established. We remark that the cluster expansion (8) corresponds to the product formula in Tutte's Theorem 4.4.

5. Conclusion

The formal power series used by Tutte as the setting for his results have certain advantages over the numerical functions utilised in this paper. For example, it is not necessary to insist on the positivity of an expression just because it is equal to the exponential of another expression. However, in practice one is especially interested in the singularities of the functions under discussion, and for this reason numerical estimates of the radii of convergence of series are important. In the case of the colouring problem, there is a series for $l(\sigma)$ in powers of z^{-1} (where z is the number of colours), and it has a radius of convergence $R(\sigma)$ which depends on σ. The problem of finding a good estimate for $R(\sigma)$ is an important one.

Another interesting problem is to enlarge the scope of the theory so that it applies to infinite graphs. The approach through linear functionals seems promising, since there are many theorems about extending such functionals.

REFERENCES

1. G. A. Baker, Linked-cluster expansion for the graph-vertex coloration problem. *J. Combinatorial Theory Ser. B* **10** (1971) 217–231.
2. N. L. Biggs, "Interaction Models," London Mathematical Society Lecture Notes Series No. 30, Cambridge Univ. Press, London and New York, 1977.
3. N. L. Biggs, On cluster expansions, *Quart. J. Math. Oxford* (2) **29** (1978) 159–173.
4. L. Comtet, "Advanced Combinatorics," Reidel Publ., Dordrecht, Netherlands, 1974.
5. W. T. Tutte, On dichromatic polynomials, *J. Combinatorial Theory* **2** (1967) 301–320.
6. H. Whitney, The coloring of graphs, *Ann. of Math.* **33** (1932) 688–718.

AMS 05C99

DEPARTMENT OF MATHEMATICS
ROYAL HOLLOWAY COLLEGE
EGHAM, SURREY, ENGLAND

Matroids, Graphs, and 3-Connectivity

ROBERT E. BIXBY†

and

WILLIAM H. CUNNINGHAM‡

1. Introduction

This paper describes results on 3-connectivity in matroids which, like much of the work of Tutte [9, 10, 11], are based on properties of cocircuits and their bridges. The results are of two kinds: those concerning general matroids, and those concerning binary matroids. The first group of results presents a recursive characterization of 3-connectivity, leading to an efficient matroid algorithm for detecting 3-connectivity or its absence. The second group of results generalizes to binary matroids theorems of Tutte on 3-connected bond-matroids; in particular, we give a characterization of those binary matroids which are graphic (that is, polygon-matroids), generalizing a theorem of Tutte on planarity.

---

† Research partially supported by National Science Foundation Grant MPS74-075-07506-A01.

‡ Research partially supported by National Science Foundation Grant MCS76-08803 to Johns Hopkins University and by a grant from Carleton University.

Terminology used but not defined here is defined by Welsh [14]. Where r is the rank function of a matroid M on E, a *separator* of M is a set $S \subseteq E$ such that $r(S) + r(E\setminus S) = r(E)$. An *elementary separator* of M is a minimal nonempty separator. A matroid is *nonseparable* if it has at most one elementary separator. A *parallel-class* of a matroid is a set consisting of an element and all elements in M parallel with it. The *simplification* of a matroid is the (simple) matroid obtained by deleting each loop and identifying the members of each parallel-class. Given a basis B of M, a *B-fundamental cocircuit* of M is a cocircuit having just one element from B. A *2-separation* of a nonseparable matroid M is a partition $\{E_1, E_2\}$ of E such that $|E_1|, |E_2| \geq 2$ and $r(E_1) + r(E_2) = r(E) + 1$. Matroid M is *3-connected* if it is nonseparable and has no 2-separation. The importance of 3-connection has been emphasized by Whitney's uniqueness theorem for graphic matroids [15], Tutte's "wheels and whirls" theorem [12], and work on matroid decomposition [1, 3, 4].

Let Y be a cocircuit of the nonseparable matroid M on E. A *bridge* of Y in M is an elementary separator of $M\setminus Y$, the matroid obtained by deleting Y from M. If Y has more than one bridge, Y is a *separating* cocircuit; otherwise Y is *nonseparating*. A *Y-component* of M is a matroid of the form $M/(E\setminus(B \cup Y))$, the contraction of M to $B \cup Y$, where B is a bridge of Y. (Notice that Y is a nonseparating cocircuit of each Y-component.) Let $B_1, B_2, \ldots, B_k$ be the bridges of Y, having corresponding Y-components $M_1, M_2, \ldots, M_k$. A B_i-*segment* of Y is a parallel-class of M_i contained in Y; let π_i denote the partition of Y into its B_i-segments, for $1 \leq i \leq k$. For $i \neq j$, B_i *avoids* B_j if there exists $S_i \in \pi_i$ and $S_j \in \pi_j$ such that $S_i \cup S_j = Y$. The *avoidance graph* of Y is the simple graph whose vertices are the bridges of Y, with two bridges adjacent if and only if they are not avoiding. All of these concepts were introduced by Tutte [11] for binary matroids, as part of his theory characterizing graphic matroids. [The reader may find it useful to apply these concepts to the polygon-matroid of a nonseparable graph G; Y will be a minimal edge-cutset of G. In this context the following observations can be made: (1) If Y is nonseparating, then it is the set of edges incident with some vertex of G; (2) for any cocircuit Y, the avoidance graph of Y is bipartite.]

The main result of the first part of the paper is the following theorem, which relates the 3-connectivity of a matroid to properties of its Y-components, for any cocircuit Y.

Theorem 1 *Let M be a nonseparable matroid which is both simple and cosimple, and let Y be a cocircuit of M. Then*

(a) *M has a 2-separation $\{E_1, E_2\}$ such that $Y \subseteq E_2$ if and only if the simplification of some Y-component has a 2-separation;*

(b) *M has a 2-separation $\{E_1, E_2\}$ such that Y meets both E_1 and E_2 if and only if the avoidance graph of Y is not connected.*

Theorem 1 suggests a recursive procedure for testing a matroid for 3-connectivity. If M is simple, cosimple, and nonseparable and the avoidance graph of Y is connected, then testing 3-connectivity of M can be accomplished by testing 3-connectivity of the simplifications of the Y-components. These are smaller than M unless Y is nonseparating; the following result provides a method of finding a separating cocircuit (or detecting that M is 3-connected).

Theorem 2 *Let M be a simple, nonseparable matroid, and let B be a basis of M. If every B-fundamental cocircuit is nonseparating, then M is 3-connected.*

A polygon of a connected graph G is defined to be *peripheral* if its edge-set is a nonseparating cocircuit of the bond-matroid of G. In an important paper [10], Tutte used peripheral polygons of 3-connected graphs to obtain new results on planar graphs (and to prove essentially all the previously known characterizations). Each of the following results on binary matroids (which are proved in Section 3) is, for the special case of bond-matroids, a theorem of [10].

Theorem 3 *Let M be a 3-connected binary matroid on E, where $|E| \geq 4$, and let $e \in E$. Then e is an element of two distinct nonseparating cocircuits of M.*

Theorem 4 *Let M be a 3-connected binary matroid. Then each cocircuit of M is a mod 2 sum of nonseparating cocircuits.*

Theorem 5 *A 3-connected binary matroid M is the polygon-matroid of a graph if and only if each element is contained in at most two nonseparating cocircuits.*

2. Recursive Characterization and Algorithm for 3-Connectivity

In this section we give proofs of Theorems 1 and 2 and show that the resulting 3-connectivity algorithm is "good," or "polynomial-time-bounded." We will use elementary results on matroids whose proofs may be found, for example, in [14]; these include the submodularity of the rank function, descriptions of separability in terms of the rank function and the circuit family, and descriptions of minors in terms of their rank functions and their circuit families.

Lemma 1 *If C_1, C_2 are complementary separators of $M \backslash Y$ and r' is the rank function of $M' = M/C_2$, then for any $A \subseteq C_1$, $r'(A) + r'((C_1 \cup Y) \backslash A) - r'(C_1 \cup Y) = r(A) + r(E \backslash A) - r(E)$.*

Proof The left hand side of the above expression is equal to
$$r(A \cup C_2) - r(C_2) + r(E \backslash A) - r(C_2) - (r(E) - r(C_2))$$
$$= r(A) + r(C_2) - r(C_2) - r(E) + r(E \backslash A)$$
$$= r(A) + r(E \backslash A) - r(E).$$ ∎

It follows that, in the context of Lemma 1, M' is nonseparable if M is; in particular, Y-components of a nonseparable matroid are nonseparable.

Lemma 2 *If $\{E_1, E_2\}$ is a 2-separation of the nonseparable matroid M and the cocircuit Y of M meets both E_1 and E_2, then $E_1 \backslash Y$ is a separator of $M \backslash Y$. In addition, if M is simple, then Y is separating.*

Proof If $r(E_1 \backslash Y) < r(E_1)$ and $r(E_2 \backslash Y) < r(E_2)$, then $r(E_1 \backslash Y) + r(E_2 \backslash Y) \leq r(E_1) + r(E_2) - 2 = r(E) - 1 = r(E \backslash Y)$. Thus we may assume that $r(E_1 \backslash Y) = r(E_1)$ or $r(E_2 \backslash Y) = r(E_2)$, say the former. Then in the matroid $M/(E_1 \backslash Y)$, Y is a cocircuit and $Y \cap E_1$ is a set of loops. But this implies that $Y \cap E_1 = \emptyset$, a contradiction. Therefore the first statement of the lemma is proved. Now, if Y is nonseparating, it must be that $Y \supseteq E_1$ (say). But then, since Y meets E_2, we have $r(E_2) = r(E)$, so $r(E_1) = 1$. If M is simple, this is not possible. ∎

Proof of Theorem 1(a) Suppose that the simplification of M_i, the Y-component corresponding to bridge B_i, is not 3-connected. Then M_i has a 2-separation $\{A_1, A_2\}$ such that $r_i(A_1), r_i(A_2) \geq 2$, where r_i is the rank function of M_i. By Lemma 2, since Y is a nonseparating cocircuit of M_i, we may assume that $Y \subseteq A_2$. By Lemma 1, $1 = r_i(A_1) + r_i(A_2) - r_i(B_i \cup Y) = r(A_1) + r(E \backslash A_1) - r(E)$. Clearly $|A_1|, |E \backslash A_1| \geq 2$, so M has a 2-separation $\{E_1, E_2\}$ with $Y \subseteq E_2$.

Now suppose that M has a 2-separation $\{E_1, E_2\}$ such that $Y \subseteq E_2$. Let $A \subseteq E_1$ be such that, for each j, $A \cap B_j = E_1 \cap B_j$ or $A \cap B_j = \emptyset$. Then $r(A) + r(E_1 \backslash A) = r(E_1)$, so $r(A) + r(E \backslash A) \leq r(A) + r(E_2) + r(E_1 \backslash A) = r(E_1) + r(E_2) = r(E) + 1$. If $|B_j \cap E_1| = 1$ for two different choices of j, we have a set A as just shown with $|A| = 2$ and $r(A) + r(E \backslash A) \leq r(E) + 1$, contradicting the fact that M is simple and cosimple. Therefore, since $|E_1| \geq 2$, there exists i such that $|E_1 \cap B_i| \geq 2$. Where $A = E_1 \cap B_i$, we have $r(A) + r(E \backslash A) = r(E) + 1$, so by Lemma 1, where r_i is the rank function of the Y-component M_i corresponding to B_i, $r_i(A) + r_i((B_i \backslash A) \cup Y) = r_i(B_i \cup Y) + 1$. It is easy to check that $r_i(A), r_i((B_i \backslash A) \cup Y) \geq 2$, so the simplification of M_i is not 3-connected. ∎

The proof of Theorem 1(b) (especially the "if" part) is somewhat more difficult. We will need the following two lemmas. The truth of the first result (Lemma 3) is, perhaps, surprising; it should be pointed out that its conclusion is not generally true for every component of the avoidance graph.

Lemma 3 *If the avoidance graph of Y is not connected, then there exists a component of the graph having vertex-set C and a partition $\{T_1, T_2\}$ of Y such that:*

T_1 *contains all but one B-segment for each bridge $B \in C$;*
T_2 *contains all but one B-segment for each bridge $B \notin C$.*

Proof We first prove *claim* 1: For the vertex-set C of any component of the avoidance graph, there exists, for each $B_i \notin C$, a set $S_i \in \pi_i$ such that, for all $B_j \in C$, there exists $S_{ij} \in \pi_j$ with $S_i \cup S_{ij} = Y$. If this is not true, then there exist adjacent bridges B_j, $B_k \in C$ and $S_j \in \pi_j$, $S_k \in \pi_k$ and distinct S_i, $S_i' \in \pi_i$ with $S_i \cup S_j = Y = S_i' \cup S_k$. But then $S_j \supseteq Y \setminus S_i \supseteq S_i' \supseteq Y \setminus S_k$, so $S_j \cup S_k = Y$, a contradiction.

Next, we prove *claim* 2: Where C and D are the vertex-sets of any two components of the avoidance graph, there exists $S_i \in \pi_i$ for all $B_i \in C \cup D$ such that, for all $B_i \in C$ and $B_j \in D$, $S_i \cup S_j = Y$. For if this is not true, by claim 1 there exist adjacent bridges B_i, $B_k \in D$ and bridge $B_j \in C$ such that $S_{ij} \neq S_{kj}$. Then $S_i \supseteq Y \setminus S_{ij} \supseteq S_{kj} \supseteq Y \setminus S_k$, so $S_i \cup S_k = Y$, a contradiction.

We now prove Lemma 3 by induction on the number k of components of the avoidance graph. Consider first the case $k = 2$, and let C, D be the vertex-sets of the components. Applying claim 2, and letting T_1 be $\cap(S_i: B_i \in D)$ and T_2 be $Y \setminus T_1$, this case is verified. Now assume that the result is true whenever the avoidance graph has fewer than $k \geq 3$ components, and suppose that the avoidance graph in question has components with vertex-sets $C_1, C_2, \ldots, C_k$. We make use of the induction hypothesis by observing that, if certain of the bridges of Y are contracted from M, then in the new matroid, Y is still a cocircuit and its bridges are just those that remain, and their segments are the same as before. Thus we may assume that C_{k-1} (say) has the property that there exists a partition $\{T_1, T_2\}$ of Y such that:

T_1 contains all but one B-segment for $B \in C_{k-1}$;
T_2 contains all but one B-segment for $B \in C_1 \cup C_2 \cup \cdots \cup C_{k-2}$.

Furthermore, since the case $k = 2$ has been proved, there exists a partition $\{T_1', T_2'\}$ of Y such that:

T_1' contains all but one B-segment for $B \in C_{k-1}$;
T_2' contains all but one B-segment for $B \in C_k$.

Now if $T_1 \cap T_1'$ contains all but one B-segment for every B in C_{k-1}, we are done, because $T_2 \cup T_2'$ contains all but one B-segment for every $B \in C_k \cup C_1 \cup \cdots \cup C_{k-2}$. Therefore, we may assume that there exists $B_i \in C_{k-1}$ such that, for distinct S_i, $S_i' \in \pi_i$, we have $S_i \cup T_1 = Y = S_i' \cup T_1'$. Then $S_i \supseteq T_2$ and $S_i' \supseteq T_2'$, so $T_2 \cap T_2' = \emptyset$; it follows that $T_1' = Y \setminus T_2' \supseteq T_2$. Thus

T_1' contains all but one B-segment for $B \in C_1 \cup \cdots \cup C_{k-1}$;
T_2' contains all but one B-segment for $B \in C_k$. ∎

Lemma 4 *Let A_1, A_2 be disjoint separators of $M \setminus Y$ and let S, $\emptyset \subset S \subset Y$, be such that the elements of S are in parallel in both $M/(E \setminus (A_1 \cup Y))$ and $M/(E \setminus (A_2 \cup Y))$. Then the elements of S are in parallel in $M/(E \setminus (A_1 \cup A_2 \cup Y))$.*

Proof Let a, b be distinct elements of S. Since a, b are in parallel in both $M/(E\backslash(A_1 \cup Y))$, $M/(E\backslash(A_2 \cup Y))$, there exist circuits C_1, C_2 of M such that $C_1 \cap (A_1 \cup Y) = \{a, b\} = C_2 \cap (A_2 \cup Y)$. If $C_1 \cap A_2 = \emptyset$, then a, b are in parallel in $M/(E\backslash(A_1 \cup A_2 \cup Y))$, and we are done. Otherwise, there exists $c \in C_1 \cap A_2$. Therefore, there is a circuit C of M such that $c \in C \subseteq (C_1 \cup C_2)\backslash\{a\}$. But then $|C \cap Y| \leq 1$, so $C \cap Y = \emptyset$. Since A_2 is a separator of $M\backslash Y$, this implies that $C \subseteq A_2$, whence $C \subset C_1$, a contradiction. The proof is complete. ∎

Proof of Theorem 1(b) Suppose that M has a 2-separation $\{E_1, E_2\}$ such that Y meets both E_1 and E_2. Then by Lemma 2, $r(E_1\backslash Y) + r(E_2\backslash Y) = r(E\backslash Y)$, so each bridge B_i is contained in E_1 or E_2. Let $T_i = E_i \cap Y$ for $i = 1$ and 2. Let r', r'' be the rank functions of $M/(E_2\backslash T_2)$, $M/(E_1\backslash T_1)$, respectively. Then $r'(T_2) + r''(T_1) = r(E_2) - r(E_2\backslash T_2) + r(E_1) - r(E_1\backslash T_1) = r(E) + 1 - (r(E) - 1) = 2$. Since $T_1, T_2 \neq \emptyset$, it follows that the elements of T_2 are in parallel in $M/(E_2\backslash T_2)$ and that the elements of T_1 are in parallel in $M/(E_1\backslash T_1)$. Therefore, for every bridge $B_i \subseteq E_2$, T_1 is contained in a B_i-segment, and for every $B_i \subseteq E_1$, T_2 is contained in a B_i-segment. Thus every B_i contained in E_1 avoids every B_i contained in E_2. Thus the avoidance graph of Y is not connected, provided that each of E_1, E_2 contains at least one bridge. Now suppose that E_1 (say) contains no B_i, so that $E_1 = T_1$. Since $T_2 \neq \emptyset$, $r(E_2) = r(E)$, so $r(E_1) \leq 1$, and, since M is simple, $\{E_1, E_2\}$ is not a 2-separation of M.

Now, suppose that the avoidance graph of Y is not connected. Then we may choose C and $\{T_1, T_2\}$ as in Lemma 3. Let $E_1 = T_1 \cup (\cup(B_i : B_i \in C))$ and let $E_2 = T_2 \cup (\cup(B_i : B_i \notin C))$. Then the elements of T_1 are in parallel in $M/(E\backslash(Y \cup B_i))$ for each $B_i \notin C$, so by Lemma 4, the elements of T_1 are in parallel in $M/(E\backslash(Y \cup (E_2\backslash T_2))) = M/(E_1\backslash T_1)$. Similarly the elements of T_2 are in parallel in $M/(E_2\backslash T_2)$. Letting r'', r' be the rank functions of these two minors, we have $2 = r''(T_1) + r'(T_2) = r(E_1) - r(E_1\backslash T_1) + r(E_2) - r(E_2\backslash T_2) = r(E_1) + r(E_2) - r(E\backslash Y) = r(E_1) + r(E_2) - r(E) + 1$. Therefore, $r(E_1) + r(E_2) = r(E) + 1$. Since $T_1, T_2 \neq \emptyset$, and since each of E_1, E_2 contains at least one B_i, we have $|E_1|, |E_2| \geq 2$, and the proof is complete. ∎

Proof of Theorem 2 Since M is nonseparable, for every partition $\{E_1, E_2\}$ of E, there exists a B-fundamental cocircuit Y meeting both E_1 and E_2. If $\{E_1, E_2\}$ were a 2-separation, then Y would be separating, by Lemma 2. ∎

We now explain how this theory provides a good matroid algorithm for testing for 3-connectivity. First, we state the algorithm.

Algorithm for 3-Connectivity *Step* 0 Check that M is nonseparable, simple, and cosimple. If it fails to satisfy any of these conditions, stop; M is not 3-connected. (There are five small exceptions to this rule, all having fewer than four elements.)

Step 1 If M has rank 2 or less, stop; M is 3-connected.

Step 2 Choose a basis B of M and find a B-fundamental cocircuit Y which is separating. If none exists, stop; M is 3-connected.

Step 3 Check that the avoidance graph of Y is connected. If it is not, stop; M is not 3-connected.

Step 4 Apply the algorithm recursively to the simplifications of the Y-components.

We remark that Step 0 need be applied only to the initial matroid M. This is because Y-components of cosimple, nonseparable matroids are themselves cosimple and nonseparable, and their simplifications are, of course, simple. The main subroutine required for implementation of the above algorithm is an algorithm for finding the elementary separators of a matroid. There is a simple and efficient algorithm for this problem [3], which was first described by Tutte [9] for binary matroids. It is based on the observation that, given any basis B, a set S is a separator if and only if every B-fundamental cocircuit is contained in S or in $E\backslash S$. Beginning with the (0, 1)-matrix whose rows are the incidence vectors of the B-fundamental cocircuits of M for some basis B of M, the amount of computation required to find the elementary separators of M is bounded by a function of the order of $r(E) \cdot |E|$. It might appear that as many as $r(M')$ applications of this subroutine could be required when Step 2 of the 3-connectivity algorithm is applied to a matroid M'. Where M is the initial matroid, this would lead to a bound of the order of $(r(M))^2$ on the total number of applications of the subroutine. However, by refining the choice of B in Step 2, we can improve this bound, using the following result.

Proposition *Let B be a basis of the nonseparable matroid M, let Y be a B-fundamental cocircuit of M, and let M_i be the Y-component of M corresponding to the bridge B_i of Y. Then $B' = B \cap (B_i \cup Y)$ is a basis of M_i, and any B'-fundamental cocircuit Y' of M_i is a B-fundamental cocircuit of M. Moreover, if Y' is separating in M', then Y' is separating in M.*

Proof Since Y contains exactly one element e of B, $B\backslash\{e\}$ is a basis of $E\backslash Y$ and so $B \cap B_i$ is a basis of B_i. Since Y is a cocircuit of M_i, $B' = (B \cap B_i) \cup \{e\}$ is a basis of M_i. Now Y' is clearly a cocircuit of M, and it has just one element from B, so it is a B-fundamental cocircuit of M. Since Y, Y' are both B'-fundamental cocircuits of M_i, it is easy to see that $Y\backslash Y'$ is a cocircuit of $M_i\backslash Y'$. Therefore, if $\{A_1, A_2\}$ is a separation of $M_i\backslash Y'$, we may assume that $A_2 \supseteq Y\backslash Y'$; moreover, $B_i\backslash Y'$ is a separator of $M\backslash(Y \cup Y')$. Using Lemma 1 with $M\backslash Y'$ in place of M, we have, where r_i is the rank function of M_i, $0 = r_i(A_1) + r_i(A_2) - r_i((B_i \cup Y)\backslash Y') = r(A_1) + r((E\backslash Y')\backslash A_1) - r(E\backslash Y')$. Therefore, Y' is separating in M, as required. ∎

We have already observed that, where M_i is a Y-component of the nonseparable matroid M, Y is a nonseparating cocircuit of M_i. Therefore, it follows from the proposition that the total number of executions of the elementary separators subroutine, during the application of the 3-connectivity algorithm to matroid M, need not exceed $r(M) + 1$. Moreover, the proposition shows that the $(0, 1)$-matrix needed for application of the subroutine to any matroid M', is a submatrix of the corresponding matrix for the initial matroid M. Therefore, assuming the existence of efficient algorithms for constructing this initial matrix, and for computing parallel classes in M and its minors, we can assert that there is a bound on the total computational effort required by the elementary separators subroutine which is of the order of $(r(M))^2 \cdot |E|$. It is not difficult to see that, under the same assumption, the work required to execute the other steps of the 3-connectivity algorithm is dominated by this bound. (However, there are some interesting aspects of implementation of the other steps, especially Step 3; these are treated in [2], where the efficiency of such steps is crucial to the bound for the whole algorithm.) As a consequence we have a computation bound for the 3-connectivity algorithm of the order of $(r(M))^2 \cdot |E|$. (An algorithm for testing for k-connectivity for any $k \geq 2$ has been given by Cunningham and Edmonds [3, 4]; it requires the order of $|E|^{k-1}$ iterations of the major step of Edmonds' matroid intersection algorithm [5].) Where M is not 3-connected, one may be interested in actually finding a 2-separation. This is straightforward, unless the algorithm terminates in Step 3. In this case the proofs of Lemma 3 and Theorem 1(b) yield an efficient algorithm for computing a 2-separation of the matroid being considered.

3. Binary Matroids and Graphs

This section contains the proofs of Theorems 3, 4, and 5 and some further discussion of these results. We will be using a number of well-known properties of binary matroids; we will also need Tutte's "wheels" theorem [12] for binary matroids. One important concept due to Tutte, which perhaps requires some explanation, is that of "line." A *line* of a matroid M is a minimal set which is the union of two distinct circuits; equivalently, it is a union of circuits having rank 2 less than its cardinality. If M is binary, a line of M contains at most three circuits. We will often be dealing with *colines*: lines of the dual matroid. We will need the following two lemmas for the proof of Theorem 3. The first (Lemma 5) is just the statement that Tutte's definition [11] of B-segment is, in the case of a binary matroid, equivalent to ours. This result is 8.52 of [11], and we reproduce Tutte's proof for completeness. Where Y is a cocircuit of a binary matroid M and Z is a cocircuit of $M_i \setminus Y$ contained in bridge B_i, then $Z \cup Y$ is a coline of M_i. Tutte defines a *primary segment determined by B_i* to be a set $P \subseteq Y$ such that for some such Z, $P \cup Z$ and $(Y \backslash P) \cup Z$ are cocircuits of M_i.

Lemma 5 *If B_i is a bridge of the cocircuit Y of the nonseparable binary matroid M, the B_i-segments are the minimal nonempty intersections of the primary segments determined by B_i.*

Proof Let T_i be a minimal nonempty intersection of primary segments determined by B_i. Each primary segment contains a cocircuit of $M_i \backslash B_i$, and the B_i-segments are the minimal nonempty intersections of such cocircuits, so T_i contains a B_i-segment S_i. Suppose that $S_i \neq T_i$. Then there exists a cocircuit X of M_i which meets T_i but does not contain it, and $X \cap B_i$ is not a cocircuit of $M_i \backslash Y$. Choose X so that $|B_i \cap X|$ is minimum. Then $M_i \backslash Y$ has a cocircuit Z such that $Z \subset B_i \cap X$, and M_i has a cocircuit X' such that $X' \cap B_i = Z$. By the choice of X, X' contains or is disjoint from T_i. Consider the mod 2 sum $X + X'$. It meets but does not contain T_i, and it is a union of disjoint cocircuits (since M_i is binary), one of which meets but does not contain T_i, a contradiction. ∎

Lemma 6 *Let A, B, C be circuits of a matroid M such that (1) $A \cup B$ and $A \cup C$ are lines; (2) $A \not\subseteq B \cup C$; (3) $B \backslash A$ and $C \backslash A$ are contained in different elementary separators of M/A. Then $B \cup C$ is a line.*

Proof Let r' be the rank function of M/A. Then

$$r(A \cup B \cup C) = r'((B \cup C) \backslash A) + r(A)$$
$$= r'(B \backslash A) + r'(C \backslash A) + r(A)$$
$$= r(A \cup B) + r(A \cup C) - r(A)$$
$$= |A \cup B| - 2 + |A \cup C| - 2 - (|A| - 1)$$
$$= |A \cup B \cup C| - 3.$$

There exists $x \in A \backslash (B \cup C)$. Any set $D \subseteq (A \cup B \cup C) \backslash \{x\}$ has $|D| - r(D) \leq 2$ because $|\cdot| - r(\cdot)$ is a nondecreasing function. Thus $r(B \cup C) = |B \cup C| - 2$. ∎

Proof of Theorem 3 We will derive Theorem 3 from the following *claim*: Let Y be a separating cocircuit containing e, and B be a bridge of Y; then there exists a cocircuit Y' such that $e \in Y'$ and for some bridge B' of Y', $B \subset B'$. Applying the claim repeatedly, we obtain a nonseparating cocircuit J_1 containing e. Since M is 3-connected, there exists another cocircuit Y containing e and $J_1 \backslash Y$ meets a bridge B of Y. Again, applying the claim repeatedly, we obtain a second nonseparating cocircuit J_2 containing e.

It remains only to prove the claim. Since M is 3-connected, it follows from Theorem 1 that the avoidance graph of Y is connected. Therefore there exists a bridge B' of Y such that B does not avoid B'. It follows from Lemma 5 that there exists a primary segment P determined by B and a primary segment P' determined by B' such that $e \in P'$, $P \nsubseteq P'$, and $P \cup P' \neq Y$. (Otherwise B does not avoid B'.) Therefore, we may choose $D \subseteq B$ and $D' \subseteq B'$ such that $C = D \cup P$ and $C' = D' \cup P'$ are cocircuits of M and $C \cup Y$, $C' \cup Y$ are colines of M. It follows from Lemma 6 that $C \cup C'$ is a coline of M, so $C \backslash C'$ is a cocircuit of $M \backslash C'$. Since $P \nsubseteq P'$, this cocircuit meets both B and Y, so $M \backslash C'$ has a bridge properly containing B; moreover, $e \in C'$. ∎

The proof of Theorem 3 given above is based on an attempt to generalize Tutte's proof [10] for bond-matroids. Theorem 4 seems to be more difficult to prove. Although the argument in [10] can be generalized to a proof of Theorem 4, a good deal of rather complicated binary-matroid theory seems to be required. In order to simplify the presentation, we have chosen instead to prove Theorem 4 by induction using the "wheels" theorem. The following two lemmas will be needed.

Lemma 7 *If e is an element of the binary 3-connected matroid M and D is a nonseparating cocircuit of M/e, then D is a nonseparating cocircuit of M or there exists a partition $\{P, Q\}$ of D such that $P \cup \{e\}$ and $Q \cup \{e\}$ are nonseparating cocircuits of M.*

Proof The set D is a cocircuit of M. If D is separating, then $M \backslash D$ is separable while $(M \backslash D)/e$ is not, so $\{e\}$ is a separator of $M \backslash D$. Therefore, $\{e\}$ is a coloop of $M \backslash D$, so $D \cup \{e\}$ is a coline of M. Since M is binary, there exists a partition $\{P, Q\}$ of D such that $P \cup \{e\}$ and $Q \cup \{e\}$ are cocircuits of M. Because e is a coloop of $M \backslash D$, we have $(M/e) \backslash D = M \backslash (D \cup \{e\}) = (M \backslash (P \cup \{e\})) \backslash Q$, so $(M \backslash (P \cup \{e\})) \backslash Q$ is nonseparable. If $P \cup \{e\}$ is a separating cocircuit of M, it must be that Q is a separator of $M \backslash (P \cup \{e\})$. Therefore, $r(Q) + r(E \backslash (D \cup \{e\})) = r(E \backslash (P \cup \{e\})) = r(E) - 1$. But, since $D \cup \{e\}$ is a coline, $r(E \backslash (D \cup \{e\})) = r(E) - 2$, so $r(Q) = 1$. If $|E| \geq 4$, then since M is 3-connected, $|Q| = 1$; but then $Q \cup \{e\}$ is a cocircuit of cardinality 2, a contradiction. Therefore, $P \cup \{e\}$ is nonseparating, and similarly for $Q \cup \{e\}$. (When $|E| \leq 3$, the truth of the lemma is easily checked.) ∎

Lemma 8 *If e is an element of the binary 3-connected matroid M and D is a nonseparating cocircuit of $M \backslash e$, then D is a nonseparating cocircuit of M, or $D \cup \{e\}$ is a nonseparating cocircuit of M, or there exists a partition $\{P, Q\}$ of D such that $P \cup \{e\}$ and $Q \cup \{e\}$ are nonseparating cocircuits of M.*

Proof Either D or $D \cup \{e\}$ is a cocircuit of M. If $D \cup \{e\}$ is, then it is clearly nonseparating. If D is a cocircuit of M and is separating, then $D \cup \{e\}$ is a coline and the lemma can be proved in a manner similar to Lemma 7. ∎

Proof of Theorem 4 The result is easily seen to be true when $|E| \leq 2$ or M is the polygon-matroid of a wheel graph [11] (indeed, of any 3-connected graph). Let us assume that it is true whenever $|E| < k \geq 3$ and suppose that M has k elements. If M is the polygon-matroid of a wheel, we are done; otherwise by the wheels theorem, there exists $e \in E$ such that $M \backslash e$ or M/e is 3-connected.

Case 1: M/e *Is 3-connected* Let $N(M)$ be the matrix whose rows are the incidence vectors of nonseparating cocircuits of M. It follows from Lemma 7 that for every vector x in the row space (over the binary field) of $N(M/e)$, $(x, 0)$ is in the row space of $N(M)$. In addition, by Theorem 3, there is a row of $N(M)$ having a 1 in the column corresponding to e. Thus $N(M)$ has rank at least the rank of $N(M/e)$ plus 1. By the induction hypothesis, $N(M/e)$ has rank $r(M/e) = r(M) - 1$. It follows that $N(M)$ has rank at least $r(M)$, and so each cocircuit is a mod 2 sum of nonseparating cocircuits, as required.

Case 2: $M \backslash e$ *Is 3-Connected* Each cocircuit D' of M which is not of the form D or $D \cup \{e\}$ for some cocircuit D of $M \backslash e$ is a mod 2 sum of two cocircuits of M of the second form. Thus it is enough to prove the result where D' is D or $D \cup \{e\}$ for some cocircuit D of $M \backslash e$. By the induction hypothesis, D is the mod 2 sum $\sum D_i$ of nonseparating cocircuits D_i of $M \backslash e$. Using Lemma 8, each D_i can be replaced in the sum by itself, or by $D_i \cup \{e\}$, or by $P_i \cup \{e\}$ and $Q_i \cup \{e\}$, where $\{P_i, Q_i\}$ is a partition of D_i, such that the new sum $\sum D_i'$ consists of nonseparating cocircuits of M. This mod 2 sum is either D or $D \cup \{e\}$, and since M is binary, it must be a disjoint union of cocircuits. Thus $\sum D_j'$ is whichever of D, $D \cup \{e\}$ is a cocircuit, that is, $\sum D_j' = D'$. ∎

Proof of Theorem 5 The condition that each element be contained in at most 2 nonseparating cocircuits is clearly necessary, because a nonseparating cocircuit of the polygon matroid of a graph G must be the set of edges incident with some vertex of G. It is easy to check that the condition is sufficient when $|E| \leq 3$. When $|E| \geq 4$, consider the matrix $N(M)$ whose rows are the incidence vectors of nonseparating cocircuits of M. By Theorem 4, M is the linear independence matroid (over the binary field) of $N(M)$. But it follows from Theorem 3 and the hypothesis, that $N(M)$ has exactly 2 ones in each column. Therefore, $N(M)$ is the incidence matrix of a graph, and so M is the polygon-matroid of that graph. ∎

We remark that both of the hypotheses (binary and 3-connected) are needed for Theorems 3 and 5, and that 3-connection is necessary in Theorem 4. There exist 3-connected nonbinary matroids having no nonseparating cocircuits at all; an example is the matroid having 5 elements, every 3-element subset being a basis. Such matroids would be counterexamples to both

Theorem 3 and Theorem 5 if the binary hypotheses were dropped. The necessity of the hypothesis of 3-connection in Theorems 3 and 4 is demonstrated by any simple nonseparable polygon-matroid having a 2-separation. Its necessity in Theorem 5 is demonstrated by taking any nongraphic binary matroid and replacing each element by 2 elements in series. (Less trivially, one can compose [1, 3] two Fano matroids to obtain a nonseparable, nongraphic binary matroid having 12 elements, for which every element is in exactly two nonseparating cocircuits.)

When only bond-matroids are considered, a result stronger than Theorem 3 can be proved. Tutte [10] shows that, for any element e of a 3-connnected bond-matroid having $|E| \geq 4$, there exist 2 nonseparating cocircuits having only e in common. This result does not extend to general binary matroids; the Fano matroid provides a counterexample. However, the method of proof of Theorem 3 does yield the following extension: For any element e of a 3-connected binary matroid having $|E| \geq 4$, there exist nonseparating cocircuits $J_1, J_2, \ldots, J_k$ such that $\cap J_i = \{e\}$.

A cocircuit of a matroid M is said to be *nodal* (or *totally bridge-separable* [11]) if its avoidance graph has no edges. It is easy to see that, in a polygon-matroid (3-connected or not), every cocircuit is a mod 2 sum of nodal cocircuits. Tamari [8] has extended this result to arbitrary binary matroids. For the case of 3-connected binary matroids, his result follows immediately from Theorem 4, since every nonseparating cocircuit is nodal. In addition, it is straightforward, using decomposition arguments [1, 3], to deduce Tamari's result itself from Theorem 4. Tamari also gives a procedure for constructing a "cocircuit basis" consisting of nodal cocircuits. Such a procedure can also be derived from our methods.

Finally, we wish to discuss the relationship of Theorem 5 to three other characterizations of graphic matroids [6, 7, 13] which it resembles. First, we note that Theorem 5 is a "good characterization," in the sense that it provides a short proof that a matroid is not graphic. (One simply exhibits an element and three nonseparating cocircuits containing it.) On the other hand, both Welsh's theorem [13] and Sachs's theorem [7] characterize "graphicness" as being equivalent to the existence of a family of cocircuits satisfying certain properties, and so these results are not good characterizations.

Fournier's theorem [6] is, however, more closely related to Theorem 5. Given three cocircuits Y_1, Y_2, Y_3 of a matroid M, we say that Y_1 *does not separate* Y_2 and Y_3 if $Y_2 \setminus Y_1$ and $Y_3 \setminus Y_1$ are contained in the same elementary separator of $M \setminus Y_1$. Otherwise, Y_1 *separates* Y_2 and Y_3. Fournier's theorem says that M is graphic if and only if for any three cocircuits having a nonempty intersection, there exists one which separates the other two. It is easy to see that the condition is necessary. Sufficiency is more difficult to prove; Fournier's proof uses Tutte's deep excluded-minor characterization of graphic matroids [11]. For the binary 3-connected case, we can show that

Fournier's theorem follows from Theorem 5: If an element is contained in three distinct nonseparating cocircuits, then they have nonempty intersection and no one separates the other two. The result can now be extended to the class of all binary matroids by decomposition arguments [1, 3]. Finally, to go from the binary to the general case, we observe that any nonbinary matroid has a coline containing more than three cocircuits [11]. It is easy to see that such a coline contains three cocircuits having a common element, and that no one of these separates the other two. Therefore, the present methods provide a more elementary proof of Fournier's theorem.

ACKNOWLEDGMENT

Theorems 3, 4, and 5 were conjectured to one of the authors by Jack Edmonds and are also strongly suggested by remarks of Tutte [10].

REFERENCES

1. R. E. Bixby, Composition and decomposition of matroids and related topics, Ph.D. Thesis, Cornell University, Ithaca, New York, 1972.
2. R. E. Bixby and W. H. Cunningham, Converting linear programs to network problems, *Math. Operations Res.*, to appear.
3. W. H. Cunningham, A combinatorial decomposition theory, Ph.D. Thesis, University of Waterloo, Ontario, Canada, 1973.
4. W. H. Cunningham and J. Edmonds, Decomposition of matroids and linear systems, to appear.
5. J. Edmonds, Matroid intersection, *Ann. Discrete Math.*, to appear.
6. J.-C. Fournier, Une relation de séparation entre cocircuits d'un matroide, *J. Combinatorial Theory Ser. B* **12** (1974) 181–190.
7. D. Sachs, Graphs, matroids, and geometric lattices, *J. Combinatorial Theory* **9** (1970) 192–199.
8. R. Tamari, Combinatorial algorithms in certain classes of binary matroids, Ph.D. Thesis, Cornell University, Ithaca, New York, 1977.
9. W. T. Tutte, An algorithm for determining whether a given binary matroid is graphic, *Proc. Amer. Math. Soc.* **11** (1960) 905–917.
10. W. T. Tutte, How to draw a graph, *Proc. London Math. Soc.* **13** (1963) 743–768.
11. W. T. Tutte, Lectures on matroids, *J. Res. Nat. Bur. Standards Sect. B* **69** (1965) 1–47.
12. W. T. Tutte, Connectivity in matroids, *Canad. J. Math.* **18** (1966) 1301–1324.
13. D. J. A. Welsh, On the hyperplanes of a matroid, *Proc. Cambridge Philos. Soc.* **65** (1969) 11–18.
14. D. J. A. Welsh, "Matroid Theory," Academic Press, New York, 1976.
15. H. Whitney, Congruent graphs and the connectivity of graphs, *Amer. J. Math.* **54** (1932) 150–168.

AMS 05B35, 68A20

Robert E. Bixby
DEPARTMENT OF INDUSTRIAL ENGINEERING
 AND MANAGEMENT SCIENCE
NORTHWESTERN UNIVERSITY
EVANSTON, ILLINOIS

William H. Cunningham
DEPARTMENT OF MATHEMATICS
CARLETON UNIVERSITY
OTTAWA, ONTARIO

GRAPH THEORY AND RELATED TOPICS

On the Mixed Achromatic Number and Other Functions of Graphs

FRED BUCKLEY

and

A. J. HOFFMAN

1. Introduction

In this paper we introduce a graph-theoretic function which we call the mixed achromatic number of a graph, since its relation to the mixed chromatic number [6] is analogous to the relation of achromatic number [1] to chromatic number. These functions and certain other graph-theoretic functions will be defined in Section 3, and the purpose of this paper is to explore a very simple question about the relative rate of growth of these functions: for two such functions f and g, we will consider whether there can exist a sequence of graphs $G_1, G_2, \ldots,$ such that $f(G_n) \to \infty$ while $\{g(G_n)\}$ is bounded. If this is false, we will say f is *bounded* (from above) by a function of g. If f is bounded from above by a function of g, and g is bounded from above by a function of f, we will say that f and g are *equivalent functions*.

The formalism for considering this question about certain graph-theoretic functions is facilitated by regarding the set $\mathscr{G}$ of all graphs as a partially ordered set, with $G \leq H$ if G is (isomorphic to) an induced subgraph of H. Indeed, the formalism is appropriate for considering similar questions about functions on any infinite partially ordered set P. This we do in Section 2, where we also define the key notion of Ramsey function on P [7]. In Section 3, we define the seven graph-theoretic functions we will examine: $\mathrm{achr}(G) \equiv$ achromatic number of G, $\mathrm{eq}(G) \equiv$ number of equivalence classes of $V(G)$, $\mathrm{bip}(G) \equiv$ smallest cardinality of a partition of $E(G)$ into complete bipartite graphs; $\mathrm{achr}^*(G) \equiv$ mixed achromatic number of G, $\mathrm{eq}^*(G) \equiv$ number of mixed equivalence classes of G, $\mathrm{bip}^*(G) \equiv$ smallest cardinality of a partition of $E(G)$ into complete bipartite graphs and cliques, $\mathrm{match}^*(G) \equiv \min(\mathrm{match}(G), \mathrm{match}(\bar{G}))$, where $\mathrm{match}(G)$ is the largest number of independent edges in G. In Section 4, we prove

Theorem 1.1 *The functions* $\mathrm{achr}(G)$, $\mathrm{eq}(G)$, *and* $\mathrm{bip}(G)$ *are Ramsey functions, and they are equivalent.*

The fact that eq is bounded by a function of achr was proved in [2], achr bounded by a function of eq is trivial, and eq and bip equivalent was shown in [3]. The work reported in this paper began when we learned about [2] at the Qualicum Beach Conference in 1976 and in subsequent conversations at the University of Victoria, and recalled relevant material in [3]. We are grateful to Brian Alspach, Pavol Hell, and Donald Miller for these contacts, and also to Paul Erdös, A. Hajnal, and Peter Hammer for additional discussions. In particular, the discussions with Hammer helped in the formulation of the material in Section 5, where we prove

Theorem 1.2 *The functions* $\mathrm{achr}^*(G)$, $\mathrm{eq}^*(G)$, $\mathrm{bip}^*(G)$, *and* $\mathrm{match}^*(G)$ *are Ramsey functions,* eq^* *and* bip^* *are equivalent functions, and* achr^* *and* match^* *are equivalent functions. Further,* eq^* *is bounded by a function of* achr^*, *but not conversely.*

Finally, we will discuss in Section 6 the relationship of these seven functions of G to the spectrum of the (0,1) adjacency matrix $A(G)$ of G. This will be mostly a restatement of material already given in [3] and [4], and inferable from Theorem 1.1, and will be of the form that f and g are equivalent for five of the seven functions f already described and certain functions g of the spectrum of $A(G)$. But we will also show that such a relation cannot hold for $\mathrm{achr}^*(G)$ or $\mathrm{match}^*(G)$, even though it holds for the five other graph theoretic functions we are discussing.

Theorem 1.3 *There exist two sequences of graphs* $G_1, G_2, \ldots,$ *and* $H_1, H_2, \ldots,$ *such that* $A(G_n)$ *and* $A(H_n)$ *have, for each n, the same spectrum, but*

$$\mathrm{achr}^*(G_n) \to \infty, \quad \mathrm{achr}^*(H_n) = 2 \quad \text{for all } n,$$
$$\mathrm{match}^*(G_n) \to \infty, \quad \mathrm{match}^*(H_n) = 1 \quad \text{for all } n.$$

2. Functions, Especially Ramsey Functions, on a Partially Ordered Set

Let P be an infinite partially ordered set with an element $0 \in P$ such that $0 \leq p$ for all $p \in P$. We will consider real functions f on P such that

$$f(0) = 0, \tag{2.1}$$

$$p \leq q \quad \text{implies} \quad f(p) \leq f(q), \tag{2.2}$$

$$\sup_{p \in P} f(p) = \infty. \tag{2.3}$$

Note that (2.1) and (2.2) imply f is nonnegative. Call a function satisfying (2.1)–(2.3) *proper*. Suppose f and g are proper. We will say $f \leq g$ (that is, f is bounded by a function of g) if, for every $Q \subset P$

$$\sup_{p \in Q} f(p) = \infty \quad \text{implies} \quad \sup_{p \in Q} g(p) = \infty. \tag{2.4}$$

Assume f proper. An *f-chain* C of P is a sequence

$$0 = p_0 < p_1 < p_2 < \cdots \tag{2.5}$$

such that

$$f(p_n) \to \infty. \tag{2.6}$$

Let C be an f-chain (2.5) and define the following function on P:

$$h_c(p) = \max\{n \mid p_n \leq p\}. \tag{2.7}$$

Note that since $0 = p_0 \leq p$, (2.2) and (2.6) imply that $h_c(p)$ is a proper function on P, and $h_c \leq f$. Also, if $f_1, \ldots, f_t$ are proper, and $f_i \leq f$ for all i, then $\max_{i=1,\ldots,t} f_i \leq f$, and is proper. Let $\mathscr{A}$ be a finite collection of f-chains $C_1, \ldots, C_t$, and let $h_\mathscr{A}(p) = \max_{i=1,\ldots,t} h_{c_i}(p)$. By the foregoing, we know $h_\mathscr{A}$ is proper, and $h_\mathscr{A} \leq f$. If, conversely, there exists a finite collection $\mathscr{A}$ of f-chains such that

$$f \leq h_\mathscr{A} \tag{2.8}$$

then f is said to be a *Ramsey function* on P [7].

This definition was given in [7], along with examples of certain P and proper f that were, and were not, Ramsey functions. In what follows, our partially ordered set will be the set $\mathscr{G}$ of all graphs, with $0 = \emptyset$, and $G \leq H$ means G is (isomorphic to) an induced subgraph of H.

3. Some Graph-Theoretic Functions and Some Chains of Graphs

We begin with some notation. If S is a set, $|S|$ is its cardinality. If G is a graph, $\bar{G}$ is the graph with $V(\bar{G}) = V(G)$, and $A(\bar{G}) + A(G) = J - I$, where

J is the matrix of all ones. If $S \subset V(G)$, $\langle S \rangle$ is the induced subgraph of G subtended by the vertices in S.

If G is a graph, its *achromatic number* $\mathrm{achr}(G)$ is the largest r such that there exists a partition

$$V(G) = S_1 \cup \cdots \cup S_r \tag{3.1}$$

with each S_i an independent set, but $S_i \cup S_j$ ($i \neq j$) is not an independent set.

Say that vertices i and j of G are *equivalent* if $i = j$ or $i \neq j$ but every vertex k is adjacent to i if and only if k is adjacent to j. Since i is not adjacent to itself, it is clear that definition yields an equivalence relation on $V(G)$ such that the corresponding equivalence classes form a partition

$$V(G) = F_1 \cup \cdots \cup F_t \tag{3.2}$$

in which each F_i is an independent set. We denote the number t of equivalence classes by $\mathrm{eq}(G)$.

Next, let $B_1, \ldots, B_u$ be complete bipartite graphs such that each $V(B_i) \subset V(G)$ and

$$E(G) = E(B_1) \cup \cdots \cup E(B_u) \tag{3.3}$$

is a partition of $E(G)$. The smallest possible u such that such a partition (3.3) exists will be denoted by $\mathrm{bip}(G)$.

Now we define the corresponding *-functions. Let r^* be the largest integer such that there exists a partition

$$V(G) = S_1 \cup \cdots \cup S_{r^*} \tag{3.1*}$$

and each S_i an independent set or $\langle S_i \rangle$ is a clique, but for $i \neq j$, $S_i \cup S_j$ is not an independent set and $\langle S_i \cup S_j \rangle$ is not a clique. Note that this implies

$$|\{i \mid |S_i| = 1\}| \leq 1. \tag{3.1*a}$$

The *mixed achromatic number* $\mathrm{achr}^*(G)$ is the number r^*.

Say that vertices i and j of G are *mixed equivalent* if $i = j$ or $i \neq j$ and every vertex $k \neq i, j$ is adjacent to i if and only if k is adjacent to j. This definition yields an equivalence relation on $V(G)$ such that the vertices F of each (mixed) equivalence class have the property that either F is an independent set or $\langle F \rangle$ is a clique. If the corresponding partition of $V(G)$ is

$$V(G) = F_1 \cup \cdots \cup F_{t^*} \tag{3.2*}$$

then we denote the number t^* of mixed equivalence classes by $\mathrm{eq}^*(G)$.

Let $\mathrm{bip}^*(G)$ be the smallest integer u^* such that there exist graphs $G_1, G_2, \ldots, G_{u^*}$ with each $V(G_i) \subset V(G)$,

$$E(G) = E(G_1) \cup \cdots \cup E(G_{u^*}) \tag{3.3*}$$

a partition of $E(G)$, each G_i a clique or a complete bipartite graph.

For match* and each of these six functions, the value of these functions for the graph $\emptyset$ is 0. If we consider each of these functions as functions on the partially ordered set $\mathcal{G}$, it is obvious that it is proper [i.e., satisfies (2.1)–(2.3)].

We now define several chains of graphs in $\mathcal{G}$, which will be needed to show that the different functions are Ramsey functions on $\mathcal{G}$. Each of these chains is of the form

$$\emptyset = G_0 < G_1 < G_2 < \cdots, \tag{3.4}$$

where we must specify each G_n. We will let $G = G_1 \cup G_2$ have the meaning: $V(G_1)$ and $V(G_2)$ are disjoint, $V(G) = V(G_1) \cup V(G_2)$, and $E(G) = E(G_1) \cup E(G_2)$. Also $G = G_1 + G_2$ has the meaning: $V(G_1)$ and $V(G_2)$ are disjoint, $V(G) = V(G_1) \cup V(G_2)$, $E(G) = E(G_1) \cup E(G_2)$ and all edges joining $V(G_1)$ with $V(G_2)$. The symbol nG means the graph formed by n disjoint copies of G.

$G_n = K_n$ (the clique on n vertices);   (3.4a)

$G_n = nK_2$ (sometimes called the *ladder graph*);   (3.4b)

G_n is the bipartite graph, with the vertices of one part denoted by $1, \ldots, n$, the vertices of the other part denoted by $1', \ldots, n'$, and i and j' adjacent if and only if $i \neq j$. Call this graph W_n;   (3.4c)

G_n is the bipartite graph, with the vertices of one part denoted by $1, \ldots, n$, the vertices of the other part denoted by $1', \ldots, n'$, and i adjacent to j' if and only if $i \leq j$. Call this graph on $2n$ vertices T_n;   (3.4d)

$G_n = \overline{nK_2}$ (sometimes called the *cocktail party graph*);   (3.4e)

$G_n = 2K_n$;   (3.4f)

$G_n = K_{n,n} = \overline{2K_n}$;   (3.4g)

$G_n = K_n \cup \overline{K}_n$;   (3.4h)

$G_n = K_n + \overline{K}_n$.   (3.4i)

Next, let G and H be two graphs such that $V(G) = V(H)$, $E(G) \cap E(H) = \emptyset$. By $G \dotplus H$ we mean the graph with $V(G \dotplus H) = V(G) = V(H)$, and $E(G \dotplus H) = E(G) \cup E(H)$.

$G_n = 2K_n \dotplus nK_2$,   (3.4j)

$G_n = 2K_n \dotplus T_n$,   (3.4k)

$G_n = (K_n \cup \overline{K}_n) \dotplus nK_2$,   (3.4l)

$G_n = (K_n \cup \overline{K}_n) \dotplus W_n$,   (3.4m)

$G_n = (K_n \cup \overline{K}_n) \dotplus T_n$.   (3.4n)

We will note at the appropriate occasions in Sections 4 and 5 which of the chains (3.4a)–(3.4n) are f-chains for each of the seven graph-theoretic functions we are considering.

4. Proof of Theorem 1.1

Lemma 4.1 *Each of the chains (3.4a)–(3.4d) is an f-chain, where f is any one of* achr(G), eq(G), bip(G).

We must show that, if $f(G)$ is any of the three functions, and G_n is given by (3.4a), ..., (3.4d), then $f(G_n) \to \infty$.

Proof Clearly, achr(K_n) $= n$, achr(W_n) $\geq n$.

Consider T_n. The independent sets (1), (2, 1'), (3, 2'), ..., (n, ($n-1$)'), (n') satisfy (3.1) and the accompanying conditions, so achr(T_n) $\geq n + 1$. We now show that achr $((\substack{!\\2})K_2) \geq t$. Let the edges of $(\substack{!\\2})K_2$ join vertices i and i', $i = 1, 2, \ldots, (\substack{!\\2})$. Denote the edges of K_t by $E_1, \ldots, E_{(\substack{!\\2})}$ and the vertices of K_t by $V_1, \ldots, V_t$. If E_k has V_i and V_j as endpoints, $i < j$, associate k with V_i and k' with V_j. The $t - 1$ numbers (primed or unprimed) associated with each V_i are all different, hence form an independent set in $(\substack{!\\2})K_2$. On the other hand there is an edge of $(\substack{!\\2})K_2$ joining any two of these t independent sets. Thus achr$((\substack{!\\2})K_2) \geq t$, whence achr($nK_2$) $\to \infty$.

Next consider eq(G). We have eq(K_n) $= n = $ eq(W_n). Also, eq(T_n) $=$ eq(nK_2) $= 2n$. Thus eq(G_n) $\to \infty$, where G_n is given by any one of (3.4a)–(3.4d).

Finally, consider bip(G). Using eigenvalue arguments, Graham and Witsenhausen have shown that bip(K_n) $= n - 1$. Without such arguments, it is easy to show that bip(K_n) $\to \infty$. For let B be a complete bipartite graph, with parts of cardinality a and b, in the partitioning. It follows that bip(K_n) $\geq 1 + \max\{\text{bip}(K_a), \text{bip}(K_b), \text{bip}(K_{n-a-b})\}$, which, together with bip(K_2) $= 1$, implies bip(K_n) $\to \infty$. Obviously bip(nK_2) $= n$. Consider next bip(W_n). We have bip(W_2) $=$ bip($2K_2$) $= 2$, so assume $n \geq 3$. Let B be a complete bipartite graph, whose edges are contained in W_n. Let $i_1, i_2, \ldots, i_r, i'_{r+1}, \ldots, i'_{r+s}$ be the vertices of B. Then all the numbers $i_1, \ldots, i_r, i_{r+1}, \ldots, i_{r+s}$ are distinct. Obviously, bip(W_n) $\geq 1 + \max\{\text{bip}(W_r), \text{bip}(W_s), \text{bip}(W_{n-r-s})\}$. It follows that bip($W_n$) $\to \infty$. Finally, bip(T_1) $=$ bip(K_2) $= 1$. If B is a complete bipartite graph whose edges are contained in T_n, let $i_1 < i_2 < \cdots < i_r \leq i'_{r+1} < i'_{r+2} < \cdots < i'_{r+s}$ be the vertices of B. Then bip(T_n) $\geq 1 + \max\{\text{bip}(T_{i_r-1}), \text{bip}(T_{n-i_r})\}$, which implies bip($T_n$) $\to \infty$. ∎

Lemma 4.2 *There exists a function $R(n, m)$ such that, if $t \geq R(n, m)$ and the edges of K_t are colored with m colors, there exists a graph $H \leq K_t$, with $|V(H)| = n$, all of whose edges have the same color. (This is Ramsey's theorem.)*

Lemma 4.3 *There exists a function $F(n)$ such that if A is a rectangular $(0,1)$ matrix with $F(n)$ rows, all different, then A contains a square submatrix of order n which, after rearranging rows and columns, is one of the following:*

$$I, \qquad (4.1a)$$

$$J - I, \qquad (4.1b)$$

$$T, \qquad (4.1c)$$

where $T = (t_{ij})$ has $t_{ij} = 1$ if and only if $i \leq j$.

This lemma is proved in [3, Lemmas 4.3 and 4.5], where it is shown that

$$F(n) \leq 3 \cdot 2^{R(n,4)-1} - 1.$$

Now to prove Theorem 1.1. Let $\mathscr{A}$ be the collection of four eq-chains (3.4a)–(3.4d). We shall first show that

$$\text{eq} \leq h_\mathscr{A}, \qquad (4.2)$$

which means that eq is a Ramsey function on $\mathscr{G}$. By Lemma 4.1,

$$h_\mathscr{A} \leq \text{achr} \qquad (4.3)$$

We will then prove

$$\text{achr} \leq \text{eq}. \qquad (4.4)$$

Taken together, (4.2)–(4.4) establish that $\text{achr} \leq \text{eq} \leq \text{achr}$, and achr is a Ramsey function. Similarly, we will show

$$\text{bip} \leq \text{eq}, \qquad (4.5)$$

which together with (4.2) and (from Lemma 4.1) $h_\mathscr{A} \leq \text{bip}$ will prove $\text{bip} \leq \text{eq} \leq \text{bip}$ and bip is a Ramsey function. So we are done if we prove (4.2), (4.4), and (4.5).

To prove (4.2), it is sufficient to show there exists a function $F_1(n)$ such that, for all G,

$$\text{eq}(G) < F_1(h_\mathscr{A}(G)). \qquad (4.6)$$

Let $R(n,m)$ be the function described in Lemma 4.2, and write $R(n)$ for $R(n, 2)$. Let $F(n)$ be the function described in Lemma 4.3. Then set

$$F_1(n) = R(\max\{1 + n, F(R(1 + n))\}). \qquad (4.7)$$

Suppose (4.6) is false, so that, for some G,

$$\text{eq}(G) \geq R(\max\{1 + h_\mathscr{A}(G), F(R(1 + h_\mathscr{A}(G)))\}).$$

Then there exists in G a set V of vertices, $|V| \geq F_1(h_\mathscr{A}(G))$, such that no two are equivalent. It follows from Lemma 4.2 that there exists a subset $W \subset V$,

$|W| = \max\{1 + h_{\mathscr{A}}(G), F(R(1 + h_{\mathscr{A}}(G))\}$, such that W is an independent set or $\langle W \rangle$ is a clique. But the latter is impossible, by (3.4a) and the definition of $h_{\mathscr{A}}(G)$. Hence G contains an independent set of vertices W, with $|W| = F(R(1 + h_{\mathscr{A}}(G)))$, no two vertices of W equivalent.

Let $X = V(G) - W$, and let A be the (0, 1) matrix whose rows correspond to vertices in W, columns to vertices in X, and $A = (a_{ij}) = 1$ if and only if i is adjacent to j. All rows of A are different. By Lemma 4.3, A contains a square submatrix of order $R(1 + h_{\mathscr{A}}(G))$, which after rearranging rows and columns is one of the matrices (4.1). Let $W' \subset W$ and $X' \subset X$ be the subsets of W and X corresponding to this square submatrix of order $R(1 + h_{\mathscr{A}}(G))$. By Lemma 4.2, X' contains a subset X'', $|X''| = 1 + h_{\mathscr{A}}(G)$, and X'' is an independent set or $\langle X'' \rangle$ is a clique. As before, the latter alternative is impossible. Let $W'' \subset W$ have cardinality $1 + h_{\mathscr{A}}(G)$ so that the submatrix of A found by the rows corresponding to W'' and columns corresponding to X'' is of the form (4.1a), (4.1b), or (4.1c). If of the form (4.1a), consider $\langle W'' \cup X'' \rangle$, which is of the form (3.4b), proving $h_{\mathscr{A}}(G) \geq 1 + h_{\mathscr{A}}(G)$, a contradiction. Similarly (4.1b) contradicts (3.4c) and (4.1c) contradicts (3.4d). Hence (4.6) is true, which proves (4.2).

To prove (4.4), let $V(G) = F_1 \cup \cdots \cup F_e$ be the partition of $V(G)$ into equivalence classes, $V(G) = S_1 \cup \cdots \cup S_r$ be a partition of $V(G)$ into independent sets such that $r = \mathrm{achr}(G)$, $S_i \cup S_j$ ($i \neq j$) is not an independent set. For each $i = 1, \ldots, r$, let $T(i) = \{j | S_i \cap F_j \neq \varnothing\}$. Observe that, if $T(i) = T(k)$, $S_i \cup S_k$ is independent. Hence the sets $T(i)$ are distinct, whence $r = \mathrm{achr}(G) \leq 2^{\mathrm{eq}(G)} - 1$, which proves (4.4).

All that remains is to prove (4.5). But it is obvious from the definition of equivalence that, if we partition $A(G)$ corresponding to the equivalence classes, the diagonal blocks are 0, the off-diagonal blocks are 0 or J. It follows that $\mathrm{bip} \leq \binom{\mathrm{eq}}{2}$, which proves (4.5). ∎

5. Proof of Theorem 1.2

Let $\mathscr{A}'$ be the finite collection of nine chains in $\mathscr{G}$ given by (3.4b)–(3.4e) and (3.4j)–(3.4n). We will first establish that

$$\text{Each chain in } \mathscr{A}' \text{ is an eq*-chain,} \tag{5.1}$$

$$\mathrm{eq}^* \leq h_{\mathscr{A}'}, \tag{5.2}$$

$$\mathrm{eq}^* \leq \mathrm{bip}^*, \tag{5.3}$$

$$\mathrm{bip}^* \leq \mathrm{eq}^*. \tag{5.4}$$

If we prove (5.1)–(5.4), this will show that eq* and bip* are Ramsey functions, and $\mathrm{eq}^* \leq \mathrm{bip}^* \leq \mathrm{eq}^*$. Now the proofs of (5.1), (5.2), and (5.4) are entirely analogous to the proofs of the corresponding statements for $\mathscr{A}$, eq, and bip given in Section 4, so we discuss only (5.3). We shall prove eq* $\leq 2^{2\text{bip}^*}$. Let bip*(G) = bip*. This implies that there are subsets $S_1, S_2, \ldots, S_{\text{bip}^*}$ of $V(G)$ and graphs $G_1, G_2, \ldots,$ such that $V(G_i) = S_i$ and

$$E(G) = E(G_1) \cup \cdots \cup E(G_{\text{bip}^*})$$

is a partition of $E(G)$, with each $\langle G_i \rangle$ a clique or a complete bipartite graph. Assume $G_1, \ldots, G_a$ are cliques, (a may be bip*), the remaining $G_{a+1}, \ldots, G_{\text{bip}^*}$ are complete bipartite graphs. Denote the vertices of the respective two parts of S_{a+i} by T_{a+2i-1} and T_{a+2i}. Thus we are led to consider

$$S_1, \ldots, S_a, T_{a+1}, T_{a+2}, T_{a+3}, T_{a+4}, \ldots, T_{2\text{bip}^*-a-1}, T_{2\text{bip}^*-a}.$$

For each vertex i of G, let $P(i) = \{j | i \in S_j \text{ or } T_j\}$. Suppose i and k are distinct vertices of G such that $P(i) = P(k)$; we will show that i and k are mixed equivalent. Let v be any vertex $\neq i, k$ which is adjacent to i. Then the edge (v, i) is in one of the cliques or one of the complete bipartite graphs. Suppose (v, i) is in clique $G_j, j \leq a$. Then $j \in P(i)$. But since $P(i) = P(k), j \in P(k)$, hence $(v, k) \in E(G_j)$, so v is adjacent to k. If (v, i) is in one of the complete bipartite graphs, then $i \in T_{a+2j-1}, v \in T_{a+2j}$ (or vice versa), and $k \in T_{a+2j-1}$ (or T_{a+2j}) establishes that k and v are adjacent. Thus, eq*$(G) \leq 2^{2\text{bip}^*-a}$, implying eq* $\leq 2^{2\text{bip}^*}$, which establishes (5.3).

We now turn to achr*(G) and match*(G). We first show eq* $\leq$ achr*. To do this, it is sufficient to show that each of the nine chains in $\mathscr{A}'$ is an achr*-chain; i.e., achr*$(G_n) \to \infty$ for each of these chains. Since no new arguments are involved, we omit the demonstration. On the other hand, from (3.4f), we see achr* is not bounded by a function of eq*. All that remains to be shown is that achr* and match* are Ramsey functions on $\mathscr{G}$ and are equivalent functions.

Let $\mathscr{A}^*$ be the collection of six chains given by (3.4b), (3.4e)–(3.4i). Clearly each is an achr*-chain and is a match*-chain as well. We will be done if we prove both of the following:

$$\text{achr*} \leq h_{\mathscr{A}^*}, \tag{5.5}$$

$$\text{match*} \leq h_{\mathscr{A}^*}. \tag{5.6}$$

Let $R(n) = R(n, 2)$ from Lemma 4.2, and $F(n)$ be the function described in Lemma 4.3. To prove (5.5) we show that if

$$\text{achr*}(G) \geq 4F(R(R(2n)))R(R(R(n)F(R(2n)))) + 1, \tag{5.7}$$

then $h_{\mathscr{A}^*}(G) \geq n$. Write $B = F(R(R(2n))), A = R(R(R(n)F(R(2n))))$.

Recalling (3.1*a), at most one of the achr*(G) subsets of $V(G)$ forming a suitable partition of $V(G)$ has cardinality one. By (5.7) there are at least $2AB$ disjoint independent subsets of $V(G)$, each of cardinality at least two, any pair of which are joined by an edge, or at least $2AB$ cliques, each of cardinality at least two, any pair of which are joined by at least one nonedge. Assume the former. The latter case is treated analogously. Indeed, note that the complementary graphs of any chain of A^* also form a chain of A^*.

Label the independent sets $S_1, S_2, \ldots, S_{2AB}$. By hypothesis there exist vertices $v_i \in S_i$ such that v_1 and v_2 are adjacent, v_3 and v_4 are adjacent, etc. Call V the set of odd-numbered v_i, the even-numbered W. Let M be the (0, 1) square matrix of order AB whose rows correspond to vertices in V, the columns to vertices in W, with a 1 in a given row and column if the corresponding vertices are adjacent. Concerning M we know

$$\text{the diagonal of } M \text{ consists entirely of ones.} \quad (5.8)$$

Either there exist at least B rows of M, no two of which are the same (case 1), or M contains at least A identical rows (case 2).

Case 1 We consider the submatrix N_1 of B rows of M, no two of which are the same. By Lemma 4.3, N_1 contains a square submatrix N_2 of order $R(R(2n))$, which is in one of the forms (4.1). By Lemma 4.2 applied to the vertices of G corresponding to the rows of N_2 there is a subset $V' \subset V$, with cardinality $R(2n)$ such that

$$V' \text{ is an independent set} \quad (5.9)$$

or

$$\langle V' \rangle \text{ is a clique.} \quad (5.10)$$

Consider the principal submatrix of N_2 with respect to V'. This square submatrix N_3 is of order $R(2n)$ and has one of the forms (4.1). By Lemma 4.2 applied to the vertices of G corresponding to the columns of N_3 there is a subset $W' \subset W$ with cardinality $2n$ such that

$$W' \text{ is an independent set} \quad (5.11)$$

or

$$\langle W' \rangle \text{ is a clique.} \quad (5.12)$$

Now we show the result of the possible combinations (with intermediate steps) in the accompanying tabulation.

MIXED ACHROMATIC NUMBER AND GRAPH FUNCTIONS 115

Combination		Result
(5.9), (5.11), and (4.1a)		(3.4b) appears
(5.9), (5.11), and (4.1b)		(3.4e) appears
(5.9), (5.11), and (4.1c)	T_{2n}	(3.4g) appears
(5.9), (5.12) or (5.10), (5.11), and (4.1a)	$(K_{2n} \cup \overline{K}_{2n}) + 2nK_2$	(3.4h) appears
(5.10), (5.11), and (4.1b)	$(K_{2n} \cup \overline{K}_{2n}) + W_{2n}$	(3.4i) appears
(5.10), (5.11), and (4.1c)	$(K_{2n} \cup \overline{K}_{2n}) + T_{2n}$	(3.4h) appears
(5.10), (5.12), and (4.1a)	$(K_{2n} \cup K_{2n}) + 2nK_2$	(3.4f) appears
(5.10), (5.12), and (4.1b)	$(K_{2n} \cup K_{2n}) + W_{2n}$	(3.4e) appears
(5.10), (5.12), and (4.1c)	$(K_{2n} \cup K_{2n}) + T_{2n}$	(3.4f) appears

Case 2 In this case M contains A identical rows. Recalling (5.8), this implies M contains a square submatrix J of order A. By Lemma 4.2 applied to the vertices corresponding to the rows of J, there is a subset $V'' \subset V$ with cardinality $R(R(n)F(R(2n)))$ such that

$$V'' \text{ is an independent set} \qquad (5.13)$$

or

$$\langle V'' \rangle \text{ is a clique.} \qquad (5.14)$$

Consider a square submatrix J'' of J with cardinality $R(R(n)F(R(2n)))$ rows corresponding to vertices of V''. By Lemma 4.2 applied to the columns of J'', there is a subset $W'' \subset W$ corresponding to columns of J'' of cardinality $R(n)F(R(2n))$ such that

$$W'' \text{ is an independent set} \qquad (5.15)$$

or

$$\langle W'' \rangle \text{ is a clique.} \qquad (5.16)$$

(5.13) and (5.15) show that (3.4g) appears.

(5.13) and (5.16), or (5.14) and (5.15) show that (3.4i) appears.

We now handle the combination of (5.14) and (5.16). From the paragraph containing statement (5.7), we know that for each $v \in V''$ there exists at least one other vertex x in the same independent S_i. For each $v \in V''$, pick one such x yielding $|X| = |V''|$. Let Z be the square (0,1) matrix whose rows correspond to vertices of V'', columns to vertices of X, with a 1 in a given row and column if and only if the corresponding vertices are adjacent. Concerning Z, we know

$$\text{the diagonal of } Z \text{ is all 0.} \qquad (5.17)$$

Now there are at least $F(R(2n))$ rows of Z which are different (case 2a), or at least $R(n)$ rows of Z are identical (case 2b).

Case 2a By Lemma 4.3 applied to the $F(R(2n))$ different rows of Z, followed by an application of Lemma 4.2 we conclude that there exists $H \leq G$ such that $|V(H)| = 4n$ and the adjacency matrix of H is one of the following:

$$\begin{bmatrix} J - I & I \\ I & J - I \end{bmatrix}, \tag{5.18a}$$

$$\begin{bmatrix} J - I & J - I \\ J - I & J - I \end{bmatrix}, \tag{5.18b}$$

$$\begin{bmatrix} J - I & T \\ T^t & J - I \end{bmatrix}, \tag{5.18c}$$

$$\begin{bmatrix} J - I & I \\ I & 0 \end{bmatrix}, \tag{5.18d}$$

$$\begin{bmatrix} J - I & J - I \\ J - I & 0 \end{bmatrix}, \tag{5.18e}$$

$$\begin{bmatrix} J - I & T \\ T^t & 0 \end{bmatrix}. \tag{5.18f}$$

(5.18a) produces a graph (3.4f), (5.18b) gives (3.4e), (5.18c) gives (3.4f), (5.18d) gives (3.4h), (5.18e) gives (3.4i), and (5.18f) gives (3.4h).

Case 2b (5.17) and Lemma 4.2 produce a graph (3.4f) or (3.4h). Thus $h_{A^*} \geq n$, and

$$h_{A^*} \leq \text{achr}^* \leq h_{A^*}. \tag{5.19}$$

The theorem will be complete if we prove (5.6), match$^* \leq h_{A^*}$. To do this we show that

if match$^*(G) \geq F(R(R(2n)))R(n(F(2n)))$ then $h_{A^*}(G) \geq n$. Let $B \equiv F(R(R(2n)))$, $A \equiv R(n(F(2n)))$. The definition of match$^*(G)$ and (5.19) imply match$(G) \geq AB$, and match$(\bar{G}) \geq AB$.

Consider a matching in G of cardinality AB. Let the matching edges be labeled $(v_1, w_1), (v_2, w_2)$, etc. Let $V = \{v_i | (v_i, w_i) \in \text{matching}\}$, $W = \{w_i | (v_i, w_i) \in \text{matching}\}$. Let M be the matrix whose rows correspond to the vertices of V, columns to vertices of W, with a 1 in a given row and column if and only if the corresponding vertices are adjacent. About M we know

the diagonal of M consists entirely of 1's. (5.20)

Either there exist B rows of M, no two of which are the same (case 1), or M contains at least A identical rows (case 2).

Case 1 This is identical to case 1 for achr*(G).

Case 2 M contains A identical rows. (5.20) implies M contains a square submatrix J of order A. By Lemma 4.2 applied to the vertices corresponding to the rows of J there is a subset, V', of these vertices of cardinality $nF(2n)$ such that

$$V' \text{ is an independent set} \qquad (5.21)$$

or

$$\langle V' \rangle \text{ is a clique.} \qquad (5.22)$$

Similarly there is a subset, W', of the vertices corresponding to the columns of J, of cardinality $nF(2n)$ such that

$$W' \text{ is an independent set} \qquad (5.23)$$

or

$$\langle W' \rangle \text{ is a clique.} \qquad (5.24)$$

Noting $nF(2n) \geq 2n$. (5.21) and (5.23) yield a graph (3.4g). (5.21) and (5.24), or (5.22) and (5.23) yield (3.4i).

Now for the last case (5.22) and (5.24). Recalling match$(\bar{G}) \geq AB$. This together with (5.22) implies that G contains a subgraph (not induced) of the form

$$(K_{nF(2n)} \cup \bar{K}_{nF(2n)}) + nF(2n)K_2 \qquad \text{(case 2a)}$$

or

$$(K_{nF(2n)} \cup \bar{K}_{nF(2n)}) + W_{nF(2n)} \qquad \text{(case 2b)}$$

with other possible adjacencies between vertices of the K and $\bar{K}$ only (i.e., the $\bar{K}$ is induced), or (3.4e) or (3.4f) appears.

Case 2a Consider the square matrix Q of order $nF(2n)$, whose rows correspond to the vertices of K and columns correspond to the vertices of $\bar{K}$. The diagonal of Q is all ones. Either n rows of Q are identical (case 2a1) or $F(2n)$ rows of Q are all different (case 2a2).

Case 2a1 Q contains a square submatrix J of order n. Thus G contains an induced subgraph (3.4i).

Case 2a2 By Lemma 4.3 applied to the rows of Q, Q contains a square submatrix of order $2n$ and form (4.1). For (4.1a), (3.4h) appears, (4.16) produces (3.4i) as does (4.1c).

Case 2b This case follows from case 2a by considering the complementary graph. This completes the proof that $h_{A^*} \geq n$. Thus

$$h_{A^*} \leq \text{match}^* \leq h_{A^*}. \qquad (5.25)$$

(5.19) and (5.25) together imply that achr* and match* are equivalent functions. ∎

6. Eigenvalues of Graphs

If G is a graph its $(0, 1)$ adjacency matrix $A(G)$ is defined by

$$A(G) = (a_{ij}) = \begin{cases} 1 & \text{if } i \text{ and } j \text{ are adjacent vertices,} \\ 0 & \text{if not.} \end{cases}$$

In particular, $a_{ii} = 0$. Since $A(G)$ is a real symmetric matrix, its eigenvalues are real, and we denote them by $\lambda_1(G) \geq \lambda_2(G) \geq \cdots$. We now introduce the following functions on a graph. For any set S of real numbers, $\cup(S)(G) = |\{i \mid \lambda_i(G) \notin S\}|$. For any pair of real numbers $x \leq y$,

$$\Lambda(x, y)(G) = |\{i \mid \lambda_i(G) \leq x\} \cup \{i \mid \lambda_i(G) \geq y\}|.$$

We also write $\Lambda(x, \infty)(G) = |\{i \mid \lambda_i(G) \leq x\}|$.

Theorem 6.1 [3]

$$\cup(\{0\}) \leq \Lambda(-1, \infty) \leq \text{eq} \leq \text{bip} \leq \text{achr} \leq \cup(\{0\}).$$

The relations

$$\cup(\{0\}) \leq \Lambda(-1, \infty) \leq \text{eq} \leq \text{bip} \leq \cup(\{0\})$$

were shown in [3]. These relations, together with Theorem 1.1, prove Theorem 6.1.

Theorem 6.2 [4].

$$\cup(\{0, -1\}) \leq \Lambda((-\sqrt{5} - 1)/2, 1)$$
$$\leq \Lambda(-2, (\sqrt{5} - 1)/2) \leq \text{eq}^*$$
$$\leq \text{bip}^* \leq \cup(\{0, -1\}).$$

The relations

$$\cup(\{0, -1\}) \leq \Lambda((-\sqrt{5} - 1)/2, 1) \leq \Lambda(-2, (\sqrt{5} - 1)/2)$$
$$\leq \text{eq}^* \leq \cup(\{0, -1\})$$

were shown in [4]. These, together with Theorem 1.2, prove Theorem 6.2.

All that remains is to prove Theorem 1.3. Let H_n be K_{1,n^2}, and let $G_n = K_{n,n} \cup \overline{K}_{(n-1)^2}$. Then

$$|V(G_n)| = |V(H_n)| = n^2 + 1, \qquad \lambda_1(G_n) = \lambda_1(H_n) = n,$$

$$\lambda_2(G_n) = \cdots = \lambda_{n^2}(G_n) = \lambda_2(H_n) = \cdots = \lambda_{n^2}(H_n) = 0,$$

$$\lambda_{n^2+1}(G_n) = \lambda_{n^2+1}(H_n) = -n.$$

Now achr*$(H_n) = 2$, while achr*$(G_n) \geq n$. Also match*$(H_n) = 1$, but match*$(G_n) \geq n$.

In the language of [5], eq, bip, achr, eq*, and bip* are spectral functions, but achr* and match* are not.

REFERENCES

1. F. Harary, G. Prins, and S. T. Hedetenimi, An interpolation theorem for graphical homomorphisms, *Portugal. Math.* **26** (1967) 435–462.
2. P. Hell and D. J. Miller, Graphs with given achromatic number, *Discrete Math.* **16** (1976) 195–207.
3. A. J. Hoffman, Eigenvalues and edge partitionings of graphs, *Linear Algebra and Appl.* **5** (1972) 137–146.
4. A. J. Hoffman, Application of Ramsey-style theorems to eigenvalues of graphs, *Combinatorics, Part 2, Proc. Advanced Study Inst. Combinatorics, Breukelen, Math. Centre Tracts* **56** (1974) 43–57.
5. A. J. Hoffman, Spectral functions of graphs, *Proc. Internat. Congr. Math., Vancouver* **2** (1974) 461–463.
6. A. J. Hoffman and L. Howes, On eigenvalues and colorings of graphs II, *Ann. New York Acad. Sci.* **175** (1970) 238–242.
7. A. J. Hoffman and P. Joffe, Nearest S-matrices of given rank and Ramsey functions of eigenvalues of bipartite S-graphs, *Proc. Internat. Colloq. Combinatorial Problems, Orsay, 1976.*

AMS 05C99

Fred Buckley
DEPARTMENT OF MATHEMATICS
CITY UNIVERSITY OF NEW YORK
NEW YORK, NEW YORK

A. J. Hoffman
IBM T. J. WATSON RESEARCH CENTER
YORKTOWN HEIGHTS, NEW YORK

On Tutte's Conjecture for Tangential 2-Blocks

BISWA TOSH DATTA

1. Introduction

Let $PG(n, 2)$ be the finite projective space of dimension n over the Galois field $GF(2)$. If Δ is a non-null subset of points in $PG(n, 2)$, then $\langle \Delta \rangle$ denotes the space generated by the points in Δ and $\dim \langle \Delta \rangle$ denotes the dimension of $\langle \Delta \rangle$. Again, if Δ and Δ' are any two subsets of points in $PG(n, 2)$, then $\Delta - \Delta'$ denotes the set of points which are in Δ, but not in Δ'; $|\Delta|$ denotes the number of points in Δ.

Let k be a positive integer. A nonempty set β of points in $PG(n, 2)$ is called a *k-block* in $PG(n, 2)$ if its dimension is at least k and $\beta \cap \Sigma_{n-k} \neq \emptyset$, for every $(n - k)$-space Σ_{n-k} in $PG(n, 2)$. A k-block β is *minimal* if no proper subset of β is a k-block. Let β be a k-block in $PG(n, 2)$ and Δ be a nonempty subset of β; if $\dim \langle \Delta \rangle \leq n - k$, we define the *tangent* of Δ in β, denoted by $t(\Delta)$, as any $(n - k)$-space, say Σ_{n-k}, such that $\Delta \subseteq \Sigma_{n-k}$ and $\beta \cap \Sigma_{n-k} \subseteq \langle \Delta \rangle$. If $\Delta = \{A_1, A_2, \ldots, A_m\}$, then $t(\Delta)$ is also denoted by $t(A_1, A_2, \ldots, A_m)$. A non-null subset β of points in $PG(n, 2)$ is said to be a *tangential k-block* if β is a k-block and every nonempty subset of β, of dimension $\leq n - k$, has a tangent in β. Let q be a positive integer. A *q-stigm* in $PG(n, 2)$ is defined to be a set of q points in a $(q - 2)$-space of $PG(n, 2)$ such that each $q - 1$ elements of

121

the set are linearly independent. An *odd stigm* is a q-stigm for which q is an odd integer ≥ 3. If β is a k-block in PG(n, 2) and St is an odd stigm contained in β, then we say that St is an odd stigm of β. Now we have the following results.

Proposition 1.1 [9] *Every tangential k-block is minimal.*

Proposition 1.2 [9] *Every k-block contains an odd stigm.*

Proposition 1.3 [9] *The odd stigm are the minimal 1-blocks.*

Proposition 1.4 [9] *Let β be a k-block of dimension d in PG(n, 2) and let Σ_{d-k+1} be a subspace of dimension $d - k + 1$. Then $\beta \cap \Sigma_{d-k+1}$ contains an odd stigm.*

There exist tangential 2-blocks of dimensions 2, 3, and 5, called Fano block, Desargues block, and Petersen block, respectively [9]. Now, Tutte's conjecture on tangential 2-blocks may be stated as follows.

Tutte's Conjecture *The only tangential 2-blocks in PG(n, 2) are the Fano, Desargues, and Petersen blocks.*

An affirmative answer to Tutte's conjecture would imply the four-color theorem, and would provide an affirmative answer to a special case of Hadwiger's conjecture [9] and Tutte's conjecture on Tait coloring [9].

2. Some Concepts and Results on Tangential 2-Blocks

In the following definitions and propositions, n denotes a positive integer ≥ 3, unless it is mentioned otherwise; also, Σ_m denotes an m-dimensional subspace in PG(n, 2), where m is a non-negative integer $\leq n$.

Definition Let Σ_k and Σ_{n-k-1} be two subspaces in PG(n, 2) such that $0 \leq k \leq n - 3$ and $\Sigma_k \cap \Sigma_{n-k-1} = \emptyset$. Let β be a non-null subset of points in PG(n, 2) such that $\Sigma_k \cap \beta = \emptyset$. Then the *projection* of β through Σ_k on Σ_{n-k-1} is denoted by $P(\beta, \Sigma_k, \Sigma_{n-k-1})$ and defined by

$$P(\beta, \Sigma_k, \Sigma_{n-k-1}) = \{A : A \in \Sigma_{n-k-1} \text{ and } \exists S \in \beta \text{ such that}$$
$$\langle \Sigma_k \cup \{S\} \rangle \cap \Sigma_{n-k-1} = \{A\}\}.$$

Σ_k is called the *vertex* and Σ_{n-k-1} the *base* of the projection. Now we have the following.

Proposition 2.1 [4] *Let β be an n-dimensional tangential 2-block in PG(n, 2). Let Σ_k and Σ_{n-k-1} be subspaces such that $\Sigma_k \cap \Sigma_{n-k-1} = \emptyset$. Assume that there exists a $(k + 2)$-space Σ_{k+2} containing Σ_k such that*

(i) $\forall \Sigma_k' \subseteq \Sigma_{k+2}$ and $\Sigma_k' \neq \Sigma_k$, $\Sigma_k' \cap \beta \neq \emptyset$; and
(ii) if Σ_{k+1}^i, $i = 1, 2, 3$, are the three $(k+1)$-spaces in Σ_{k+2} and on Σ_k, then $\dim \langle \beta \cap \Sigma_{k+1}^i \rangle = k + 1$, $\forall i \in \{1, 2, 3\}$.

Then $P(\beta, \Sigma_k, \Sigma_{n-k-1})$ is an $(n - k - 1)$-dimensional tangential 2-block in Σ_{n-k-1}.

Corollary 2.2 [4] *Let β be an n-dimensional tangential 2-block in PG(n, 2). Let Σ_2 be a plane in PG(n, 2) containing six points of β. Let $Z \in \Sigma_2 - \beta$ and Σ_{n-1} be any $(n - 1)$-space in PG(n, 2) such that $Z \notin \Sigma_{n-1}$. Then $P(\beta, \{Z\}, \Sigma_{n-1})$ is an $(n - 1)$-dimensional tangential 2-block in Σ_{n-1}.*

Corollary 2.3 [4] *Let β be an n-dimensional tangential 2-block in PG(n, 2). (i) If $(n - 1)$-dimensional tangential 2-blocks do not exist, then there cannot be any plane with six points of β. (ii) If $(n - 2)$-dimensional tangential 2-blocks do not exist, then there can be at most one plane with six points of β.*

Definition Let β be an n-dimensional tangential 2-block in PG(n, 2). Let k and l be two non-negative integers such that $0 \leq k \leq n - 3$ and $k \leq l \leq n - 1$. A subspace Σ_k is called an *attenuation space* for β with respect to Σ_l if

(i) $\Sigma_k \subseteq \Sigma_l$,
(ii) $\dim \langle \beta \cap \Sigma_k \rangle < k$, and
(iii) $\forall \Sigma_{k+1}$ in PG(n, 2), $\Sigma_{k+1} \cap \Sigma_l = \Sigma_k \Rightarrow |(\Sigma_{k+1} - \Sigma_k) \cap \beta| \leq 1$.

In the above definition, condition (iii) implies that $S_1 + S_2 \notin \Sigma_k$, for any two points S_1 and S_2 of $\beta - \Sigma_l$. Now we have the following theorem.

Proposition 2.4 [4] *Let β be an n-dimensional tangential 2-block in PG(n, 2). Let Σ_k and Σ_l be two subspaces in PG(n, 2) such that $0 \leq k \leq n - 3$, $k + 2 < l \leq n - 1$, $\Sigma_k \subseteq \Sigma_l$, and $\dim \langle \beta \cap \Sigma_k \rangle < k$. Then Σ_k is an attenuation space for β with respect to Σ_l provided there exists a $(k + 2)$-space Σ_{k+2} and a nonempty subset Δ of β such that*

(i) $\dim \langle \Delta \rangle \leq n - 3$, $\langle \Delta \rangle \subseteq \Sigma_l$, and $\langle \Delta \rangle \cap \Sigma_{k+2} = \emptyset$,
(ii) $\Sigma_k \subseteq \Sigma_{k+2} \subseteq \Sigma_l$, and
(iii) $\forall \Sigma_k' \subseteq \Sigma_{k+2}$, $\Sigma_k' \neq \Sigma_k \Rightarrow (\langle \Sigma_k' \cup \Delta \rangle - \langle \Delta \rangle) \cap \beta \neq \emptyset$.

Definition Let β be an n-dimensional tangential 2-block in PG(n, 2). Let Σ_m be an m-space in PG(n, 2), where $0 \leq m \leq n - 1$. A point X of Σ_m is said to be an *attenuation point* for β with respect to Σ_m if (i) $X \notin \beta$ and (ii) $X \neq S_1 + S_2$, for every pair of points S_1 and S_2 belonging to $\beta - \Sigma_m$. Clearly, an attenuation point is an attenuation space of dimension 0.

Proposition 2.5 [4] *Let β be an n-dimensional tangential 2-block in PG(n, 2). Let Σ_m be an m-space in PG(n, 2), where $1 \leq m \leq n - 1$, and let*

$X \in \Sigma_m$. Then X is an attenuation point for β with respect to Σ_m provided there exists a nonempty subset Δ of $\beta \cap \Sigma_m$ such that

(i) $\dim \langle \Delta \rangle \leq n - 3$;
(ii) $X \in t(\Delta)$, but $X \notin \langle \Delta \rangle$; and
(iii) $X \in t(\Delta \cup \{S\})$, $\forall S \in \beta - \Sigma_m$.

Definition Let Σ_m be an m-space, where $1 \leq m \leq n - 1$ and β is an n-dimensional tangential 2-block in $PG(n, 2)$. Let X be an attenuation point in Σ_m and Δ be a nonempty subset of $\beta \cap \Sigma_m$. Then Δ is said to *induce* the attenuation point X with respect to Σ_m if Δ and X satisfy the three conditions (i), (ii), and (iii) of Proposition 2.5.

Definition Let β be an n-dimensional tangential 2-block and Σ_{n-1} an $(n-1)$-space in $PG(n, 2)$. Let $\emptyset \neq \Delta \subseteq \beta \cap \Sigma_{n-1}$ and $\dim \langle \Delta \rangle = d$, where $0 \leq d \leq n - 3$. Then Δ is called a *polarizing set* for β with respect to Σ_{n-1} if there exist three $(d + 1)$-spaces, denoted by $\rho^i_{d+1}(\Delta)$, $i = 1, 2, 3$, such that

(i) $\rho^i_{d+1}(\Delta) \supseteq \Delta$, $\forall i \in \{1, 2, 3\}$, and
(ii) $\beta - \Sigma_{n-1} \subseteq \bigcup_{i=1}^{3} \rho^i_{d+1}(\Delta)$.

From the preceding definition it follows that (i) if S and $S' \in \rho^i_{d+1}(\Delta)$, where S and $S' \in \beta - \Sigma_{n-1}$ and $i \in \{1, 2, 3\}$, then $S + S' \in \langle \Delta \rangle$; (ii) if $\emptyset \neq \Delta \subseteq \beta \cap \Sigma_{n-1}$, $\dim \langle \Delta \rangle = n - 3$, and $t(\Delta) \not\subseteq \Sigma_{n-1}$, then Δ is a polarizing set for β with respect to Σ_{n-1}. Now we have the following proposition.

Proposition 2.6 [4] *Let β be an n-dimensional tangential 2-block and Σ_{n-1} an $(n-1)$-space in $PG(n, 2)$. Let $\emptyset \neq \Delta \subseteq \beta \cap \Sigma_{n-1}$. Then Δ is a polarizing set with respect to Σ_{n-1} provided*

(i) Δ *induces a set μ of attenuation points with respect to Σ_{n-1} such that* $\dim \langle \Delta \cup \mu \rangle = n - 3$, *and*
(ii) $t(\Delta) \not\subseteq \Sigma_{n-1}$.

Definition Let β be an n-dimensional tangential 2-block in $PG(n, 2)$. If $g_i(P) = \{P, A_i, B_i\}$, is a line of β, $\forall i \in \{1, 2, 3\}$, then the three concurrent lines $g_1(P)$, $g_2(P)$, and $g_3(P)$ are called *generators* of a cone in $C_3(P)$, where $C_3(P)$ is the 3-space spanned by the generators; and the line $\{P, A_1 + A_2 + A_3, A_1 + A_2 + B_3\}$ is called the *axis* of the cone; P is called the *vertex* of the cone.

Let $g_i(P) = \{P, A_i, B_i\}$ be a line of β, where $i = 1, 2, 3$, $C_3(P) = \langle \bigcup_{i=1}^{3} g_i(P) \rangle$, $A(P) = \{P, A_1 + A_2 + A_3, A_1 + A_2 + B_3\}$, and $V_2(P) = \langle \{P, A_1 + A_2, A_1 + A_3\} \rangle$. With the above notations, we have the following result.

Proposition 2.7 [4] (i) *If l is a line of $C_3(P)$ through a point of $A(P) - \{P\}$, then l must meet one of the generators of $g_1(P)$, $g_2(P)$, and $g_3(P)$.*
(ii) $A(P) \cap \beta = \{P\}$.
(iii) *If $n \geq 4$ and $\emptyset \neq \mu \subseteq \beta - C_3(P)$, $\dim\langle\mu\rangle \leq n - 2$, and $\langle\mu\rangle \cap C_3(P) = \emptyset$, then $t(\mu) \cap A(P) = \emptyset$.*
(iv) *If $n \geq 4$ and $\emptyset \neq \mu \subseteq \beta - C_3(P)$, $\dim\langle\mu\rangle \leq n - 2$, and $\langle\mu\rangle \cap C_3(P) \neq \emptyset$, then $\langle\mu\rangle \cap C_3(P) = A(P) - \{P\}$.*
(v) *$V_2(P)$ does not contain any odd stigm of β.*
(vi) *If $n \geq 4$ and $X \in A(P) - \{P\}$, then X is an attenuation point for β with respect to $C_3(P)$.*

Let β be an n-dimensional tangential 2-block, where $n \geq 2$. In our investigation of a hypothetical tangential 2-block β, we use the following notations.

Let $P \in \beta$ and $\pi(P)$ denote a tangent of P. Then $\pi(P)$ is an $(n - 2)$-space in $PG(n, 2)$. There are exactly three $(n - 1)$-spaces on $\pi(P)$ and we denote them by $\tau_1(P)$, $\tau_2(P)$, and $\tau_3(P)$. Then $\forall i \in \{1, 2, 3\}$, $\tau_i(P)$ has an odd stigm of β containing P; we denote this odd stigm of β in $\tau_i(P)$ by $St_i(P)$, $i = 1, 2, 3$. We denote the space $\langle St_i(P)\rangle \cap \pi(P)$ by $\lambda_i(P)$, $i = 1, 2, 3$.

Definition Let β be an n-dimensional tangential 2-block, where $n \geq 2$. Let $P \in \beta$ and $St_i(P)$ be an r_i-stigm, where r_i is a positive odd integer and $r_i \geq 3$, $i = 1, 2, 3$. Then $St_1(P)$, $St_2(P)$, and $St_3(P)$ are said to constitute a (r_1, r_2, r_3)-*tangential stigm system* with respect to P.

Proposition 2.8 [4] (i) *If $P \in \beta$, then P has an (r_1, r_2, r_3)-tangential stigm system for some odd integers r_1, r_2 and r_3, where $r_i \geq 3$, $i = 1, 2, 3$.*
(ii) *Let σ be a permutation on $\{1, 2, 3\}$ and r_i be an odd integer ≥ 3, for $i = 1, 2, 3$. Then the existence (or nonexistence) of an (r_1, r_2, r_3)-tangential stigm system with respect to a point P of β implies the existence (or nonexistence) of an $(r_{\sigma(1)}, r_{\sigma(2)}, r_{\sigma(3)})$-tangential stigm system with respect to the point P.*

Proposition 2.9 [4] *Let β be an n-dimensional tangential 2-block in $PG(n, 2)$, where $n \geq 4$. Let Σ_{n-1} be any $(n - 1)$ space in $PG(n, 2)$. Then $|\beta - \Sigma_{n-1}| \geq 5$.*

Proposition 2.10 [4] *Let β be an n-dimensional tangential 2-block and Σ_{n-2} be an $(n - 2)$-space in $PG(n, 2)$. Let τ_1, τ_2, and τ_3 be the three $(n - 1)$-spaces which are on Σ_{n-2} and in $PG(n, 2)$. Let $P \in \beta \cap \Sigma_{n-2}$ and r_1 and r_2 be two positive odd integers such that $5 \leq r_i \leq 7$, $i = 1, 2$. Let St_i be an r_i-stigm of β in τ_i such that $St_i \cap \Sigma_{n-2} = \{P\}$, $\forall i \in \{1, 2\}$. If $\lambda_i = \langle St_i\rangle \cap \Sigma_{n-2}$, $i = 1, 2$, then $\lambda_i \not\subseteq \lambda_j$, $\{i, j\} = \{1, 2\}$.*

Proposition 2.11 [6] *Let β be an n-dimensional tangential 2-block in $PG(n, 2)$ such that $n \geq 4$, and there exists at most one plane with six points of β.*

If there exists a (3, 3, 3)-tangential stigm system with respect to each point of β, then $\dim\langle\beta\rangle \geq 8$.

Proposition 2.12 [4] *Let β be an n-dimensional tangent 2-block. Let m be an integer such that $3 \leq m \leq n$. Let $\mathrm{St}_1 = \{P, A_1, A_2, \ldots, A_{m-1}\}$, $\mathrm{St}_2(P) = \{P, B_1, B_2\}$, and $\mathrm{St}_3(P) = \{P, C_1, C_2\}$ be any three distinct odd stigms of β, where $\langle \mathrm{St}_i \rangle \not\subseteq \langle \mathrm{St}_1 \rangle$, $\forall i \in \{2, 3\}$. Then $\dim\langle \bigcup_{i=1}^{3} \mathrm{St}_i \rangle = m$.*

Proposition 2.13 *Let β be an n-dimensional tangential 2-block in $\mathrm{PG}(n, 2)$, where $n \geq 4$, and m be a positive integer such that $m = n$ if n is even and $m = n - 1$ if n is odd. Let $P \in \beta$, $\mathrm{St}_1(P) = \{P, A_1, A_2, \ldots, A_m\}$, $\mathrm{St}_2(P) = \{P, B_1, B_2\}$, and $\mathrm{St}_3(P) = \{P, C_1, C_2\}$.*
Then

  (i) *n is odd;*
  (ii) *the only line of β in $\tau_i(P)$ is $\mathrm{St}_i(P)$, $\forall i \in \{2, 3\}$;*
  (iii) *if there is no tangential 2-block of dimension $n - 1$, then $\beta \cap \tau_1(P) \subseteq \langle \mathrm{St}_1(P) \rangle$.*

Proof (i) The proof of (i) follows immediately from Proposition 2.12. (ii) Suppose that (ii) is false. Without loss of generality assume that $\{P, D_1, D_2\}$ be a line of β in $\tau_2(P)$, which is different from $\mathrm{St}_2(P)$. By Proposition 2.12, it follows that $B_1 + D_1$ and $C_1 + D_1 \notin \langle \mathrm{St}_1(P) \rangle$. But by (i), $\dim\langle \mathrm{St}_1(P)\rangle = n - 2$; hence $B_1 + C_1 \in \langle \mathrm{St}_1(P) \rangle$, which leads to a contradiction by Proposition 2.12. Thus (ii) is proved. (iii) Suppose that (iii) is false. So, $\exists S \in \langle \lambda_1(P) \cup \{B_1 + C_1\}\rangle - \{P\}$. Because of our hypothesis and Corollary 2.3, $B_i + C_j \notin \beta$, $\forall i, j \in \{1, 2\}$. So, $S = B_1 + C_1 + X$, where $X \in \lambda_1(P) - \{P\}$. Without loss of generality, assume that $S = B_1 + C_1 + \sum_{i=1}^{2k} A_i$, where k is a positive integer and $2 \leq 2k \leq n - 3$. Then $\langle\{P, B_1, C_1\}\rangle \cap \langle S, A_1, A_2, \ldots, A_{2k-1}, A_{2k+2}, \ldots, A_{n-1}\rangle = \emptyset$, which is a contradiction. Thus (iii) is proved. This completes the proof. ∎

Let β be an n-dimensional tangential 2-block in $\mathrm{PG}(n, 2)$, where $n \leq 3$. Let $P \in \beta$, $\mathrm{St}_1(P) = \{P, A_1, A_2, \ldots, A_{2m}\}$, $2 \leq 2m \leq n$, $\mathrm{St}_2(P) = \{P, B_1, B_2\}$, and $\mathrm{St}_3(P) = \{P, C_1, C_2\}$, and $I_k = \{1, 2, 3, \ldots, k\}$, where k is a positive integer. Let

$$\mu_i(U_t) = \left\{ \sum_{j \in J} A_j + U_t : J \subseteq I_{2m} \text{ and } |J| = i \right\}, \quad t \in I_2, \quad U \in \{B, C\},$$

$$i \in I_{2m-1}$$

$$\nu(U) = \beta \cap \{X + U_t : X \in \langle \mathrm{St}_1(P)\rangle, t \in I_2\}, \quad U \in \{B, C\}.$$

Using the above notations, we have the following result.

Proposition 2.14 (i) $|\beta \cap \mu_i(U_t)| \leq 1$, $\forall i \in I_{2m-1}$, $\forall t \in I_2$, and $\forall U \in \{B, C\}$.

(ii) If $2m \geq 4$ and $\beta \cap \mu_i(U_t) = \{\sum_{j \in J} A_j + U_t\}$, $J \subseteq I_{2m}$, $|J| = i$, $i \in I_{2m-1}$, $t \in I_2$, then $\forall k \in I_{2m-i-1}$, $\beta \cap \mu_{i+k}(U_t) \subseteq \{\sum_{j \in J} A_j + \sum_{j \in J'} A_j + U_t\}$ for some $J' \subseteq I_{2m} - J$, $|J| = k$.

(iii) Let there exist two lines of β through each point of β. If $2m + 1 = n$ and no proper subset of $St_1(P)$ is a 3-stigm, then there exist permutations σ and σ' on I_{2m} such that

$$v(B) = \left\{\sum_{i=1}^{k} A_{\sigma(i)} + B_1 : k \in I_{2m-1}\right\},$$

and

$$v(C) = \left\{\sum_{i=1}^{k} A_{\sigma'(i)} + C_1 : k \in I_{2m-1}\right\}.$$

Proof (i) Suppose that (i) is false. Without loss of generality assume that $|\beta \cap \mu_i(B_1)| \leq 2$, $i \in I_{2m-1}$. Without loss of generality assume that $\sum_{j=1}^{i} A_j + B_1$ and $\sum_{j \in J} A_j + B_1 \in \beta$, where $J \subseteq I_{2m}$, $|J| = i$, and $J \neq I_i$. Now, $\sum_{j \in J} A_j + B_1 = \sum_{j \in J'} A_j + B_2$, where $J' = I_{2m} - J$. Clearly, $J' \cap I_i \neq \emptyset$. Let $k \in J' \cap I_i$ and $J'' = \{t : t \in J' \cup I_i - \{k\}\}$ and $\Delta = \{A_j : j \in J''\}$. Then $t(\Delta) \cap \langle\{A_k, B_1, B_2\}\rangle = \emptyset$, which is a contradiction. Thus the proof of (i) follows.

(ii) Let $\{\beta \cap \mu_i(U_t) = \sum_{j=1}^{i} A_j + U_t\}$, $t \in I_2$, $U \in \{B, C\}$. Suppose that $\beta \cap \mu_{i+k}(U_t) \neq \emptyset$, $k \in I_{2m-i-1}$. If possible, let

$$\beta \cap \mu_{i+k}(U_t) = \left\{\sum_{j \in J} A_j + U_t\right\},$$

where $J \subseteq I_{2m}$, $|J| = i + k$, and $I_i \not\subseteq J$. Now, $\sum_{j \in J} A_j + U_t = \sum_{j \in J'} A_j + U_{t'}$, where $J' = I_{2m} - J$, $\{t, t'\} = I_2$. Clearly, $I_i \cap J' \neq \emptyset$. Let $r \in I_i \cap J'$, $J'' = I_i \cup J' - \{r\}$, and $\Delta = \{A_j : j \in J''\}$. Then $t(\Delta) \cap \langle\{A_r, U_1, U_2\}\rangle = \emptyset$, which leads to a contradiction. Thus (ii) is proved.

(iii) Since no proper subset of $St_1(P)$ is a 3-stigm and there are two lines of β through each point of β, the proof of (iii) follows from (ii) and Proposition 2.13. This completes the proof. ∎

In order to illustrate the applications of the previous results we present a short proof of the following result, which was proved in [9].

Proposition 2.15 *Four-dimensional tangential 2-blocks do not exist.*

Proof If possible, let β be a 4-dimensional tangential 2-block in $PG(4,2)$. Let $P \in \beta$. By Propositions 2.10 and 2.12, assume that $St_i(P) = \{P, A_i, B_i\}$, $i = 1, 2, 3$. Let $C_3(P) = \langle \bigcup_{i=1}^{3} St_i(P) \rangle$ and $V_2(P) = \langle \{P, A_1 + A_2,$

$A_1 + A_3\}\rangle$. Assert that $\beta \cap V_2(P) - \{P\} \neq \emptyset$. If possible, let $V_2(P)$ be a tangent of P. Let τ_1', τ_2', and τ_3' be the three 3-spaces on $V_2(P)$, where without the loss of generality, $\tau_1' = C_3(P)$. By Propositions 1.4, 2.10, and 2.12, τ_i' contains a 3-stigm, say, $\{P, S_i, S_i'\}$, $\forall i \in \{2, 3\}$. Since $C_3(P) - V_2(P) \cup \{A_1 + A_2 + A_3, A_1 + A_2 + B_3\} \subseteq \beta$, it follows from Proposition 2.7 that $\langle\{P, S_2, S_3\}\rangle$ is a Fano block. This is a contradiction by Proposition 1.1. So, the assertion is true. Without loss of generality assume that $A_1 + A_2 \in \beta$. Since $St_i(A_1 + A_2)$ is a line of β, $\forall i \in \{1, 2, 3\}$, it follows from Proposition 2.7 that $\exists i \in \{1, 2, 3\}$ such that $St_i(A_1 + A_2) = \{A_1 + A_2, S_1, S_2\}$, where S_1 and $S_2 \in \beta - C_3(P)$. Now $|\beta - C_3(P)| \geq 5$, by Proposition 2.9. Let $S_i \in \beta - C_3(P) \cup \{S_1, S_2\}$, $i = 3, 4, 5$. Now, $t(A_3) \cap \langle\{P, A_1, A_2\}\rangle = \{A_1 + B_2\}$; whence we deduce that $\{A_3\}$ is a polarizing set with respect to $C_3(P)$. Considering the distribution of S_1, S_2, S_3, S_4, and S_5 among $\rho_1^i(A_3)$, $i = 1, 2, 3$, we deduce that $\exists i \in \{1, 2\}$ and $j \in \{3, 4, 5\}$ such that $S_i + S_j = A_3$. Without loss of generality assume that $S_1 + S_3 = A_3$. Hence $S_2 + S_3 = A_1 + A_2 + A_3$, which is a contradiction by Proposition 2.7(vi). This completes the proof. ∎

Proposition 2.16 *Let β be a 7-dimensional tangential 2-block in* PG(7, 2). *Let P be a point of β such that* (3, 3, 3)-*tangential stigm system does not exist with respect to P. If there exist two lines of β through each point of β, then P cannot possess a* (7, 3, 3)-*tangential stigm system.*

Proof If possible, let there exist two lines of β through each point of β and $St_1(P) = \{P, A_1, A_2, A_3, A_4, A_5, A_6\}$, $St_2(P) = \{P_1, B_1, B_2\}$ and $St_3(P) = \{P, C_1, C_2\}$. By Proposition 2.13(iii), $\langle \lambda_1(P) \cup \{B_1 + C_1\}\rangle$ is another tangent of P, say $\pi'(P)$. Let $\tau_1'(P), \tau_2'(P)$, and $\tau_3'(P)$ be the three 6-spaces on $\pi'(P)$, where, without loss of generality, $\tau_1' = \tau_1'(P)$ and $St_i(P) \subseteq \tau_2'$, $\forall i \in \{2, 3\}$. By using notations of Proposition 2.14, we have, without loss of generality, $v(B) = \{A_1 + B_1, A_2 + B_2, A_1 + A_3 + B_1, A_1 + A_3 + A_4 + B_1, A_2 + A_5 + B_2\}$ and $v(C) = \{A_{\sigma(1)} + C_1, A_{\sigma(2)} + C_2, A_{\sigma(1)} + A_{\sigma(3)} + C_1, A_{\sigma(1)} + A_{\sigma(3)} + A_{\sigma(4)} + C_1, A_{\sigma(2)} + A_{\sigma(5)} + C_2\}$, where σ is a suitable permutation on I_6, by Proposition 2.14. Now, $\tau_3'(P)$ must contain an odd stigm, say $St_3'(P)$, by Proposition 1.4. By Lemma 1.1 in [6] and Proposition 2.13 we conclude that $St_3'(P)$ is a 5-stigm. By Proposition 2.14 $St_3'(P) - \{P\} \subseteq \tau_3'(P) \cap (v(B) \cup v(C))$. Also, by Proposition 2.14 we may assume, without loss of generality, that $v(B) = \{A_1 + B_1, A_2 + B_2, A_1 + A_3 + B_1, A_1 + A_3 + A_4 + B_1, A_2 + A_5 + B_2\}$ and $v(C) = \{A_i + C_1, A_j + C_2, A_1 + A_i + C_1, A_1 + A_2 + A_i + C_1, A_j + A_k + C_2\}$, $\{i, j, k\} \subseteq I_6 - I_2$, $i \neq j \neq k \neq i$, and $St_3' = \{P, A_1 + B_1, A_2 + B_2, A_i + C_1, A_1 + A_2 + A_i + C_1\}$. Now, assert that $\beta \cap \langle St_1(P)\rangle = St_1(P)$. Since P does not possess a (3, 3, 3)-tangential stigm system, $P + A_i \notin \beta$, $\forall i \in I_6$. If possible, let $A_r + A_s + A_t \in \beta$, for some r, s, and $t \in I_6$. Since P has no (3, 3, 3)-tangential stigm system and there are two lines of β through each point of β, it follows that $\exists X \in \tau_2(P) \cap (v(B) \cup \{B_1, B_2\})$, $X' \in \tau_2(P) \cap v(C)$, $Y \in \tau_3(P) \cap v(B)$, and $Y' \in \tau_3(P) \cap (v(C) \cup \{C_1, C_2\})$ such that $X + Y = X' + Y' = A_r + A_s + A_t$, whence we deduce that $X + Y \in \{A_2 + A_5 + A_6, A_4 + A_5 + A_6, A_3 + A_4 + A_6\}$ and $X' + Y' \in \{A_2 + A_k + A_l, A_u + A_v + A_j\}$, where k, l, u, and $v \in \{3, 4, 5, 6\} - \{i, j\}$. Suppose that $X + Y = A_2 + A_5 + A_6$. Then $X = B_2$, $Y = A_1 + A_2 + A_4 + B_1$, $X' = A_j + C_2$, $Y' = A_1 + A_i + C_1$, $\{i, j\} = \{3, 4\}$, and $\{k, l\} = \{5, 6\}$. Now, $A_1 + A_2 + A_3 + A_4$ is a point on the axis of the cone whose generators are $\{B_2, B_1, P\}$, $\{B_2, A_2, A_2 + B_2\}$, and $\{B_2, A_1 + A_3 + A_4 + B_1, A_2 + A_5 + A_6\}$; but $A_1 + A_2 + A_3 + A_4 \in t(A_1 + A_2 + A_i + C_1, A_j, C_1)$ and hence $t(A_1 + A_2 + A_i + C_1, A_j, C_1)$ contains a point on the generators of the cone, and this leads to a contradiction. So, $X + Y \ne A_2 + A_5 + A_6$. Suppose that $X + Y = A_4 + A_5 + A_6$. Then $X = A_1 + A_3 + B_1$, $Y = A_2 + B_2$, $X' = A_1 + A_2 + A_i + C_1$, $Y' = C_2$, and $X' + Y' = A_u + A_v + A_j$, where $\{u, v, j\} = \{4, 5, 6\}$. So, $i = 3$. Note that $A_1 + A_2 + A_3 + A_j$ is a point on the axis of the cone whose generators are $\{C_2, C_1, P\}$, $\{C_2, A_j, A_j + C_2\}$ and $\{C_2, A_1 + A_2 + A_3 + C_1, A_4 + A_5 + A_6\}$. Now, $j = 4 \Rightarrow A_1 + A_2 + A_3 + A_4 \in t(A_1 + A_3 + A_4 + B_1, A_2, B_1)$, which leads to a contradiction. Again, $j = 5 \Rightarrow A_1 + A_2 + A_3 + A_5 \in t(A_2 + A_5 + B_2, A_1, A_3, B_2)$, which leads to a contradiction. Also, $j = 6 \Rightarrow A_1 + A_2 + A_3 + A_6 = P + A_4 + A_5 \in t(A_2 + A_5 + B_2, A_2, A_4, B_1)$, which leads to a contradiction. So, $X + Y \ne A_4 + A_5 + A_6$. Similarly, we can show that $X + Y \ne A_3 + A_4 + A_6$. Thus we arrive at a contradiction. So, $\beta \cap \langle \text{St}_1(P) \rangle = \text{St}_1(P)$. Hence it follows that $\{B_1, B_2, P\}$ and $\{B_1, A_1, A_1 + B_1\}$ are the only two lines of β, each of which contains B_1. By Lemma 1.1 in [6] and Proposition 2.13 we deduce that there exists a tangent, say π, such that one of the three 6-spaces on π contains neither $\{B_1, B_2, P\}$ nor $\{B_1, A_1, A_1 + B_1\}$, but contains a 5-stigma, say St', of β. Without loss of generality, we assume that $\pi(B_1) = \pi$, A_1 and $B_2 \notin \tau_3(B_1)$, and St' $\subseteq \tau_3(B_1)$. Since $A_1 \notin \tau_3(B_1)$, we deduce that St' $\not\subseteq v(B) \cup \text{St}_1(P)$. So, St' $\cap v(C) \ne \varnothing$; hence it follows that $|\text{St}' \cap v(B)| = 1$ and $|\text{St}' \cap \text{St}_1(P)| = 1$. Hence $\exists X \in \text{St}_1(P)$, $Y \in v(B)$, and Z and $W \in v(C)$ such that St' $= \{B_1, X, Y, Z, W\}$. But it is easy to check that such an odd stigm St' does not exist in $\tau_3(B_1)$, which leads to a contradiction. Hence the proof of the lemma follows. ∎

Proposition 2.17 *Tangential 2-blocks of dimensions 6 and 7 do not exist.*

The proof can be found in [5] and [6], and is therefore omitted.

Proposition 2.18 *Let β be an n-dimensional tangential 2-block in $\text{PG}(n, 2)$, where $2 \le n \le 7$. Let r be an odd integer such that $r = n$ if n is odd and $r = n + 1$ if n is even. If there exists a $(r, 3, 3)$-tangential stigm system with respect to each point of β, then β is either a Fano, Desargues, or Petersen block.*

Proof First, suppose that n is even. By Proposition 2.13, it follows that $n = 2$. Let $P \in \beta$. Because of Proposition 2.8, we observe that P has a $(3, 3, 3)$-tangential stigm system; hence we conclude that β contains all the seven points of $PG(2, 2)$. Clearly, these seven points of β form a tangential 2-block, called the Fano block.

Next assume that $n \in \{3, 5, 7\}$. Assert that $n \neq 7$. If possible, let $n = 7$. By Corollary 2.3 and Propositions 2.11 and 2.17, there exists a point $P \in \beta$ such that P has no $(3, 3, 3)$-tangential stigm system. Hence by Proposition 2.16, P has no $(7, 3, 3)$-tangential stigm system. But this contradicts our hypothesis. Thus $n \neq 7$. So, $n \in \{3, 5\}$. Let $n = 3$. Because of Proposition 2.8, we conclude that there exists a $(3, 3, 3)$-tangential stigm system with respect to each point of β. Let $P \in \beta$ and $St_i(P) = \{P, A_i, B_i\}, i = 1, 2, 3$. By Proposition 2.7(ii), $A_1 + A_2 + A_3$ and $A_1 + A_2 + B_3 \in \beta$. Since there are three lines of β through A_1, assume, without loss of generality, that $A_1 + A_2, A_1 + A_3$, and $A_2 + X_3 \in \beta$, for some $X_3 \in \{A_3, B_3\}$. But $\{A_1 + A_2, A_1 + A_3\} \subseteq \beta \Rightarrow \beta \cap \{A_1 + B_2, A_1 + B_3\} = \emptyset$, by Proposition 2.14. Again, there are three lines of β through $A_2 + X_3$; hence it follows easily that $X_3 = A_3$. So, $\beta = (\bigcup_{i=1}^{3} St_i(P)) \cup \{A_1 + A_2, A_1 + A_3, A_2 + A_3\}$. Note that β is the complement of the 5-stigm $\{A_1 + B_2, A_1 + B_3, A_2 + B_3, A_1 + A_2 + A_3, A_1 + A_2 + B_3\}$ in $PG(3, 2)$. Now, it is easy to check that the ten points in β form a 3-dimensional tangential 2-block; and since these ten points lie in threes on ten lines in a Desargues configuration, we refer to it as the Desargues block.

Finally, we assume that $n = 5$. Since 4-dimensional tangential 2-block does not exist, it follows from Corollary 2.3 and Proposition 2.11 and 2.16 that $\exists P \in \beta$ such that P has no $(3, 3, 3)$-tangential stigm system. By our hypothesis, we can assume that $St_1(P) = \{P, A_1, B_1, C_1, D_1\}, St_2(P) = \{P, A_2, B_2\}$, and $St_3(P) = \{P, A_3, B_3\}$, where $P + X_1 \notin \beta, \forall X_1 \in \{A_1, B_1, C_1, D_1\}$. By Propositions 2.13 and 2.14(iii) and by rearranging the letters and suffixes, assume, without loss of generality, that

$$\beta = \left(\bigcup_{i=1}^{3} St_i(P)\right) \cup \{A_1 + A_2, B_1 + B_2, A_1 + C_1 + A_2, X_1 + A_3,$$
$$Y_1 + B_3, X_1 + Z_1 + A_3\},$$

where X_1, Y_1, and Z_1 are three distinct elements of $\{A_1, B_1, C_1, D_1\}$. By Proposition 2.13, $\langle \lambda_1(P) \cup \{A_2 + A_3\}\rangle$ is another tangent of P, say π'. Let τ_1', τ_2', and τ_3' be the three 4-spaces on π'. Without loss of generality, assume that $\tau_1' = \tau_1(P)$ and $St_i(P) \subseteq \tau_3', \forall i \in \{2, 3\}$. By Proposition 1.4 and 2.13(ii), we conclude that τ_2' contains a 5-stigm, say St'. Since $A_1 + C_1 + A_2$ and $X_1 + Z_1 + A_3 \in \tau_3'$, it follows that $St' = \{P, A_1 + A_2, B_1 + B_2, X_1 + A_3, Y_1 + B_3\}$ whence we conclude that, without loss of generality,

$X_1 = C_1$ and $Y_1 = D_1$. So, $Z_1 \in \{A_1, B_1\}$. Since we can interchange A_1 and B_1, assume, without loss of generality, that $Z = A_1$. Hence

$$\beta = \left(\bigcup_{i=1}^{3} \text{St}_i(P)\right) \cup \{A_1 + A_2, B_1 + B_2, A_1 + C_1 + A_2, C_1 + A_3,$$

$$D_1 + B_3, A_1 + C_1 + A_3\}.$$

Now, it can be shown by direct verification that these 15 points of β do form a 5-dimensional tangential 2-block, which is called the Petersen block because of its relation to the Petersen graph through embedding [9]. This completes the proof. ∎

REFERENCES

1. C. Berge, "The Theory of Graphs and its Applications," Wiley, New York, 1962.
2. H. R. Brahana, The four color problem, *Amer. Math. Monthly* **30** (1923).
3. R. D. Carmichael, "Introduction to the Theory of Groups of Finite Order," Dover, New York, 1937.
4. B. T. Datta, On tangential 2-blocks, *Discrete Math.* **15** (1976) 1–22.
5. B. T. Datta, Nonexistence of six-dimensional tangential 2-blocks, *J. Combinatorial Theory Ser. B* **21** (1976) 171–193.
6. B. T. Datta, Nonexistence of seven-dimensional tangential 2-blocks, *Discrete Math.*, to appear.
7. G. A. Dirac, A theorem of R. L. Brooks and conjecture of H. Hadwiger, *Proc. London Math. Soc.* **7** (1957) 161–195.
8. J. Petersen, Sur le théorème de Tait, *Intermédiaire des Mathematiciens* **5** (1898) 225–227.
9. W. T. Tutte, On the algebraic theory of graph colorings, *J. Combinatorial Theory* **1** (1966) 15–50.
10. W. T. Tutte, Lectures on matroids, *J. Res. Nat. Bur. Standards Sect. B* **69** (1965) 1–47.
11. O. Veblen, An application of modular equations in analysis situs, *Ann. of Math.* **14** (1912) 86–94.
12. O. Veblen and J. W. Young, "Projective Geometry," Vol. 1, Ginn, Boston, Massachusetts, 1910.

AMS 05B25, 05B25

DEPARTMENT OF MATHEMATICS
THE OHIO STATE UNIVERSITY
LIMA, OHIO

Intersection and Distance Patterns

M. DEZA

and

I. G. ROSENBERG

1. Introduction

For m fixed and finite sets $A_0, \ldots, A_{m-1}$, the $m \times m$ matrices $(|A_i \cap A_j|)$ and $(|A_i \backslash A_j|)$ are called the intersection and distance matrices of $(A_0, \ldots, A_{m-1})$. If we arrange the entries into a vector and eliminate duplications we obtain the intersection and distance patterns. The characterization of the set Y of the intersection matrices is a rather difficult combinatorial problem which is fully understood only for $m \leq 4$. In this paper we survey the known results and various connections (e.g., quadratic forms, convex cones, graph theory, etc.) and discuss the integer programming aspects. The dual cone of conv Y and the facets of conv Y are studied. In particular we prove that there are facets of the intersection matrices which correspond to unique sets $A_0, \ldots, A_{m-1}$ (unique up to renaming of elements). Finally we study special distance patterns which we call elliptic and extensions of strong Δ-systems. Some special subproblems, related problems, and generalizations not discussed here can be found [5, 8–10].

2. Intersection Patterns and Quadratic Forms

Let $m \in \mathbb{P} = \{1, 2, \ldots\}$ be fixed. Set $k = \frac{1}{2}m(m + 1)$. Given finite sets $A_0, \ldots, A_{m-1}$ set $ij\tilde{} = |A_i \cap A_j|$ and call the vector

$$(00\tilde{}, 01\tilde{}, \ldots, (0, m - 1)\tilde{}, 11\tilde{}, \ldots, (m - 1, m - 1)\tilde{})$$

the *intersection* or *realizable pattern* of $A_0, \ldots, A_{m-1}$. Let Y be the set of all k-vectors obtained in this way. Sometimes the numbers $ij\tilde{}$ are arranged into the $m \times m$ $(ij\tilde{})$ called the *intersection* matrix of $A_0, \ldots, A_{m-1}$. Let Z denote the set of intersection matrices. Obviously each intersection matrix is a symmetric matrix over $\mathbb{N} = \{0, 1, \ldots\}$ in which each diagonal entry dominates all entries in the row passing through it. However, the characterization of Z is a rather difficult problem. For $m = 4$, the set Z has been described as the set of integer non-negative solutions of 40 inequalities ([2, 3] and rediscovered [14]; the number of inequalities may sound worse than it really is because, translated into the language of Hamming distances, 30 of them are the obvious triangle inequalities and the remaining 10 are pentagon inequalities introduced in [3]). Before proceeding further we briefly mention some other interpretations of Y. Setting $A = A_0 \cup \cdots \cup A_{m-1} = \{a_0, \ldots, a_{n-1}\}$ we can characterize each A_i by its characteristic zero–one n-vector $x_i = (x_{i0}, \ldots, x_{i,n-1})$ having $x_{ij} = 1$ if $a_j \in A_i$ and $x_{ij} = 0$ otherwise. Then clearly $ij\tilde{} = |A_i \cap A_j| = x_i x_j^T$ (scalar product). Thus the intersection matrix is $(ij\tilde{}) = XX^T$, where X is the $m \times n$ zero–one matrix having rows $x_0, \ldots, x_{m-1}$. Hence the set Z of intersection matrices is

$$Z = \{XX^T : X \text{ zero–one } m \times n \text{ matrix}, n = 1, 2, \ldots\}$$

or, equivalently, each intersection matrix is the matrix of the scalar products of m binary n-vectors ($n = 1, 2, \ldots$). We can relax the restriction X binary in several ways. For $S \subseteq \mathbb{R}$ (=reals) set $T_S = \{XX^T : X \ m \times n \text{ matrix over } S\}$. Thus T_S consists of the matrices of inner products of m row-vectors from S^n (whose determinants are exactly Gram's determinants defining the volume of the parallelotops determined by the row-vectors). A generalization to Hilbert spaces and complex numbers is in [11]. We have the obvious inclusions

$$\left. \begin{array}{c} Z = T_{[0, 1]} \subseteq T_\mathbb{N} \subseteq T_{\mathbb{Q}_+} \subseteq T_{\mathbb{R}_+} \\ \cap \qquad \cap \\ T_{[-1, 1]} \subseteq T_\mathbb{Z} \subseteq T_\mathbb{Q} \end{array} \right\} \subseteq T_\mathbb{R}$$

($\mathbb{Z}$ denotes the integers, $\mathbb{Q}$ the rationals, and $\mathbb{Q}_+$ and $\mathbb{R}_+$ the non-negative rationals and reals.) The number n of columns in the definition of T_S is not necessarily completely unrestricted, for example by a result of Mordell (see, e.g. [16]) each integer positive definite matrix has a representation XX^T where the rational matrix X has at most $m + 3$ columns.

The matrices of the form XX^T, X square rational, are studied in Hasse-Minkowski's theory of the congruences of quadratic forms (see, e.g. [13, Sect. 10.4]). In particular, the decomposition of $(r - \lambda)I + \lambda J$ into XX^T, X square rational, was applied by Bruck, Ryser, and Chowla (see, e.g. [13, Sect. 10.3]) as a necessary condition for the existence of block designs.

The sets $T_{[0, 1]}$ and $T_\mathbb{N}$ are submonoids of $(\mathbb{N}^{n^2}, |)$ while $T_{[-1, 1]}$ and $T_\mathbb{Z}$ are submonoids of $(\mathbb{Z}^{n^2}, +)$ (in both cases + denotes the vector or componentwise addition). Further $T_{\mathbb{Q}_+}$ and $T_{\mathbb{R}_+}$ are pointed convex cones over $\mathbb{Q}$ and $\mathbb{R}$ while $T_\mathbb{Q}$ and $T_\mathbb{R}$ are subspaces of the vector space $\mathbb{R}^{n^2}$. The set $T_{\mathbb{R}_+}$ agrees with the set of completely positive matrices [13, Sect. 16.2] and $T_\mathbb{R}$ is nothing but the set of positive semidefinite matrices. The dual cone to $T_{\mathbb{R}_+}$ is the set of $m \times m$ copositive matrices [13, Sect. 16.2] while the dual of $T_\mathbb{R}$ is $T_\mathbb{R}$ itself. The defining properties of these two dual cones may be viewed as special instances of copositivity with respect to a given cone $D \subseteq \mathbb{R}^{n^2}$ (the cone being $\mathbb{R}_+^{n^2}$ and $\mathbb{R}^{n^2}$, respectively; see, e.g. [1, Chap. 3, Sect. 9]).

The dual cone to $Y = T_{[0, 1]}$ is described later. The dual cone to $T_{[-1, 1]}$ is used to describe the covariance functions of unit stationary processes [17].

3. Distance Patterns

Another equivalent formulation is based on symmetric differences of sets. Given sets $A_0, \ldots, A_{m-1}$ choose $A_m = \emptyset$ and for all $i, j \in \mathbf{m} + 1 = \{0, 1, \ldots, m\}$ and set $ij^* = |(A_i \backslash A_j) \cup (A_j \backslash A_i)|$. Then $ij^* = ii^\sim + jj^\sim - 2ij^\sim$ where, as before, k stands for $\frac{1}{2}m(m + 1)$. We may form the k-dimensional vector $(01^*, \ldots, 0m^*, 12^*, \ldots, 1m^*, \ldots, (m - 2, m - 1)^*)$ and call it the *distance pattern* (the name is explained below). The corresponding $(m + 1) \times (m + 1)$ matrix (ij^*), called the *distance matrix*, is clearly a symmetric matrix over N with zero diagonal, which explains the fact that the distance matrix carries the same information as the smaller intersection matrix). The name is justified by the fact that ij^* is nothing else than the Hamming distance between the characteristic vectors x_i and x_j introduced earlier. Recall that the Hamming distance of $x, y \in \{0, 1\}^n$ is the square of their Euclidean distance or, equivalently, $\sum_{i=0}^{n-1} |x_i - y_i|$, i.e., it equals the number of places where x differs from y. The intersection and distance patterns are in 1–1 correspondence, which, in fact, is a linear nonsingular operator on the vector space $\mathbb{R}^k$. The formula inverse to $ij^* = ii^\sim + jj^\sim - 2ij^\sim$ is $ij^\sim = \frac{1}{2}(im^* + jm^* - ij^*)$.

4. Integer Programming Aspect

We describe an integer linear programming aspect of intersection patterns. Let $Q = \{1, 2, \ldots, l\}$ (where $l = 2^m - 1$). For $q \in Q$ let $q = q_{(0)} + 2q_{(1)} + \cdots + 2^{m-1}q_{(m-1)}$ be the binary expansion of q (i.e., all

$q_{(i)} \in \{0, 1\}$) and let $\|q\| = q_{(0)} + \cdots + q_{(m-1)}$ (the number of ones in the binary expansion of q). Let $R = \{q \in Q: \|q\| \leq 2\} = \{r_0, \ldots, r_{k-1}\}$ where $r_0 < r_1 < \cdots < r_{k-1}$. Let $E = (e_{rq})$ be the $k \times l$ binary matrix whose rows and columns are indexed by R and Q and defined by setting (i) $e_{rq} = 1$ if in the binary expansions of r and q we have $r_{(0)} \leq q(0), \ldots, r_{(m-1)} \leq q_{(m-1)}$ and (ii) $e_{rq} = 0$ otherwise (e.g., $e_{15} = 1$ and $e_{34} = 0$). Denoting by x_q the frequency of the column $(q_{(0)}, \ldots, q_{(m-1)})^T$ in X we have

Proposition 1 *A vector $y \in \mathbb{N}^k$ is an intersection pattern if and only if $y = Ex$ for some $x \in \mathbb{N}^l$.*

Corollary 1 *The set Y of intersection patterns is a submonoid of $(\mathbb{N}^k, +)$ (componentwise addition).*

In particular, Y contains all non-negative integer multiples of its elements.

5. Convex Cones and Dual Cones of Intersection Patterns

It is natural to ask about the convex hull, conv Y, of Y (the least convex subset of $\mathbb{R}^k$ containing Y) and the set $I = \mathbb{N}^k \cap \text{conv } Y$ called the set of *pseudo-realizable vectors* [4, 7]. Clearly conv $Y = E\mathbb{R}_+^l$. Also we have

Proposition 2

$$I = \mathbb{N}^k \cap \left(\bigcup_{n \in \mathbb{P}} n^{-1} Y \right).$$

The set I may not agree with Y. For one thing there may be $x', x'' \in \mathbb{N}^l$ and a rational $0 < \lambda < 1$ such that $x = \lambda x' + (1 - \lambda)x''$ is not an integer itself but Ex is. This happens already for $m = 5$ ([3] rediscovered in [14]). This shows that, in general, Y cannot be described as the set of integer points of a cone determined by a system of linear inequalities (but, in principle, may be described by linear inequalities and number-theoretical congruences). In this respect our problem differs from several other combinatorial problems. However, even the knowledge of conv Y would provide a good start for the characterization of Y. Since conv Y is a pointed polyhedral cone with apex 0 there is a unique set $D^{\sim} \subseteq \mathbb{R}^{1k}$ such that for every $d \in D^{\sim}$ we have $dy \geq 0$ whenever $y \in Y$. The set $D^{\sim}$ is the *dual* or *polar cone* to conv Y and corresponds to the set of all homogeneous linear inequalities valid for Y. Thus an economical description of D provides a good description of conv Y. We have

Proposition 3 *Let $d \in \mathbb{R}^{1k}$. Then $d \in D^{\sim}$ if and only if dE is non-negative.*

For $d \in D^{\sim}$ let F_d stand for $\{y \in Y: dy = 0\}$.

Corollary 2 For $d \in \tilde{D}$
$$F_d = \{Ex: x \in \mathbb{N}^l, x_i = 0 \text{ whenever } dE_{*i} > 0\}.$$
Let $M = \{(i, j): 0 \leq j \leq m - 1\}$.

Proposition 4 The following statements are equivalent for $c = (c_{00}, \ldots, c_{0,m-1}, \ldots, c_{m-1,m-1}) \in \mathbb{R}^k$:

(A) $c \in \tilde{D}$,
(B) $\sum_{(i,j) \in B^2 \cap M} c_{ij} \geq 0$ for every $B \subseteq \mathbf{m}$.
(C) The quadratic form $f(z) = \sum_{(i,j) \in M} c_{ij} z_i z_j$ is non-negative on $\{0, 1\}^m$.

The last condition stipulates that the values of the form are non-negative on the vertices of the m-dimensional unit hypercube.

We can relate the condition (B) of Proposition 4 to the following combinatorial problem which may be of some interest on its own. Let K_m^+ be the complete (unoriented) graph with loops on $\mathbf{m} = \{0, \ldots, m - 1\}$. The condition (B) in Proposition 4 means that c_{ij} is an edge-valuation by reals (i.e., a map $M \to \mathbb{R}$) such that the sum of the valuations of edges having both ends in B is non-negative for all $B \subseteq \mathbf{m}$.

For $b \in \mathbb{Z}^m$ the inequality

$$\sum_{0 \leq i \leq j \leq m} b_i b_j \widetilde{ij} + \frac{1}{2} \sum_{i \in \mathbf{m}} b_i(b_i - 1) \widetilde{ii} \geq 0 \tag{1}$$

is a valid inequality on Y ([2, 3], see also [14]). We refer to it as the *regular inequality induced by* b.

Example 1 Let $0 \leq n \leq m$. Set $b_i = -1$, for $i \in \mathbf{n}$ and $b_i = 1$ otherwise, and obtain

$$\sum_{0 \leq i \leq j < n} \widetilde{ij} + \sum_{n \leq i < j < m} \widetilde{ij} - \sum_{0 \leq i < n \leq j < m} \widetilde{ij} \geq 0. \tag{2}$$

We are interested in the tersest possible description of conv Y, provided by the facets of conv Y. The facets among the regular inequalities are the following. For $q \in Q$ set $v_q = \sum_{i \in \mathbf{m}} q_{(i)} b_i$. Further let $V = \{q \in Q : v_q \in \{0, 1\}\}$ and let $E^{(b)}$ be the submatrix of E consisting of the columns of E indexed by $q \in V$. The regular inequality induced by b is a facet of conv Y iff rank $E^{(b)} = k - 1$. In particular, if $|V| < k - 1$ then automatically it is not a facet of conv Y.

Proposition 5 Let $m > 3$, $b_0 = \cdots = b_{f-1} = -1$, and $b_f = \cdots = b_{m-1} = 1$. Then the regular inequality corresponding to b is a facet of conv Y if and only if $1 < f < m - 1$.

Write $Y_m (\tilde{D}_m)$ to denote more specifically the set of intersection patterns (its dual) of m sets. Each element $d \in \tilde{D}_m$ yields a valid inequality from $\tilde{D}_{m'}$

for all $m' > m$. Indeed define $c' \in \mathbb{R}^{k'}$ by setting $c'_{ij} = c_{ij}$ if $0 \leq i \leq j < m$ and $c'_{ij} = 0$ otherwise. We have

Lemma 1 *If $c \in \tilde{D}_m$, then $c' \in \tilde{D}_{m'}$. Moreover, if $cy = 0$ is a facet of* conv Y_m, *then $c'y' = 0$ is a facet of* conv $Y_{m'}$.

6. Dual Cone of Distance Patterns

Let Y^* be the set of distance patterns and let D^* be the dual cone of conv Y^*. Although D^* can be derived from D, we can give the following direct characterization. For $B \subseteq \mathbf{m}$ let $B' = \{0, \ldots, m\} \setminus B$.

Proposition 6 *The following statements are equivalent for $c = (c_{01}, \ldots, c_{0m}, c_{12}, \ldots, c_{1m}, \ldots, c_{m-1, m}) \in \mathbb{R}^k$:*

(A) $c \in D^*$,

(B) $\displaystyle\sum_{i \in B, j \in B'} c_{ij} + \sum_{i \in B', j \in B} c_{ij} \geq 0$ for all $B \subseteq \mathbf{m}$,

(C) *the quadratic function*

$$f(z) = \sum_{i \in m+1} z_i \left(\sum_{0 \leq j < i} c_{ji} + \sum_{i < j \leq m} c_{ij} \right) - 2 \sum_{0 \leq i < j \leq m} c_{ij} z_i z_j$$

($z \in \mathbb{R}^{m+1}$) *is non-negative on* $\{0, 1\}^{m+1}$.

The vector c may be regarded as a real valuation of the edges of the complete graph K_{m+1} on $\mathbf{m} + 1$ (unoriented and loopless). The condition (B) of Proposition 6 stipulates that for each cut the sum of the values on its edges is non-negative.

The translation of the regular inequalities is the following. Let $(b_0, \ldots, b_m) \in \mathbb{Z}^{m+1}$ be such that $b_0 + \cdots + b_m = 1$. The inequality

$$-\sum_{0 \leq i < j \leq m} b_i b_j ij^* \geq 0$$

is valid for all distance patterns and will be termed *hypermetric* [2–4, 14, 15].

Example 2 Let $m = 2n$. Set $b_i = -1$ for $0 \leq i < n$, $b_i = 1$ for $n \leq i \leq 2n$. The corresponding inequality is

$$\sum_{0 \leq i < n \leq j \leq 2n} ij^* - \sum_{0 \leq i < j < n} ij^* - \sum_{n \leq i < j \leq 2n} ij^* \geq 0.$$

Its left side is always even because translated into intersection patterns it equals twice the left side of (2).

Choosing $n = 1$ we get the triangle inequality. For $n = 2$ we get the inequality

$$02\tilde{*} + 03\tilde{*} + 04\tilde{*} + 12\tilde{*} + 13\tilde{*} + 14\tilde{*} - 01\tilde{*} - 23\tilde{*} - 24\tilde{*} - 34\tilde{*} \geq 0$$

called the *pentagon inequality* [3].

7. Reduction to Inequalities

The system $y = Ex$ can be solved for x_R (where x_R is the vector of all x_i from x indexed by $r \in R = \{q \in Q : \|q\| \leq 2\}$) obtaining $x_R = Fx_{Q\setminus R} + Gy$. It turns out that both F and G are integer matrices; hence x_R is integer whenever both $x_{Q\setminus R}$ and y are. Postulating the non-negativity of x_R we can convert $y = Ex$ into a system of inequalities:

Proposition 7 [3] *Let* $y = (00\tilde{\ }, \ldots, (m-1, m-1)\tilde{\ }) \in \mathbb{N}^k$. *Then* $y \in Y$ *if and only if the system of inequalities*

$$\sum_{q \in Q \setminus R} q_{(i)} q_{(j)} x_q \leq i\tilde{j} \qquad (0 \leq i < j \leq m-1) \qquad (3)$$

$$\sum_{q \in Q \setminus R} q_{(i)}(2 - \|q\|) x_q \leq 2i\tilde{i} - \sum_{j \in m} i\tilde{j} \qquad (i \in \mathbf{m}) \qquad (4)$$

is solvable in non-negative integers x_q $(q \in Q \setminus R)$.

From the elimination we obtain the formula

$$x_0 + \cdots + x_{l-1} = \sum_{0 \leq i < m} i\tilde{i} - \sum_{0 \leq i < j < m} i\tilde{j} + \sum_{q \in Q \setminus R} \binom{\|q\| - 2}{2} x_q. \qquad (5)$$

For $y \in Y$ there may be several $x \in \mathbb{N}^l$ satisfying $y = Ex$. Following [2, 3, 4] the *content* $c(y)$ is defined as the least value of $x_0 + \cdots + x_{l-1}$ over all $x \in \mathbb{N}^l$ satisfying $y = Ex$. In other words, $c(y)$ is the minimum size of $A_0 \cup \cdots \cup A_{m-1}$ over the families $\{A_0, \ldots, A_{m-1}\}$ having y as their intersection pattern. It follows from Corollary 1 that c is a subadditive function from Y into $\mathbb{N}$ [i.e., $c(y_1 + y_2) \leq c(y_1) + c(y_2)$]. For a fixed $y \in Y$ the determination of $c(y)$ is a typical integer programming problem:

$$\text{minimize} \quad x_0 + \cdots + x_{l-1} \quad \text{subject to} \quad Ex = y, \quad x \in \mathbb{N}^l. \qquad (P)$$

To get a lower bound for $c(y)$ we relax this problem, i.e., replace the restriction $x \in \mathbb{N}^l$ by $x \in \mathbb{R}_+^l$. The dual problem is maximize $z \cdot y$ subject to $z \cdot E \leq (1, \ldots, 1)^T$, $z \in \mathbb{R}^k$. If both the relaxed and dual problems have solutions (that is, are feasible) and are bounded, then the optimal value $z^0 \cdot y$ of the dual does not exceed $c(y)$.

Let $d(y)$ be the optimal solution of the following problem

$$\text{minimize} \sum_{q \in Q \backslash R} \binom{\|q\| - 1}{2} x_q \quad \text{(P')}$$

subject to (3), (4), and $x_q \in \mathbb{N}$ for all $q \in Q \backslash R$. From Proposition 7 and (5) the constant $c(y)$ satisfies

$$c(y) = d(y) + \sum_{i \in m} y_{2i} - \sum_{0 \leq i < j < m} y_{2i+2j}.$$

Thus instead of relaxing (P) we can relax (P') and obtain an upper bound for $c(y)$ from the dual of (P').

The following simple bounds can be obtained directly [by multiplying both sides of $y = Ex$ by $(1, \ldots, 1)$]

$$(y_0 + \cdots + y_{l-1}) \bigg/ \binom{m+1}{2} \leq c(y) \leq y_0 + \cdots + y_{l-1}.$$

It is shown in [2] that the content $c^*(y)$ of distance patterns satisfies

$$(y_0 + \cdots + y_{l-1}) \bigg/ \left[\binom{m+1}{2}^2\right] \leq c^*(y) \leq (y_0 + \cdots + y_{l-1})/m.$$

8. Rigid Faces

Let $c \in \tilde{D}$. We say that the inequality $cy \geq 0$ is *rigid* if for each $y \in \mathbb{N}^k$ satisfying (i) $cy = 0$ and (ii) all regular inequalities, there is precisely one $x \in \mathbb{N}^l$ such that $Ex = y$.

In other words for a rigid c and y satisfying (i) and (ii) there is exactly one family $A_0, \ldots, A_{m-1}$ having y as its intersection pattern (uniqueness being interpreted as up to renaming of the elements of the sets). We exhibit a rigid regular facet of conv Y for every m.

Proposition 8 *Let $n = m - 2$. The regular inequality*

$$\sum_{0 \leq i \leq j < n} ij\tilde{} + (n, n+1)\tilde{} - \sum_{i \in n} ((i, n)\tilde{} + (i, n+1)\tilde{}) \geq 0$$

corresponding to $(-1, -1, \ldots, -1, 1, 1)$ *is a rigid facet of* conv Y.

9. Elliptic Patterns

Next we consider special distance patterns. Set $q_{(m)} = 0$ for each $q \in Q$. For each $i \in m$ we have

$$\tfrac{1}{2}(0i^* + im^* - 0m^*) = \sum_{q \in Q} (1 - q_{(0)}) q_{(i)} x_q.$$

A distance pattern is *elliptic* if $0i^* + im^* - 0m^*$ has the same value for all $i = 1, \ldots, m - 1$. The points 0 and m are called the *foci* of an elliptic pattern. Denote the common value of $0i^* + im^* - 0m^*$ by $2\delta_y$. The content of an elliptic pattern is bounded as follows:

Proposition 9 *The content of an elliptic pattern satisfies*

$$0m^* \leq c(y) \leq 0m^* + (m-1)\delta_y.$$

The elliptic pattern y will be called *singular* if $\delta_y = 0$. For singular elliptic patterns we have $c(y) = 0m^*$.

Example 3 Let $\langle x_0, \ldots, x_{2^n-1} \rangle$ be the sequence of all zero–one n-vectors in their natural order. Let y_n be its distance pattern (e.g., $y_2 = \langle 1, 1, 2, 2, 1, 1 \rangle$). Now y_n is elliptic and singular; hence the content $c(y_n) = x_0 x^*_{2^n-1} = n$ proving (not surprisingly) that the distance pattern of all zero–one n vectors can be realized by no set of zero–one vectors in a smaller dimension. Due to the symmetries of the hypercube any pair of opposite vertices can serve as the pair of foci.

Suppose $y = (01^*, \ldots, (m-1, m)^*)$ is a distance pattern and let $n \in \mathbb{N}$ be at least as big as all $1m^*, \ldots, (m-1, m)^*$. Form the vector

$$y_{(n)} = (y_{01}, \ldots, y_{0,m+1}, y_{11}, \ldots, y_{1,m+1}, \ldots, y_{m,m+1})$$

having $y_{ij} = ij^*$ for all $0 \leq i < j \leq m$ and $y_{0,m+1} = n$, $y_{i,m+1} = n - im^*$ ($0 < i \leq m$). Although $y_{(n)}$ is non-negative, it need not be a distance pattern. However, if it is, then $y_{(n)}$ is elliptic and singular. We have

Proposition 10 *Let* $y = (01^*, \ldots, (m-1, m)^*)$ *be realizable and let* $n \in \mathbb{N}$ *be at least as big as* $1m^*, \ldots, (m-1, m)^*$. *Then* $y_{(n)}$ *is realizable if and only if* $n \geq c(y)$.

Although for $n < c(y)$ the patterns $y_{(n)}$ are not realizable, they may be pseudo-realizable. Thus we define the *weak content* $w(y)$ of y as the least integer [not less than $1m^*, \ldots, (m-1, m)^*$] such that $y_{(n)}$ is pseudo-realizable. Again $w(y)$ is a subadditive function (from $\mathbb{N}^k$ into $\mathbb{N}$) [by definition $w(y) \leq c(y)$]. We note that all the known examples of pseudo-realizable but not realizable patterns are of the form $y_{(n)}$ where y is realizable and $w(y) \leq n < c(y)$.

Example 4 Let $p \in \mathbb{N}$ and let $y = (2p, \ldots, 2p)$; then y is realizable. It is shown in [5] that $c(y) \leq mp$ and $c(y) = mp$ for all $m \geq p^2 + p + 3$. Consider the simplest case $p = 1$ and $m = 5$. Then $c(y) = 5$ and it can be shown that $w(y) = 4$. Thus $y_{(4)} = (2, 2, 2, 2, 2, 4, 2, \ldots, 2)$ is pseudo-realizable (its double is realizable) but not realizable.

10. Extension of Strong Δ-Systems

A family of sets $\{A_0, \ldots, A_{m-1}\}$ is said to be a Δ-*system* corresponding to r and λ if $|A_i \cap A_j| = \lambda$ for all $0 \le i < j < m$ and $|A_i| = r$ for all $i \in \mathbf{m}$. A Δ-system $\{A_0, \ldots, A_{m-1}\}$ is a *strong Δ-system* (sΔs for short) if $A_0 \cap A_1 = A_0 \cap \cdots \cap A_{m-1}$ (see, e.g. [6]). An *extended* sΔs is a sequence $\{A_0, \ldots, A_m\}$ of sets such that $\{A_0, \ldots, A_{m-1}\}$ is an sΔs. The purpose of this paragraph is to characterize the intersection patterns of extended sΔs's. Let $k = \frac{1}{2}(m + 1) \times (m + 2)$. We have

Proposition 11 *The vector* $(y_{00}, \ldots, y_{0m}, y_{11}, \ldots, y_{1m}, \ldots, y_{mm})$ *is the intersection pattern of an extended strong Δ-system if and only if there are integers $0 \le \lambda \le r$ such that*

$$y_{ii} = r \quad \text{for all} \quad i \in \mathbf{m},$$

$$y_{ij} = \lambda \quad \text{for all} \quad 0 \le i < j < m,$$

$$\sum_{i \in \mathbf{m}} y_{im} \le y_{mm} + (m - 1)\lambda, \quad \sum_{i \in \mathbf{m}} y_{im} \le y_{mm} + (m - 1)y_{jm} \quad \text{for all} \quad j \in \mathbf{m},$$

$$y_{im} + r - \lambda \le y_{jm} \quad \text{for all} \quad i, j \in \mathbf{m},$$

$$y_{im} \le r \quad \text{for all} \quad i \in \mathbf{m},$$

Note that the second to the last inequality is the regular inequality corresponding to $(1, 1, -1)$ (setting $r = y_{ii}$ and $\lambda = y_{ij}$).

11. The Fourier–Motzkin Method

The dual cone D in principle can be obtained by a modification of the Fourier–Motzkin elimination. The idea is to eliminate successively the variables x in $Fx_{Q \setminus R} - Gy \ge 0$ until we obtain an equivalent system of inequalities involving y only. The general obstacle in this method is the very rapidly increasing number of inequalities. To keep it down at each step we should try to eliminate redundant inequalities. Denote by $\mathscr{P}_p$ the polyhedron determined by the inequalities obtained as the result of the elimination of the first p variables. The most economical description of $\mathscr{P}_p$ is the list of facets. To give this list may be difficult in general but it is possible in our case (by a relatively simple testing of the ranks of certain matrices). So far integrality has been ignored but for a complete solution we need a variant of the Fourier–Motzkin method (based on congruence and branching [18]).

REFERENCES

1. J. A. Berman, "Cones, Matrices and Mathematical Programming," Lecture Notes in Economics and Mathematical Systems No. 79, Springer-Verlag, Berlin and New York, 1973.
2. M. Deza (Tylkin), On Hamming geometry of unitary cubes (in Russian), *Dokl. Akad. Nauk SSSR* **134**, No. 5 (1960) 1037–1040
3. M. Deza (Tylkin), Realizability of matrices of distances in unitary cubes (in Russian), *Problemy Kibernet.* **7** (1962) 31–42.
4. M. Deza, Linear metric properties of binary codes (in Russian), *Proc. Soviet Union Confer. Coding Theory Transmission Information, 4th, Moscow–Tashkent* (1969) 77–85.
5. M. Deza, Une propriété extrémale des plans projectifs dans une classe de codes équidistants, *Discrete Math.* **6** (1973) 343–352.
6. M. Deza, Solution d'un problème de Erdös–Lovász, *J. Combinatorial Theory Ser. B* **16** (1974) 166–167.
7. M. Deza, Matrices des formes quadratiques non-négatives pour des arguments binaires, *C. R. Acad. Sci. Paris, Ser. A–B* **277** (1973) 893–875.
8. M. Deza, Isometries of the hypergraphs, *Proc. Internat. Confer. Theory Graphs, Calcutta, 1976*.
9. M. Deza and P. Eades, On circulant codes with prescribed distances, *Bull. Austral. Math. Soc.*, to appear.
10. M. Deza and I. Rosenberg, Generalized Intersection Patterns and Two Symbols Balanced Arrays, Preprint Publ. CRM-718, Univ. of Montreal, Montreal, 1977.
11. Ky Fan, On positive definite sequences, *Ann. of Math.* **47**, No. 3 (1946) 593–607.
12. A. J. Goldman and A. W. Tucker, Polyhedral convex cones, *in* "Linear Inequalities and Related Systems," pp. 19–40. Princeton Univ. Press, Princeton, New Jersey, 1956.
13. M. Hall, "Combinatorial Theory," Ginn (Blaisdell), Boston, Massachusetts, 1967.
14. J. B. Kelly, Products of zero-one matrices, *Canad. J. Math.* **20** (1968) 248–329.
15. J. B. Kelly, Hypermetric spaces, *in* "Proceedings of a Conference on the Geometry of Metric and Linear Spaces," Lecture Notes in Mathematics No. 490, pp. 17–31, Springer-Verlag, Berlin and New York, 1975.
16. Chao Ko, On the representation of a quadratic form as a sum of squares of linear forms, *Quart. J. Math. Oxford Ser.* **8**, No. 30 (1937) 81–98.
17. B. McMillan, History of a problem, *J. Soc. Ind. Appl. Math.* **3**, No. 3 (1965) 119–128.
18. H. P. Williams, Fourier–Motzkin elimination extension to integer programming problems, *J. Combinatorial Theory Ser. A* **21** (1976) 118–123.

AMS 05C10, 05B99

M. Deza
CNRS PARIS
3 RUE DE DURAS
PARIS, FRANCE

I. G. Rosenberg
CENTRE DE RECHERCHES MATHÉMATIQUES
UNIVERSITÉ DE MONTRÉAL
MONTRÉAL, QUÉBEC

Regular Groups of Automorphisms of Cubic Graphs

DRAGOMIR Ž. DJOKOVIĆ†

and

GARY L. MILLER‡

Background

Tutte [14] made the following definition which is crucial in understanding symmetric cubic graphs. We should call a pair consisting of a connected cubic graph G and a group of automorphisms of G an *object* denoted (G, A).

Definition Given an object (G, A) we say A is *s-transitive* (*s-regular*) over G if A is (sharply) transitive over paths of length s.

In [14] Tutte showed that:

Theorem 1 *If (G, A) is an object and A is s-transitive but not $(s + 1)$-transitive then A is s-regular.*

The other major result of Tutte, on which all that is to follow is based, is

---

† The work of the first author was supported in part by NRC Grant A-5285.
‡ The work of the second author was supported in part by NRC Grant A-5549.

Theorem 2 [13] *If (G, A) is as in Theorem 1 and A is s-regular then $1 \leq s \leq 5$.*

A very natural infinite cubic graph is the infinite tree of valence 3 which we will denote by Γ_3. The full automorphism group of Γ_3 is ω-transitive but we shall see that there are s-regular groups acting on Γ_3 for $1 \leq s \leq 5$. An interesting corollary to Tutte's two theorems is

Corollary *If (Γ_3, A) is an object and A is 6-transitive then A is ω-transitive.*

1. Vertex Fixers and Edge Stabilizers

Let (G, A) be an s-regular object, $u_0, \ldots, u_s$ a simple path in G, $A(u_0, \ldots, u_j)$ the subgroup of A which fixes $u_0, \ldots, u_j$, $0 \leq j \leq s$. The s-regularity of A implies that $|A(u_0, \ldots, u_j)| = 2^{s-j}$ for $1 \leq j \leq s$ and $|A(u_0)| = 3 \cdot 2^{s-1}$. Thus the cardinality of $A(u_0)$ is dependent only on s. In fact, up to isomorphism $A(u_0)$ is only dependent on s.

Theorem 3 *If (G, A) is an s-regular object and u is a vertex of G then $A(u)$ is unique independent of G and A.*

Proof See Propositions 2–5 in [6]. ∎

We shall denote this group by X_s.

As a corollary to Theorem 3 we get that for each s the edge fixers are unique. Let H_s denote the edge fixers. Let $\{u, v\}$ be an edge in G, (G, A) as before, and let $A[u, v]$ be the subgroup of A which need only stabilize $\{u, v\}$. Since $s \geq 1$, $A(u, v)$ is a subgroup of index 2 in $A[u, v]$ and there exist elements in $A[u, v]$ which flip the edge $\{u, v\}$. We say that A is of *type s'* if there exists an involution $\alpha \in A$ which flips an edge. Otherwise we say A is of *type s''*.

Theorem 4 *If (G, A) is an s-regular object and if s is odd then the edge stabilizer is unique and of type s'; on the other hand, if s is even then the edge stabilizer is either of type s' or type s'' and otherwise unique.*

Proof See Propositions 2–5 in [6]. ∎

Let $Y_{s'}$ ($Y_{s''}$) denote the edge stabilizer of type s' (s''). The group $Y_{s''}$ is defined only when s is even.

2. Amalgams and Amalgamated Products

An amalgam is an ordered pair (X, Y) consisting of two groups X and Y, such that multiplication in X and Y coincide on $X \cap Y$ and $X \cap Y$ is a group. Given an arc (u, v) in G we can in a natural way form the amalgam

$(A(u), A[u, v])$ where $A(u) \cap A[u, v] = A(u, v)$. By the proof of the last three theorems we get

Theorem 5 *The amalgam of an arc is unique up to s and the type.*

Proof See Proposition 10 in [6]. ∎

We shall denote this amalgam by $(X_a, Y_{a'})$ or $(X_a, Y_{a''})$, depending on the type.

For each amalgam we can construct a unique group the amalgamated product, i.e.,
$$A_{s'} = X_s \underset{H_s}{*} Y_{s'}, \qquad A_{s''} = X_s \underset{H_s}{*} Y_{s''}.$$
The importance of A_s is twofold. Not only shall we see that A_s acts s-regularly on Γ_3 but if (G, A) is an s-regular object and A_s is of the right type, then there exists a natural surjective homomorphism from A_s to A.

Given A_s and X_s, Y_s we can construct a graph $G_s = (V, E)$ as follows:
$$V = A_s/X_s \quad \text{(the left cosets of } X_s\text{)},$$
$$E = \{\{ux_s, vx_s\} \mid u^{-1}v \in X_s y X_s\},$$
where $y \in Y_s - X_s$.

We list a few facts:

Theorem 6 (1) G_s *is* Γ_3; (2) A_s *acts on* G_s *by left multiplication; and* (3) (Γ_3, A_s) *is an s-regular object.*

Proof See Proposition 11 in [6]. ∎

Let (G, A) be an s-regular object and A_s be of the same type as A and g the canonical group homomorphism from A_s to A then we can define a graph homomorphism $f: \alpha X_s \mapsto g(\alpha)u$, $\alpha \in A_s$ and u the implicit vertex in G. Now the pair (f, g) satisfies certain properties which we now define.

Definition A covering morphism is a pair $(f, g): (G, A) \to (G', A')$ such that:

(1) $g: A \to A'$ onto group homomorphism;
(2) $f: G \to G'$ onto graph homomorphism;
(3) f is locally 1–1 (neighbors of a vertex are sent to distinct vertices); and
(4) the diagram

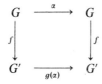

commutes, i.e., $g(\alpha) \cdot f = f \cdot \alpha$ for all $\alpha \in A$.

With one more definition we can state the key idea for the diagrams. A subgroup of A_s is said to be *small* (in A_s) if $K \lhd A_s$, $K \cap X_s = K \cap Y_s = 1$, and $(A_s : KX_s) > 2$.

Theorem 7 *Let G be a connected cubic graph and A an s-regular subgroup of $\operatorname{Aut}(G)$, $1 \le s \le 5$. Let $A_s = A_s'$ or A_s'' be of the same type as A. Then there exists a covering morphism $(f, g): (\Gamma_3, A_s) \to (G, A)$ and $\ker g$ is small in A_s.*

Conversely, let K be a small subgroup of A_s and put $A_K = A_s/K$. Let G_K be the graph whose vertex-set is A_s/KX_s and in which two vertices uKX_s and vKX_s are adjacent iff $u^{-1}v \in KX_s yX_s$. Then G_K is a connected cubic graph, A_K is an s-regular subgroup of $\operatorname{Aut}(G_K)$, and A_K is of the same type as A_s. There is a covering morphism $(f, g): (\Gamma_3, A_s) \to (G_K, A_K)$ where $g: A_s \to A_K$ is the canonical map and $f: \Gamma_3 \to G_K$ is defined by $f(uX) = uKX$.

Proof See Theorem 1 in [6]. ∎

If (f, g) is a covering morphism the kernel of (f, g) will be defined to be the kernel of g. We have from Theorem 6 that every s-regular object generates a small normal subgroup of A_s, i.e., the $\ker(f, g)$, and every small normal subgroup of A_s generates s-regular objects. The next two theorems tell us how many normal subgroups correspond to a given cubic graph.

Theorem 8 *Let (G, A) be an s-regular object, $s = 3$ or 5; then all covering morphisms $(f, g): (\Gamma_3, A_s) \to (G, A)$ have the same kernel.*

Proof See Theorem 2 in [6]. ∎

Theorem 9 *Let (G, A) be an s-regular object, $s = 1, 2, 4$, and (Γ_s, A_s) being the same type as (G, A); then over all covering morphisms from (Γ_s, A_s) to (G, A) there are either one or two kernels and if there is exactly one kernel then G is $(s + 1)$-transitive.*

Proof See Theorem 8 in [6]. ∎

Before presenting the diagram, we need a few more results about the subgroup structure of the A_s's.

Theorem 10 *The subgroup structure for the A_s's is indicated by the following diagram:*

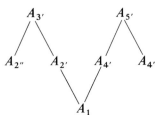

3. Some Subgroups of A_3

Some subgroups of $A_{3'}$ are shown in Fig. 1 and we shall describe them now. The lines indicate normal subgroups and the number indicates the index. Starting with A_3 it has three subgroups of index 2, namely, $A_{2'}$ and $A_{2''}$ and the even subgroup of $A_{3'}$, the subgroup which preserves the bipartition on Γ_3 denoted A_3^+. The group $A_{2'}$ contains two copies of A_1 and its even subgroup A_2^+ which equals $A_{2'} \cap A_3^+ \cap A_{2''}$. The even subgroup of $A_{1'}$ is $A_{1'} \cap A_2^+$ and $A_{1'}$ contains a normal subgroup of index 3, $A_{0'}$ which is vertex-regular on Γ_3. Now A_1^+ contains two copies of two even subgroups of vertex-regular groups on Γ_3. The subgroups $K_{4'}$, $K_{4''}$ and $Q_{3'}$, $Q_{3''}$ correspond to the two copies of the graph K_4 and the cube, respectively. The graph $K_{3,3}$ corresponds to a unique normal subgroup of A_3 since it is 3-regular. Now N corresponds to a 3-regular graph which is a 12-fold covering of Q_3 and a 16-fold covering of $K_{3,3}$. The subgroup $P \triangleleft A_{3'}$ corresponds to Petersen's graph; D', $D'' \triangleleft A_{2'}$ correspond to dodecahedron; and $Z \triangleleft A_3$ corresponds to Desargue's graph. Finally $K \triangleleft A_{3'}$ corresponds to the vertex primitive 3-regular graph on 28 vertices $G(28)$.

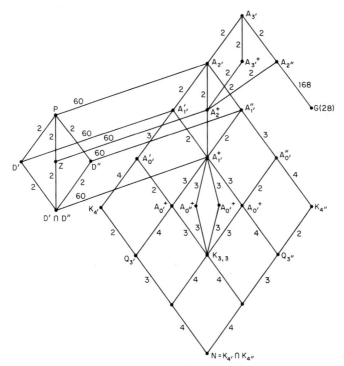

Figure 1 Partial subgroup structure of A_3.

By simply noting where the subgroup of a graph lies in the diagram a large amount of information can be read off. As an example, $K_{3,3}$ is contained in $A_{2'}$ and $A_{2''}$; therefore, Aut($K_{3,3}$) contains subgroups of type 2' and type 2". Similarly, Aut($K_{3,3}$) contains two 1-regular subgroups and at least two 0-regular subgroups: $K_{3,3}$ must be bipartite. Now Petersen's graph has no type 2" or type 1' actions since it is not contained in $A_{2''}$ nor $A_{1'}$. The fact that Petersen's graph contains no 1-regular subgroup follows by the fact that Petersen's graph is not bipartite and the following proposition:

Proposition *Let (G, A) be a 3-regular object; then the following hold*:

(1) *if A contains a 1-regular subgroup then G is bipartite and A contains two 2-regular subgroups and two 1-regular subgroups;*

(2) *if A contains two 2-regular subgroups then G is bipartite.*

Proof See Proposition 26 in [6]. ∎

4. Some Subgroups of $A_{5'}$

Some subgroups of $A_{5'}$ are shown in Fig. 2, and we shall describe them now. We have $A_4^+ = A_{4'} \cap A_{4''}$. T is the normal subgroup of $A_{5'}$ which

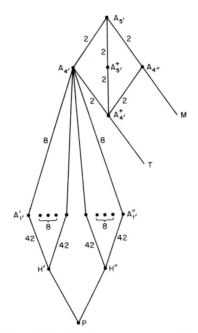

Figure 2 Partial subgroup structure of A_5.

corresponds to Tutte's 8-cage. As is well known, we have $A_{4'}/T \approx A_6$, $A_{5'}^+/T \approx S_6$, and $A_{5'}/T \approx \text{Aut}(S_6)$. Now $M \triangleleft A_s$ corresponds to the unique finite primitive 5-regular object $(G(234), \text{Aut}(SL_3(3)))$. We have $A_{4''}/M \approx SL_3(3)$ and $A_{5'}/M \approx \text{Aut}(SL_3(3))$.

Now $A_{4'}$ contains 16 copies of $A_{1'}$. They are all conjugate in $A_{5'}$ and fall into two conjugacy classes of size 8 in $A_{4'}$. Let $A_{1'}$ and $A_{1''}$ denote members of each of the two conjugacy classes. The intersection of the $A_{1'}$'s and the intersection of the $A_{1''}$'s correspond to the two copies of Heawood's graph.

Theorem 11 *Heawood's graph is the unique minimal graph which is 4-regular and whose automorphism group contains a 1-regular subgroup. So, every 4-regular object with a 1-regular subgroup is a covering of Heawood's graph.*

Proof See Proposition 29 of [6]. ■

Finally, $P = H' \cap H''$ corresponds to the unique minimal 5-regular graph such that $\text{Aut}(G)$ contains a 1-regular group.

REFERENCES

1. N. Biggs, "Algebraic Graph Theory," Cambridge Univ. Press, London and New York, 1974.
2. N. Biggs, Three remarkable graphs, *Canad. J. Math.* **25** (1973) 397–411.
3. J. Dieudonné, "La Géométrie des Groupes Classiques," Springer-Verlag, Berlin and New York, 1963.
4. D. Ž. Djoković, Automorphisms of graphs and coverings, *J. Combinatorial Theory Ser., B* **16** (1974) 243–247.
5. D. Ž. Djoković, On regular graphs V, *J. Combinatorial Theory Ser B*, to appear.
6. D. Ž. Djoković and G. L. Miller, Regular Groups of Automorphisms of Cubic Graphs, TR20, Computer Science Department, University of Rochester, Rochester, New York, 1977, to appear in *J. Combinatorial Theory Ser. B*.
7. D. Gorenstein, "Finite Groups," Harper, New York, 1968.
8. B. Huppert, "Endliche Gruppen I," Springer-Verlag, Berlin and New York, 1970.
9. R. C. Miller, The trivalent symmetric graphs of girth at most six, *J. Combinatorial Theory Ser. B* **10** (1971) 163–182.
10. M. Newman, "Integral Matrices," Academic Press, New York, 1972.
11. J.-P. Serre, Arbres, amalgames, SL_2, *Astérisque*, No. 46 (1977).
12. J. Tits, Sur le groupe des automorphismes d'un arbre, *in* "Essays on Topology and Related Topics, Volume Dedicated to G. de Rham," pp. 188–211. Springer-Verlag, Berlin and New York, 1970.
13. W. T. Tutte, "Connectivity in Graphs," Univ. of Toronto Press, Toronto, 1966.
14. W. T. Tutte, A family of cubical graphs, *Proc. Cambridge Philos. Soc.* **43** (1947) 459–474.

15. W. T. Tutte, On the symmetry of cubic graphs, *Canad. J. Math.* **11** (1959) 621–624.
16. R. M. Weiss, Über s-reguläre Graphen, *J. Combinatorial Theory Ser. B* **16** (1974) 229–233.
17. R. M. Weiss, Eckenprimitive Graphen vom Grad drei, *Abh. Math. Sem. Univ. Hamburg* **41** (1974) 172–178.

AMS 05C25

Dragomir Ž. Djoković
DEPARTMENT OF PURE MATHEMATICS
UNIVERSITY OF WATERLOO
WATERLOO, ONTARIO

Gary L. Miller
DEPARTMENT OF MATHEMATICS
MASSACHUSETTS INSTITUTE OF TECHNOLOGY
CAMBRIDGE, MASSACHUSETTS

GRAPH THEORY AND RELATED TOPICS

Problems and Results in Graph Theory and Combinatorial Analysis

PAUL ERDÖS

I published several papers with similar titles. One of my latest ones [13] (also see [16] and the yearly meetings at Boca Raton or Baton Rouge) contains, in the introduction, many references to my previous papers.

I discuss here as much as possible new problems, and present proofs in only one case. I use the same notation as used in my previous papers. $G^{(r)}(n;l)$ denotes an r-graph (uniform hypergraph all of whose edges have size r) of n vertices and l edges. If $r = 2$ and there is no danger of confusion. I omit the upper index $r = 2$. $K^{(r)}(n)$ denotes the complete hypergraph $G^{(r)}(n; \binom{n}{r})$. $K(a, b)$ denotes the complete bipartite graph ($r = 2$) of a white and b black vertices. $K_l^{(r)}(t)$ denotes the hypergraph of lt vertices $x_i^{(j)}$, $1 \leq i \leq t, 1 \leq j \leq l$, and whose $\binom{l}{r}t^r$ edges are $\{x_{i_1}^{(j_1)}, \ldots, x_{i_r}^{(j_r)}\}$ where all the i's and all the j's are distinct. $e(G(m))$ is the number of edges of $G(m)$ (graph of m vertices), the girth is the length of a smallest circuit of the graph.

1

Hajnal, Szemerédi, and I have the following conjecture: There is a function $f(m)$ tending to infinity so that if $G(n)$ is a graph of n vertices and if every subgraph $G(m)(1 \leq m \leq n)$ of it can be made bipartite by the omission of at most $f(m)$ edges, then $\chi(G(n)) \leq 3$ (or, in a slightly weaker form, its chromatic number is bounded). It is surprising that we have been able to make no progress with this extremely plausible conjecture.

It seems very likely that $f(m)$ can be taken as $c \log m$. Gallai [21] constructed a four chromatic graph $G(n)$ the smallest odd circuit of which has length at least $\sqrt{n}$, and it is easy to see that every subgraph $G(m)$ of this graph can be made bipartite by omitting $\sqrt{m}$ edges. Thus $f(m) \geq \sqrt{m}$.

Gallai and I conjectured that for every r there is an r-chromatic graph $G(n)$ the smallest odd circuit of which has length at least $n^{1/(r-2)}$. By a refinement of a method used by Hajnal and myself [17] (using Borsuk's theorem) Lovász proved this conjecture during this conference. The result of Lovász will imply that our $f(n)$ (if it exists) must be $o(n)$.

I stated that I can prove that Gallai's theorem [21] is best possible; that is, that if $G(n)$ is such that all odd circuits of $G(n)$ are longer than $cn^{1/2}$ then $\chi(G(n)) \leq 3$. This is almost certainly correct, but I have not been able to reconstruct my proof (which very likely was not correct). In fact I cannot even prove it if $cn^{1/2}$ is replaced by εn.

Another problem of Hajnal, Szemerédi, and me states: Let $\chi(G) = \aleph_1$. Is it true that, for every c, G has a subgraph of m vertices which cannot be made bipartite by the omission of cm edges? On the other hand we believe that for every $\varepsilon > 0$ there is a G with $\chi(G) = \aleph_1$ so that for every $m < \aleph_0$ every subgraph of G of m vertices can be made bipartite by the omission of fewer than $m^{1+\varepsilon}$ edges.

2

Let $|S| = n$, $A_k \subset S$, $|A_k| = cn$, $1 \leq k \leq n$. Denote by $f(n; c, \varepsilon)$ the largest integer so that there are at least $f(n; c, \varepsilon)$ sets, any pair of which have more than εn elements in common. It easily follows from Ramsey's theorem that, for every $i > 0$ and sufficiently small ε, $f(n; c, \varepsilon) > Cn^{1/(c+1)}$. I would like to prove

$$f(n; c, \varepsilon) > Cn^{1/2} \tag{1}$$

but I cannot even disprove $f(n; c, \varepsilon) > \eta n$. (1) would imply that every $G(n, [c_1 n^2])$ has $c_2 n^{1/2}$ vertices, any pair of which can be joined by vertex disjoint paths of length at most two. I can prove this with two replaced by four.

The following question is also of interest. Define a graph of n vertices as follows: The vertices are the n sets A_k. Join two of them if they have fewer than εn elements in common.

Estimate from above and below the largest possible value of the chromatic number of this graph. (1) would clearly follow if the chromatic number were less than $cn^{1/2}$.

The following question can now be posed: Let $|S| = 2n + k$. Define $G_{(n,k,l)}$ as follows: Its vertices are the $\binom{2n+k}{n}$ n-tuples of S. Two n-tuples A_i and A_j are joined if $|A_i \cap A_j| \leq l$. Estimate or determine $h(n; k, l) = \chi(G_{(n,k,l)})$.

A well-known conjecture of Kneser stated $h(n; k, 0) = k + 2$. This conjecture was recently proved by Bárány and Lovász.

A further complication can be introduced as follows: Let $h_m(n; k, l)$ be the largest integer for which $G_{(n,k,l)}$ has a subgraph $G(m)$ with $\chi(G(m)) = h_m(n; k, l)$. Determine or estimate $h_m(n, k, l)$.

The papers of Lovász and Bárány will appear soon.

3

A graph G_1 is said to be a *unique subgraph* of G if G_1 is a uniquely induced (or spanned) subgraph of G. Entringer and I [7] proved that there is a $G(n)$ which has more than

$$2^{\binom{n}{2}} \exp -(Cn^{(3/2)+\varepsilon}) \qquad (1)$$

unique subgraphs. (1) was improved to $2^{\binom{n}{2}} \exp -(cn \log n)$ by Harary and Schwenk [22]. Finally Brouwer [3] improved (1) to

$$\frac{2^{\binom{n}{2}}}{n!} e^{-Cn}. \qquad (2)$$

(2) is not far from being best possible since Pólya proved that the number of nonisomorphic graphs of n vertices is $(1 + o(1))2^{\binom{n}{2}}/n!$.

I always assumed that (2) is best possible, except for the value of C, but during this conference I had a discussion with J. Spencer who thought it quite possible that there is a graph $G(n)$ which has more than

$$\varepsilon \frac{2^{\binom{n}{2}}}{n!} \qquad (3)$$

unique subgraphs where $\varepsilon > 0$ is independent of n. I do not believe this to be true but could not disprove it and offer 100 dollars for a proof and 25 for a disproof of (3).

In trying to decide about (3) perhaps the following idea could be helpful. I had no time to think it over carefully and have to apologize to the reader if it turns out to be nonsense: Put $t_n = n \log n/\log 2$. Consider the random graph $G(n; \binom{n}{2} - t_n)$. Is it true that almost all of them have a positive fraction of their subgraphs as unique subgraphs? Is there at least one such graph with this property? More generally, determine or estimate the largest $l_n = l_n(c)$ for which there is a $G(n; l_n)$ which has more than $c2^{l_n}$ unique subgraphs, i.e., a positive proportion of the subgraphs is unique.

4

Some problems on random graphs. Is it true that almost all graphs $G(n; Cn)$ contain a path of length cn? ($c = c(C) > 0$.) I conjectured this in 1974 [15] but Szemerédi strongly disagreed. He believes that for every fixed C the longest path contained in almost all $G(n; Cn)$ is $o(n)$. On second thought I think Szemerédi may very well be right, but at the moment nothing is known.

Let $G(n; t_n)$ be a graph of n vertices and t_n edges. Is it true that for almost all such graphs the edge connectivity is equal to the minimum degree (or valency) of the graph? Rényi and I [18] proved this if $t_n < \frac{1}{2}n \log n + cn \log \log n$. In this case the minimum degree is almost surely bounded. But the conjecture should hold for a much more extended range of t_n and perhaps in fact for all values of t_n.

The probability method easily gives that for every $\eta > 0$, $0 < c < \frac{1}{2}$, and and $n > n_0(\eta, c)$ there is a $G(n; [n^{1+c}])$ which has no triangle and, for every $m > \eta n$, every induced subgraph $G(m)$ satisfies

$$(1 - \eta)n^{1+c}\left(\frac{m}{n}\right)^{1+c} < e(G(m)) < (1 + \eta)n^{1+c}\left(\frac{m}{n}\right)^{1+c}.$$

In other words the graph behaves like a random graph.

Does this remain true for $\frac{1}{2} \leq c < 1$. It certainly fails for graphs $G(n; cn^2)$. See [19, 24].

5

It is well known that every graph with "many" edges contains a large complete bipartite graph. To fix our ideas consider a $G(n; [n^2/4])$. It is easy to see that it contains a $K(r, s)$ for $s = (1 + o(1))n/2^r$ as long as $r2^r \leq (1 + o(1))n$. Is this result best possible? By a simple computation using the probability method Simonovits, Lovász, and I showed that this result is indeed best possible as long as $2^r = o(n/(\log n)^2)$. In other words we can prove that there is a $G(n; (n^2/4))$ which contains no $K(r, (1 + \varepsilon)n/2^r)$ as long as $2^r < \eta(\varepsilon)n/(\log n)^2$.

What happens if $2^r > c(n/(\log n)^2)$? More precisely: Let $G(n; [n^2/4])$ be any graph of n vertices and $[n^2/4]$ edges. For every r satisfying $r2^r < n$ consider the largest s for which our graph contains a complete bipartite $K(r, s)$. Put $f(G) = \max_r s/(n/2r)$ and let $A(n)$ be the minimum of $f(G)$ extended over all $G(n; [n^2/4])$. How does $A(n)$ behave? On the one hand it could be $1 + o(1)$ and on the other hand it could tend to infinity.

One could further ask: Colour the edges of $K(n)$ by two colours, and let $K(r, s)$ be the largest monochromatic bipartite graph for $r2^r \leq n$. Define $f_1(G)$ and A_1 as before. Conceivably $A_1 > A$. It even could be that $A \to 1$, $A_1 \to \infty$. For $2^r = o(n/(\log n)^2)$, $s < (1 + o(1))n/2^r$ holds here too.

Similarly problems can be posed for $G(n; cn^2)$, $c \neq \frac{1}{4}$, but these can easily be formulated by the interested reader.

Every $G(n; [n^2/4])$ contains a $K(r, r)$ for $r2^r < \varepsilon n$ (i.e., if $r < (1 - \varepsilon)\log n/\log 2$) and the probability method gives that it does not have to contain a $K(l, l)$ for $l > (2 + \varepsilon)\log n/\log 2$. The exact order of magnitude will probably be difficult to determine. Exactly the same question occurs if we colour a $K(n)$ by two colours and ask for the largest monochromatic $K(l, l)$. This difficulty is of course familiar in Ramsey theory.

See [23].

6

It is well known that for $n > n_0(\varepsilon, t)$ every $G^{(3)}(n; [\varepsilon n^3])$ contains a $K_3(t, t, t)$. Is it true that a $G^{(3)}(3n; n^3 + 1)$ contains a $G^{(3)}(9; 28)$? In particular does it contain a $K_3(3, 3, 3)$ and one more triple? (This conjecture is stated incorrectly in [13, p. 11].)

Very little is known about extremal problems on r-graphs (uniform hypergraphs where the edges have size r). Turán's classical problem of determining $f(n; K^{(3)}(4))$ is of course still open. I recently offered 500 dollars—in Turán's memory—for the determination of (or an asymptotic formula for) $f(n; K^{(3)}(4))$. As far as I know $f(n; G^{(3)}(4; 3))$ is also not yet known. The largest $G^3(n)$ without a $G^{(3)}(4; 3)$ I can construct it as follows: Put $n = a + b + c$ with a, b, c being as nearly equal as possible. Consider a $K_3(a, b, c)$ and put $a = a_1 + a_2 + a_3$, $b = b_1 + b_2 + b_3$, $c = c_1 + c_2 + c_3$ and iterate the preceding process. This gives a

$$G\left(n; (1 + o(1))\frac{n^3}{24}\right)$$

which contains no $G^{(3)}(4; 3)$ and which may be extremal.

Here I restate an old problem of mine which I consider very attractive. Let $G^{(r)}(n_i)$, $n_i \to \infty$, be a sequence of r-graphs of n_i vertices. We say that the family has subgraphs of edge density $\geq \alpha$ if, for infinitely many n_i, $G(n_i)$ has a subgraph $G(m_i)$, $m_i \to \infty$, so that $G(m_i)$ has at least $(\alpha + o(1))\binom{m_i}{r}$ edges. The theorem of Stone and myself implies that every $G^{(2)}(n; (n^2/2)(1 - (1/l) + \varepsilon))$ contains a subgraph of density $1 - [1/(l + 1)]$, and it is easy to see that this is best possible. If α is the largest such number then the $G(m_i)$ are called a family of subgraphs of maximal density.

Thus for $r = 2$ the possible maximal densities of subgraphs are of the form $1 - (1/l)$, $1 \leq l < \infty$. Now I conjecture that for $r > 2$ there are also only a denumerable number of possible values for the maximal densities α. I offer 500 dollars for the determination of these values for all $r > 2$—or also for a refutation of my conjecture.

The simplest unsolved problem here is as follows: Prove that there is an absolute constant $c > 0$ so that for every $\varepsilon > 0$ if

$$G^{(3)}\left(n_i; \left[\frac{n_i^3}{27}(1 + \varepsilon)\right]\right)$$

is a family of 3-graphs there is a family of subgraphs of edge density greater than $\frac{2}{9} + c$. I offer 250 dollars for a proof or disproof of this conjecture. Probably $(n^3/27)(1 + \varepsilon)$ can in fact be replaced by $(n^3/27) + n^{3-\eta_n}n$ where $\eta_n \to 0$ as $n \to \infty$.

Similar unsolved problems on the possible maximal densities arise in multigraphs and digraphs as stated in a paper of Brown et al. [4].

Let $G^{(r)}(n)$ be an r-graph. Its edge graph is the ordinary graph (i.e., $r = 2$) whose vertices are the vertices of our $G^{(r)}(n)$. Two vertices in the edge graph are joined by an edge if they are contained in one of the r-tuples of $G^{(r)}(n)$. Simonovits and I observed that the edge graph of a $G^{(r)}(n; [(1 + \varepsilon)\binom{l}{r}n^r/l^r])$ contains a $K_{l+1}^{(r)}(t)$ for every $n > n_0(t)$. The edge graph of $K_l^r([n/l])$ shows that this result is essentially best possible.

See [1, 2, 5, 11, 20].

7

In a previous paper [14] I stated the following problem: "A problem in set theory led R. O. Davies and myself to the following question: Denote by $f(n, k)$ the largest integer so that if there are given in k-dimensional space n points which do not contain the vertices of an isosceles triangle, then they determine at least $f(n, k)$ distinct distances. Determine or estimate $f(n, k)$. In particular is it true that

$$\lim_{n \to \infty} \frac{f(n, k)}{n} = \infty ? \tag{1}$$

(1) is unproved even for $k = 1$. Straus observed that if $2^k \geq n$ then $f(n, k) = n - 1$."

I perhaps should have given some explanation why (1) is difficult even for $k = 1$. Observe that $\lim_{n \to \infty} f(n, 1)/n = \infty$ implies Roth's theorem: $r_3 = o(n)$, where $r_3(n)$ is the smallest integer so that if $1 \leq a_1 < \cdots < a_k \leq n, k \geq r_3(n)$ then the a's contain an arithmetic progression (i.e., an isosceles triangle). The converse does not seem to be true, i.e., $r_3(n)/n \to 0$ does not imply (1).

Is it true that if $a_1, \ldots, a_m$ is a set of integers which does not contain an arithmetic progression of three terms then for $m > m_0(c)$ there are more than cm distinct integers of the form $a_j - a_i$? This is a special case of (1) if $k = 1$. At the moment I do not see how to prove this but perhaps it will not be very hard.

The following further questions are perhaps of some interest: Let there be given n points in k-dimensional space. Assume that every set of four of them determine at least five different distances. Is it then true that the n points determine at least $c_k n^2$ distinct distances. This is trivial for $k = 1$ since if four points on the line determine five distances three of the points must form an arithmetic progression and the fourth point is in general position. From this remark it easily follows that if there are n points on the line and any four determine at least five distances then the n points determine at least $(n^2/2) - cn$ distances. For $k > 1$ I do not know what happens, but again I am not sure if the question is really difficult.

Let $X_1, \ldots, X_n$ be n points in k-dimensional space. Denote by $A_k(X_1, \ldots, X_n)$ the number of distinct distances the n points determine. What is the set of possible values of $A_k(X_1, \ldots, X_n)$? I previously considered the smallest possible value of $A_k(X_1, \ldots, X_n)$ and, for $k > 1$, this is no doubt a very difficult question, but perhaps it is possible to make some nontrivial statement about the possible set of values of $A_k(X_1, \ldots, X_n)$, e.g., perhaps in a certain range it can take all values.

8

Let G be a graph of m edges. Denote by $f(m)$ the largest integer so that G always contains a bipartite graph of $f(m)$ edges. Edwards [6] and I proved that $f(m) > (m/2) + c\sqrt{m}$ and that in general this result is best possible.

Edwards in fact determined $f(m)$ explicitly.

Assume now that G has m edges and girth r (i.e., the smallest circuit of G has r edges). Denote by $f_r(m)$ the largest integer so that G always contains a bipartite graph of $f_r(m)$ edges. Lovász and I proved that

$$\frac{m}{2} + c_2 m^{1-c_r''} < f_r(m) < \frac{m}{2} + c_1 m^{1-c_r'}, \tag{1}$$

where c_r' and c_r'' are greater than $\frac{1}{2}$ and less than one and tend to one as r tends to infinity.

It seems certain that there is an absolute constant c_r so that

$$\frac{m}{2} + m^{c_r - \varepsilon} < f_r(m) < \frac{m}{2} + m^{c_r + \varepsilon}, \tag{2}$$

but at the moment we cannot prove (2). If (2) holds then the next step would be to get an asymptotic formula for $f_r(m) - (m/2)$.

Now we outline the proof of (1). The upper bound follows by the probability method and we do not give it here (an outline has already been published in Hungarian) since it is fairly standard. We prove the lower bound in some detail but for simplicity assume $r = 4$ (i.e., G has m edges and no triangle). The proof of the general case is not really different. We need two lemmas.

Lemma 1 *Let G have m edges and chromatic number k where $k = 2r$ or $2r - 1$. Then G contains a bipartite subgraph of at least $m(r/2r - 1)$ edges.*

If $m = \binom{s}{2}$ then $K(s)$ shows that the lemma is best possible.

Since G has chromatic number k one can decompose its vertex set into k independent subsets S_i, $1 \le i \le k$. Decompose the index set $\{i: 1 \le i \le k\}$ in all possible ways into two disjoint sets $A_u, B_u, A_u = [k/2], B_u = [(k+1)/2]$. The number of these decompositions clearly equals $\binom{k}{[k/2]}$. Consider all the $\binom{k}{[k/2]}$ bipartite subgraphs of G_u, $1 \le u \le \binom{k}{[k/2]}$, where in G_u a vertex is white if it belongs to an S_i, $i \in A_u$, and is black otherwise. Every edge of G clearly occurs in exactly $2\binom{k-2}{[k/2]-2}$ of the bipartite graphs G_u. Thus if $e(G_u)$ denotes the number of edges of G_u we have

$$\sum e(G_u) = 2m \binom{k-2}{[k/2]-1}$$

or if $k = 2r$ or $2r - 1$

$$\max_u e(G_u) \ge m \cdot 2 \binom{k-2}{[k/2]-1} = m \frac{r}{2r-1}$$

which proves the lemma. ∎

Lemma 2 *Let G have m edges and no triangle. Then its chromatic number is less than*

$$c_1 \left(\frac{m \log \log m}{\log m} \right)^{1/3} = t_m.$$

Graver and Yackel proved that a graph of chromatic number t_m has at least

$$\frac{c_2 t_m^2 \log t_m}{\log \log t_m}$$

vertices. Clearly each vertex can be assumed to have valency (or degree) at least t_m. Thus a t_m-chromatic graph has at least

$$\frac{c_2 t_m^3 \log t_m}{\log \log t_m}$$

edges, which proves our lemma. ∎

The lemma is not very far from being best possible. I showed that there is a graph of m edges which has no triangle and whose chromatic number is greater than $m^{1/3}/\log m$.

Lemmas 1 and 2 immediately give that

$$f_4(m) > \frac{m}{2} + cm^{2/3} \left(\frac{\log m}{\log \log m}\right)^{1/3}$$

which completes the proof of our theorem. The proof if $r > 4$ is clearly almost identical.

The upper bound given by the probability method for $f_4(m)$ is very much worse than $m^{2/3}$—we have no guess for the correct exponent.

See [9, 10].

9

In 1972 Faber, Lovász, and I stated the following conjecture: Let $|A_k| = n$, $1 \le k \le n$. Assume that any two of these sets have at most one element in common. Is it then true that one can colour the elements of $\bigcup_{k=1}^{n} A_k$ by n colours so that every set contains elements of all colours?

I consider this conjecture very attractive—also it does not seem to be easy and I offer 250 dollars for a proof or disproof.

Several mathematicians reformulated the conjecture as follows: Define $G(A_1, \ldots, A_n)$ as follows: The vertices are the elements of $\bigcup_{i=1}^{n} A_i$. Two vertices are joined if they belong to the same A_i. Prove $\chi(G(A_1, \ldots, A_n)) = n$. A theorem of de Bruijn and myself implies that it contains no K_{n+1}.

One advantage of this definition is that one can easily ask new questions. Let $h(m)$ be the smallest integer for which there is a set system $A_1, \ldots, A_{h(m)}$ ($|A_i| = n$, $|A_i \cap A_j| \le 1$) for which $\chi(A_1, \ldots, A_{h(m)}) = m$.

Many further generalisations are possible, e.g., $|A_i \cap A_j| \le 1$ can be replaced by $|A_i \cap A_j| \le l$. Also $|A_i \cap A_j| \le 1$ can be further strengthened in various ways. We could assume, say, that among any three sets A_i, A_j, A_l at least two are disjoint, etc.

One more new problem: In all the cases I know $G(A_1, \ldots, A_{h(m)})$ contains a $K(m)$. That this must be the case is not hard to show for fixed n and m sufficiently large. Is it true for every n and m?

See [5a].

10

Tutte (and later independently Ungár, Zykov, and Mycielski) was the first to construct graphs of arbitrarily large chromatic number having no triangles. Then I [15] and later Lovász proved that for every r there is a graph of girth r and chromatic number k.

Now Hajnal and I asked: Is there an $A(k, r)$ so that every G with $\chi(G) \geq A(k, r)$ contains a subgraph of girth r and chromatic number k? Rödl recently proved that $A(k, 4)$ exists for every k, but his upper bound for $A(k, 4)$ is probably very poor. I think $A(k, 4) < ck$ is true. It would be very interesting if one could prove that $A(k, r)$ exists for every k and r and if we could obtain some knowledge of the order of magnitude of $A(k, r)$, e.g., is it true that $\lim_{k \to \infty} [A(k, r + 1)/A(k, r)] = \infty$?

REFERENCES

1. B. Bollobás and P. Erdös, On the structure of edge graphs, *Bull. London Math. Soc.* **15** (1973) 317–321.
2. B. Bollobás, P. Erdös, and M. Simonovits, On the structure of the edge graphs II, *J. London Math. Soc.* **12** (1976) 219–224.
3. A. E. Brouwer, On the number of unique subgraphs of a graph, *J. Combinatorial Theory Ser. B* **18** (1975) 184–185.
4. W. G. Brown, P. Erdös, and M. Simonovits, Extremal problems for directed graphs, *J. Combinatorial Theory Ser. B* **15** (1973) 77–93.
5. W. G. Brown, P. Erdös, and V. T. Sós, Some extremal problems on r-graphs, *in* "New Directions in the Theory of Graphs (Proceedings of the Third Ann Arbor Conference on Graph Theory)" (F. Harary and E. M. Palmer, eds.), pp. 53–63. Academic Press, New York, 1973; see also Erdös and Spencer [20].
5a. N. G. de Bruijn and P. Erdös, On a combinatorial problem, *Indag. Mat.* **10** (1948) 421–423.
6. C. S. Edwards, An improved lower bound for the number of edges in a largest bipartite subgraph, *Recent Advances Graph Theory, Proc. Symp. Prague* (1974) 167–181; Some extremal properties of bipartite subgraphs, *Canad. J. Math.* **25** (1973) 475–485.
7. R. C. Entringer and P. Erdös, On the number of unique subgraphs of a graph, *J. Combinatorial Theory Ser. B* **13** (1972) 112–115.
8. P. Erdös, Graph theory and probability, *Canad. J. Math.* **11** (1959) 34–38.
9. P. Erdös, Graph theory and probability II, *Canad. J. Math.* **13** (1961) 346–352.
10. P. Erdös, On even subgraphs of graphs (in Hungarian), *Mat. Lapok* **18** (1964) 283–288.
11. P. Erdös, On some extremal problems on r-graphs, *Discrete Math.* **1** (1971) 1–6.

12. P. Erdös, On a problem of Grünbaum, *Canad. Math. Bull.* **15** (1972) 23–26.
13. P. Erdös, Problems and results in combinatorial analysis, *Accad. Naz. Lincei Colloq. Internat. Teorie Comb.*, Rome **2** (1973) 3–17.
14. P. Erdös, Problems and results on combinatorial number theory, *in* "A Survey of Combinatorial Theory," pp. 117–136, 136. North-Holland Publ., Amsterdam, 1973.
15. P. Erdös, Some new applications of probability methods to combinatorial analysis and graph theory, *Proc. Southeastern Confer. Combinatorics, 5th, Graph Theory Comput.*, Boca Raton (1974) 39–51.
16. P. Erdös, Problems and Results in Graph Theory and Combinatorial Analysis, *Proc. British Combinatorial Conf.*, 5th (1975) 169–192.
17. P. Erdös and A. Hajnal, Kromatikus grafókról (in Hungarian), *Mat. Lapok* **18** (1967) 1–4.
18. P. Erdös and A. Rényi, On the strength of connectedness of a random graph, *Acta Math. Acad. Sci. Hungar.* **12** (1961) 261–267; see also "The Art of Counting," pp. 574–624. MIT Press, Cambridge, Massachusetts.
19. P. Erdös and J. Spencer, "Probabilistic Methods in Combinatorics," Academic Press, New York, 1974.
20. P. Erdös and A. Stone, On the structure of linear graphs, *Bull. Amer. Math. Soc.* **52** (1946) 1087–1091.
21. T. Gallai, Kritische Graphen, *Publ. Math. Inst. Hungar. Acad. Sci.* **8** (1973) 165–192; see especially pp. 172–173, 186–189.
22. F. Harary and A. J. Schwenk, On the number of unique subgraphs, *J. Combinatorial Theory Ser. B* **15** (1973) 156–160.
23. T. Kövári, V. T. Sós, and P. Turán, On a problem of K. Zarankiewicz, *Collog. Math.* **3** (1954) 50–57; see also Erdös and Spencer [20].
24. L. Pósa, Hamiltonian circuits in random graphs, *Discrete Math.* **14** (1976) 359–364.

AMS 05C35

MATHEMATICS INSTITUTE
HUNGARIAN ACADEMY OF SCIENCES
BUDAPEST, HUNGARY

Strong Independence of Graphcopy Functions

PAUL ERDÖS, LÁSZLÓ LOVÁSZ,
and
JOEL SPENCER

Let H be a finite graph on v vertices. We define a function c_H, with domain the set of all finite graphs, by letting $c_H(G)$ denote the fraction of subgraphs of G on v vertices isomorphic to H. Our primary aim is to investigate the behavior of the functions c_H with respect to each other. We show that the c_H, where H is restricted to be connected, are independent in a strong sense. We also show that, in an asymptotic sense, the c_H, H disconnected, may be expressed in terms of the c_H, H connected.

In 1932, Whitney [1] proved that the functions c_H, H connected, were algebraically independent. Our results may be considered an extension of this work.

Notations and Conventions

All graphs G shall be finite, without loops of multiple edges. $\mathscr{G}$ denotes the family of all finite graphs. $V(G)$, $E(G)$ denote the vertex and edge sets of G.

For purposes of counting all graphs may be considered labelled. A map

$$\psi: V(H) \to V(G)$$

is called a *homomorphism* if $\{x, y\} \in E(H)$ implies $\{\psi_x, \psi_y\} \in E(G)$. If furthermore $x \neq y$ implies $\psi_x \neq \psi_y$ then ψ is a *monomorphism*. For $W \subseteq V(G)$, the restriction of G to W, denoted by $G|_W$, is that graph with vertex set W and $\{x, y\} \in E(G|_W)$ iff $\{x, y\} \in E(G)$, $x, y \in W$. Let I_s, K_s denote the empty and complete graphs respectively on s elements.

Let $H, G \in \mathscr{G}$, $|V(H)| = t$, $|V(G)| = n$. Define

$$A_H(G) = |\{\psi: V(H) \to V(G), \text{homomorphism}\}|,$$
$$B_H(G) = |\{\psi: V(H) \to V(G), \text{monomorphism}\}|,$$
$$C_H(G) = |\{W \subseteq V(G): |W| = t, G|_W \cong H\}|,$$
$$a_H(G) = A_H(G)/n^t,$$
$$b_H(G) = B_H(G)/(n)_t \quad [(n)_t = n(n-1)\cdots(n-t+1)],$$
$$c_H(G) = C_H(G)/\binom{n}{t}.$$

The lower case functions give the fraction of homomorphism, monomorphism, and copies, respectively.

Throughout this paper let k denote a fixed integer $k \geq 3$. Let $H_1, \ldots, H_m$ be all connected graphs (up to isomorphism) with $|V(H_i)| \leq k$. Let A_i, B_i, C_i, a_i, b_i, c_i denote the functions $A_{H_i}, \ldots, c_{H_i}$ for convenience. Define vector valued functions **a**, **b**, **c** by

$$\mathbf{a}(G) = (a_1(G), \ldots, a_m(G)),$$
$$\mathbf{b}(G) = (b_1(G), \ldots, b_m(G)),$$
$$\mathbf{c}(G) = (c_1(G), \ldots, c_m(G)).$$

Our object is to study the possible values for $\mathbf{a}(G)$, $\mathbf{b}(G)$, $\mathbf{c}(G)$. We wish, however, to avoid exceptional values taken by small G.

Definition S_a, S_b, S_c are defined as the sets of limit points of $\{\mathbf{a}(G)\}$, $\{\mathbf{b}(G)\}$, $\{\mathbf{c}(G)\}$, respectively. More precisely, for $x = a, b, c$

$$S_x = \{\mathbf{v} \in R^m; \exists \text{ sequence } G_n, |V(G_n)| \to \infty, \mathbf{x}(G_n) \to \mathbf{v}\}.$$

Example 1 Let $k = 3$. Let H_1 be the 2-point, 1-edge graph; H_2 the 3-point 2-edge (vee) graph, H_3 the 3-point 3-edge (triangle) graph. Then

$$\mathbf{c}(H_2) = (\tfrac{2}{3}, 1, 0) \notin S_c.$$

We show later that $S_a = S_b$ and that S_c is a nonsingular linear transformation of S_a. Though the set S_c has perhaps the greatest natural interest, we shall prove theorems for S_a, where the technical problems are minimal.

Three Constructions

Let H be a graph on vertex set $V(H) = \{1, \ldots, t\}$. Let $x_1, \ldots, x_{t+1} \geq 0$, integral. We define a new graph $H^* = H(x_1, \ldots, x_t : x_{t+1})$ as follows:

$$V(H^*) = S_1 \cup \cdots \cup S_t \cup S_{t+1},$$

where S_i are disjoint sets $|S_i| = x_i$

$$E(H^*) = \{\{\alpha, \beta\} : \alpha \in S_i, \beta \in S_j, \{i, j\} \in E(H)\}.$$

Intuitively, we have blown up the ith vertex into an independent set of size x_i and added x_{t+1} isolated points.

Notation $H(x_1, \ldots, x_t) = H(x_1, \ldots, x_t : 0)$.

$\phi : H(x_1, \ldots, x_i) \to H$ is the canonical homomorphism defined by $\phi(\alpha) = i$ iff $\alpha \in S_i$,

$$sH = H(x_1, \ldots, x_t) \quad \text{where} \quad x_1 = \cdots = x_t = s,$$

$\prod_{i=1}^{t} G_i$ is the graph consisting of vertex disjoint copies of G_i,

$\prod_{i=1}^{t} a_i G_i$ is the graph consisting of vertex disjoint copies of $a_i G_i$.

Lemma 1 *If G has no isolated points*

$$A_G(H(x_1, \ldots, x_t : x_{t+1})) = \sum_{\psi} \prod_{i \in V(G)} x_{\psi(i)},$$

where ψ ranges over all homomorphisms $\psi : G \to H$. Also

$$a_G(H(x_1, \ldots, x_t : x_{t+1})) = \sum_{\psi} \prod_{i \in V(G)} P_{\psi(i)},$$

where we define $P_j = x_j / \sum_{i=1}^{t+1} x_i$.

Proof As G has no isolated points any homomorphism is into $H(x_1, \ldots, x_t)$. For each homomorphism $\lambda : G \to H(x_1, \ldots, x_t)$, $\psi = \phi\lambda$ is a homomorphism and for each ψ there are precisely $\prod_{i \in V(G)} x_{\psi_i}$ homomorphisms λ with $\psi = \phi\lambda$. The second equality follows from division. ∎

Lemma 2

$$A_H(sG) = A_H(G) s^{|V(H)|},$$

$$a_H(sG) = a_H(G).$$

Lemma 2 is only a special case of Lemma 1.

Corollary 1 $\mathbf{a}(G) \in S_a$ *for all G.*

Proof Let $G_n = nG$ in the definition of S_a. ∎

Example 1 shows the Corollary 1 does not hold for S_c; nor does it hold for S_b.

Lemma 3 *If H is connected*

$$A_H\left(\sum_{i=1}^{t} a_i G_i\right) = \sum_{i=1}^{t} A_H(G_i) a_i^{|H|},$$

$$a_H\left(\sum_{i=1}^{t} a_i G_i\right) = \sum_{i=1}^{t} a_H(G_i) p_i^{|H|},$$

where

$$p_i = a_i |V(G_i)| \Big/ \sum_{j=1}^{t} a_j |V(G_i)|.$$

Proof The first formula follows from Lemma 2 and the observation that the range of any homomorphism ψ shall be connected and hence lie in some $a_i G_i$. The second formula follows from division. ∎

Let

$$G = \sum_{i=1}^{t} a_i G_i + I_{a_{t+1}}$$

and set $p_i = a_i |V(G_i)|/|V(G)|$ for $1 \le i \le t$. A simple calculation gives

Lemma 4 *For H connected and G given as above*

$$a_H(G) = \sum_{i=1}^{t} a_H(G_i) p_i^{|H_i|}.$$

Dimensions

Now we are able to state our main result.

Theorem 1 *There exist $\mathbf{z} \in R^m$, $\varepsilon > 0$, so that $B(\mathbf{z}, \varepsilon) \subseteq S_a$. Here $B(\mathbf{z}, \varepsilon)$ is the ball of radius ε about $\mathbf{z}$.*

We require a preliminary lemma.

Lemma 5

$$\{\mathbf{a}(G)\}_{G \in \mathscr{G}} \text{ span } R^m \text{ (as a vector space)}.$$

STRONG INDEPENDENCE OF GRAPHCOPY FUNCTIONS 169

Proof We use an indirect argument. If the lemma is false there exist $c_1, \ldots, c_m$ not all zero, so that

$$\sum_{i=1}^{m} c_i a_i(G) = 0 \quad \text{for all} \quad G \in \mathcal{G}.$$

From $\{H_i : c_i \neq 0\}$ select from among the H with any particular number v of vertices an H with the minimal number of edges. Let $G = H(x_1, \ldots, x_v : x_{v+1})$. From Lemma 1, $a_i(G)$ is a polynomial in $p_1, \ldots, p_v$ where $p_i = x_i/|V(G)|$. The coefficient of $p_1 \cdots p_v$ in $a_i(G)$ is the number of *bijective* homomorphisms $\psi : H_i \to H$. By the minimality of H this coefficient is nonzero *iff* $H = H_i$. Thus $\sum c_i a_i(G)$ is not the zero polynomial—but then there exist rational values $p_1, \ldots, p_v \geq 0$, $\sum_{i=1}^{v} p_i < 1$ for which the polynomial is nonzero. We may find $x_1, \ldots, x_{v+1} \in Z$ yielding these p_i, contradicting our assumption. ∎

Proof of Theorem 1 Let $G_1, \ldots, G_m$ be such that $\{\mathbf{a}(G_i)\}$ span R^m. Set

$$a_j(G_i) = a_{ij}, \quad 1 \leq i, j \leq n$$

so that the matrix $[a_{ij}]$ is nonsingular. Let $p_1, \ldots, p_m, p_{m+1} \geq 0$ and rational, with

$$\sum_{i=1}^{m+1} p_i = 1.$$

Let $D \in Z$, $D > 0$ so that all $Dp_i/|V(G_i)| \in Z$. Set

$$G = \sum_{i=1}^{m} \left[\frac{(Dp_i)}{|V(G_i)|}\right] G_i + I_{Dp_{m+1}}.$$

Then, by Lemma 4,

$$a_j(G_i) = \sum_{i=1}^{m} a_{ij} p_i^{|H_j|}, \quad 1 \leq j \leq m. \quad (\bigstar)$$

We consider $(\bigstar)$ as a map $\Psi : R^m \to R^m$ transforming coordinates $p_1, \ldots, p_m$ to $\eta_1, \ldots, \eta_m$. G is defined for all $0 \leq p_1, \ldots, p_m \leq m^{-1}$. Then

$$S_a \supseteq \{\Psi(p_1, \ldots, p_m) : 0 \leq p_1, \ldots, p_m \leq m^{-1}, p_i \in Q\}.$$

Since Ψ is continuous and S_a closed

$$S_a \supseteq \{\Psi(p_1, \ldots, p_m) : 0 \leq p_1, \ldots, p_m \leq m^{-1}\}.$$

Now we calculate

$$\frac{\partial \eta_j}{\partial p_i} = a_{ij} |H_j| p_i^{|H_j|-1}$$

and $Jac(\Psi)$ is a polynomial in $p_1, \ldots, p_m$. At $p_1 = \cdots = p_m = 1$

$$Jac(\Psi) = \det[a_{ij}|H_j|] = \left[\prod_{j=1}^{m}|H_j|\right]\det[a_{ij}] \neq 0,$$

so $Jac(\Psi)$ is not the zero polynomial. Hence there exist $0 < p_1, \ldots, p_m < m^{-1}$ for which $Jac(\Psi) \neq 0$. Setting $z = \Psi(p_1, \ldots, p_m)$, S_a contains a ball about z. ∎

Equivalence of Formulations

Theorem 2 $S_a = S_b$.

Proof Fix H, $|V(H)| = i$, and G, $|V(G)| = n$. There are less than $i^2 n^{i-1}$ set mappings $\psi: V(H) \to V(G)$ which are not monomorphisms. Thus

$$A_H(G) - i^2 n^{i-1} \leq B_H(G) \leq A_H(G)$$

and hence

$$a_H(G)[n^i/(n)_i] - i^2 n^{i-1}/(n)_i \leq b_H(G) \leq a_H(G)[n^i/(n)_i].$$

Now $\lim_n n^i/(n)_i = 1$ and $\lim_n i^2 n^{i-1}/(n)_i = 0$. If G_n is a sequence with $|V(G_n)| \to \infty$, then $a_H(G_n) \to \alpha$ iff $b_H(G_n) \to \alpha$. From the definition, $S_a = S_b$. ∎

Theorem 3 S_c *is a nonsingular linear transformation of* S_b *(and hence* S_a*)*.

Proof Let $|V(H)| = i$, $|V(G)| = n$. We claim

$$B_H(G) = \sum_{H_1} B_H(H_1) C_{H_1}(G),$$

where H_1 ranges over all graphs, $|V(H_1)| = |V(H)|$, which contain H as a subgraph. For if $\psi: H \to G$ is a monomorphism $\psi(V(H)) = H_1$, a graph containing H as a subgraph. Conversely, for each H_1 there are $C_{H_1}(G)$ copies of H_1 in G and $B_H(H_1)$ maps ψ into each copy. Dividing by $(n)_i$:

$$b_H(G) = \sum_{H_1} [B_H(H_1)/i!] c_{H_1}(G),$$

giving an explicit transformation from **c** to **b** for any fixed k. If we order $\{H_1, \ldots, H_m\}$ by number of edges (arbitrarily among graphs with the same number of edges) the coefficient matrix becomes upper triangular with diagonal terms $B_H(H)/i! \neq 0$ and hence the transformation is nonsingular. ∎

Topological Properties

Theorem 4 S_a is arcwise connected (and hence, by Theorems 2 and 3, so are S_b and S_c).

Proof Let $|V(H_i)| = \alpha_i$ for $1 < i \leq m$. Let $\mathbf{x} = (x_1, \ldots, x_m) \in S_a$. We claim that for $0 \leq p \leq 1$, $(x_1 p^{\alpha_1}, \ldots, x_m p^{\alpha_m}) \in S_a$. We use that, by Lemma 4

$$a_i(vG + I_w) = a_i(G)r^{\alpha_i} \quad \text{where} \quad r = v|V(G)|/[v|V(G)| + w].$$

We may find G with $|a_1(G) - x_i|$ arbitrarily small for all i and thence find v, w so that $|p - r|$ is arbitrarily small, thus making $a_i(vG + I_w)$ arbitrarily close to $x_i p^{\alpha_i}$. This completes the claim. The theorem follows as we have given an arc between an arbitrary $\mathbf{x} \in S_a$ and $\mathbf{0}$. ∎

Dependence of Disconnected Graphs

We have shown that the functions a_H, where H runs over connected graphs of size $\leq m$ are strongly independent.

Observation Let $G = \sum_{i=1}^{s} G_i$. Then

$$a_G = \prod_{i=1}^{s} a_{G_i}.$$

That is, the functions a_H, H disconnected, are dependent on the a_H, H connected.

Theorem 5 *Suppose $H_1, \ldots, H_n$ represent all graphs on $\leq k$ vertices, define $\mathbf{a}(G) = (a_1(G), \ldots, a_n(G))$, S_a as before. Then S_a has dimension m equal to the number of connected graphs on $\leq k$ vertices.*

Observation Let $H^* = H + I_s$. Then $a_{H^*} = a_H$.

Theorem 6 *Suppose $H_1, \ldots, H_n$ represent all graphs on exactly k vertices, define $\mathbf{a}(G) = (a_1(G), \ldots, a_n(G))$, S_a as before. Then S_a has dimension m equal to the number of connected graphs on $\leq k$ vertices.*

The previous theorems also hold for S_b, S_c by equivalence theorems.

Example 2 For $h = 3$, $\dim(S_c) = 3$.

Comments

1. The domain of our functions has been the class $\mathscr{G}$ of all finite graphs. Let S_a^* be the set analogous to S_a if we restrict the domain to connected graphs. We claim $S_a^* = S_a$ (and similarly $S_b^* = S_b$, $S_c^* = S_c$).

Clearly $S_a^* \subseteq S_a$. Let $\mathbf{a} \in S_a$ and fix a sequence G_n, $|V(G_n)| \to \infty$, $\mathbf{a}_n = \mathbf{a}(G_n) \to \mathbf{a}$. To each G_n add at most $|V(G_n)| - 1$ edges to form a connected graph G_n^*. One can easily show that, in an asymptotic sense, almost none of the k vertex subgraphs contain any edges of $G_n^* - G_n$ so that $\mathbf{a}(G_n^*) \to \mathbf{a}$.

We may restrict our domain to doubly connected graphs, or similar restrictions, with identical results.

2. It is not known if S_a is locally arcwise connected. In general, the topological nature of S_a is not understood.

3. In the definition of limit points it was required to find a sequence G_n with $|V(G_n)| \to \infty$. It can be shown, using probabilitic methods, that for all sequences G_n, $|V(G_n)| \to \infty$, $\mathbf{a}(G_n) \to \mathbf{a}$, there exists a sequence G_n^* (of which G_n is a subsequence) so that $|V(G_n^*)| = n$ and $\mathbf{a}(G_n^*) \to \mathbf{a}$.

4. The sets S are, in general, not convex. For example—let $k = 3$ and H_i denote the 3-point $(i - 1)$-edge graph, $1 \leq i \leq 4$, then

$$(1, 0, 0, 0), (0, 0, 0, 1) \in S_c \quad \text{(by } I_s \text{ and } K_s\text{)}$$

but

$$(\tfrac{1}{2}, 0, 0, \tfrac{1}{2}) \notin S_c.$$

5. A complete description of convex hull (S) would settle several long standing questions in graph theory. For example, for $k \geq 4$, an ancient conjecture of Erdös is that

$$\min_{\mathbf{c} \in S_c} [c_{I_k} + c_{K_k}] = 2^{1-k}.$$

6. A complete description of S for $k = 3$ appears very difficult.

REFERENCE

1. H. Whitney, The coloring of graphs, *Ann. of Math.* **33** (1932) 688–718.

AMS 05C99, 05A05

Paul Erdös
MATHEMATICS INSTITUTE
HUNGARIAN ACADEMY OF SCIENCES
BUDAPEST, HUNGARY

László Lovász
BOLYAI INTÉZET
JÓZSEF ATTILA UNIVERSITY
SZEGED, HUNGARY

Joel Spencer
MATHEMATICS DEPARTMENT
STATE UNIVERSITY OF NEW YORK
AT STONY BROOK
STONY BROOK, NEW YORK

Squaring Rectangles and Squares

A Historical Review with Annotated Bibliography

P. J. FEDERICO

The subject of this historical note is the dissection of rectangles and squares into unequal squares, with ancillary material. The treatment is first a brief general historical review, followed and supplemented by an annotated bibliography. Some terminology customarily used in this field is initially set down.

A rectangle (square) dissected into squares is called a *squared rectangle* (*square*). The component squares are the *elements* and the number of elements is the *order*. The dissection is *perfect* if no two elements are equal, otherwise it is *imperfect*. The dissection is *simple* if it does not contain any subset of elements (more than one and less than all) arranged in a rectangle (square), otherwise it is *compound*. Note that reference to a simple or compound squared rectangle does not imply that it is necessarily perfect. Figures 3 and 5–9 illustrate some compound perfect squares and Fig. 10 a simple perfect square; these will be referred to later.

The first explicit mention of division into unequal squares that has been found was in 1925. Prior to this time Dehn had published, in 1903, a study of the division of a rectangle into rectangles [1]. The treatment is analytical with the only noteworthy result, for which the paper has been frequently cited, that a rectangle cannot be divided into squares unless its sides are commensurate. There is no mention of division into unequal squares. Perhaps the puzzlists Sam Loyd and H. E. Dudeney thought of dividing a square into unequal squares. Both, between 1903 and 1925, presented the patchwork quilt puzzle, the solution of which required the division of a square into squares, not all unequal [2, 3, 53]. Dudeney also presented a puzzle of dividing a given square into unequal squares and one (given) rectangle [4].

The first examples of rectangles divided into unequal squares were given by Moroń in 1925 [5]. He first states that S. Ruziewicz had proposed the problem, "Can one assemble a rectangle from different squares?" (No publication of this question has been located.) The two examples "of rectangles which may be made up of different squares" are given, without indicating how they were obtained, to answer this question in the affirmative. Moroń's figures are reproduced here in Fig. 1, numbered I and II for later reference. Rectangle I is 33 × 32 in size and is divided into nine unequal squares; rectangle II is 65 × 47 and has ten squares. Moroń's figures introduce two conventions which have been followed; expressing the sides of the component squares in integers without any common divisor (unless some reason requires otherwise), and writing the length of the side of a square inside the square.

Other perfect squared rectangles can be formed from a given one by adding a square of the same length of side to either side, and continuing this process alternately. Moroń gives a general formula for the ratio of sides of the rectangles in this infinite series of trivially compounded squared rectangles [this ratio, incidentally, approaches τ (sometimes also called ϕ), the golden ratio, as a limit].

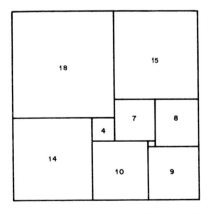 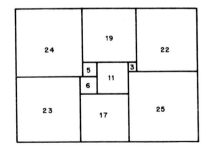

Figure 1 Moroń's two rectangles.

Moroń raises the question "For what rectangles is it possible to dissect them into squares?" He then observes "if there exists a rectangle (of different sides) for which there are two dissections R_1 and R_2 such that (1) in neither of these dissections does there appear a square equal to the smaller side of the rectangle and (2) each square of dissection R_1 is different from each square in dissection R_2, then the square is dissected into squares, all different, as shown in the following figure." Moroń's figure is reproduced here as Fig. 2. Either or both R_1 and R_2 could be compound, subject to Moroń's first condition.

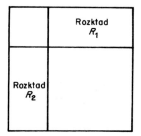

Figure 2 Moroń's figure.

Moroń's paper was practically unknown for some time. In 1935 the problem "Can a rectangle be divided into different squares?" appeared in a German journal. The answer, published in 1937, called attention to Moroń's paper and gave his two figures [10]. Steinhaus cites Moroń in the first edition (1938) of his *Mathematical Snapshots* [12]. In the meantime Kraitchik (1930) published the proposition, communicated to him by the Russian mathematician N. N. Lusin, that it was not possible to divide a square into a finite number of different squares [6]. A Japanese mathematician, M. Abe, was active in 1931–1932. His first paper, described under [8], was in Japanese and was unknown to other workers in the field until just a few years ago. His second paper [9] was in English and was known. It gives a simple perfect squared rectangle 195 × 191 and shows how an infinite series of compound squared rectangles can be built up from it with the ratio of sides approaching one in the limit.

Activity in Germany in 1937–1939 is shown by some problems [11, 19], a thesis [14], and three papers by Sprague. Sprague produced a perfect squared square, published in 1939 [15], the first one published. The square is made up by first forming two compound perfect squared rectangles with the same sides and then joining them as shown in Fig. 2. The structure of the square is shown in Fig. 3. Rectangles I and II are Moroń's two simple perfect rectangles; these are combined with an added 33 square and enlarged 29 times as shown

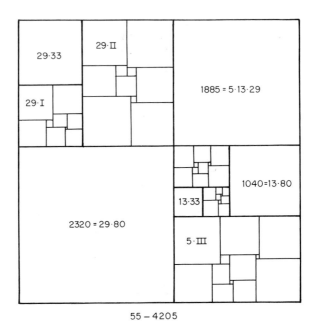

Figure 3 Structure of Sprague's square.

at the upper left. The same combination, magnified 13 times, is used again in the lower right. Rectangle III is a simple perfect rectangle of order 12 and sides 377 × 256 (number 633-b in [25] and 633-2 in [38]). The resultant compound perfect square has 55 elements and its side is 4205. Sprague had constructed a number of simple perfect rectangles in addition to the few already known (this fact is referred to in [14] and [48]) and the ingenuity of the construction of the square suggests numerous trials.

Additional results of the German activity are in Sprague [17, 18] and Reichardt and Toepken [19] and are commented on in the notes.

During the years 1936–1938, four students at Trinity College, Cambridge (R. L. Brooks, C. A. B. Smith, A. H. Stone, and W. T. Tutte) were working on the problem. The resulting paper [21] is a classic. A highly interesting account of their research, by Tutte, appeared in 1958 [36]. This account should be read not only by those having some interest in the present problem but also for its own sake, as an adventure in mathematics.

At first the group constructed simple squared rectangles empirically [36]. How this was done is as follows. Draw a rough sketch of a rectangle divided into rectangles, subject only to the condition that no two of these rectangles have an entire side in common with each other, and then imagine that each of the rectangular elements is a square. The geometry of the figure and elementary algebra enable the relative sizes of the assumed squares to be calculated. The result is a squared rectangle, perfect if none of the elements come out zero. This first stage was replaced by a theoretical treatment inspired by a wholly novel idea, which was the key to the topological aspects of the problem.

Only the highlights of the resultant paper [21] can be noted. The basic concept is the equating of a simple squared rectangle to an electrical network of a particular type. This is illustrated in Fig. 4. The squared rectangle on the left (1) is represented by the network on the right (2). Each horizontal line segment of (1) is represented by a node and each square is represented by a branch connecting the two nodes of the top and bottom lines of the square. The nodes a and b are the poles of the network and the resistance of each branch is unity. Ignoring the figures written in (2), if an unknown current is assumed entering at a and leaving at b, the Kirchhoff laws provide just enough equations to calculate the current in each branch in terms of the unknown current, the value of which is then taken so as to make the currents all integers without any common factor. These are the numbers written in (2), which are the sides of the component squares of (1). Thus electrical theory is brought into the picture and utilized, and graph theory as well, and the paper makes contributions to each.

The network of Fig. 4(2) was called a p-net (from polar net). If the two nodes a and b are connected by a new branch (in which the battery is located) the net is completed and was called a c-net. The c-nets are a special type of graph; they are all 3-connected planar graphs. The basic theorem demonstrated in the paper is that every simple rectangle can be derived from a c-net (3-connected planar graph) in the manner described. If the c-net has n edges, n p-nets are produced by removing each edge in turn, and hence n squared rectangles of order $n - 1$. The dual of a c-net results in the same rectangles, rotated 90°, and their p-nets are corresponding polar duals. The

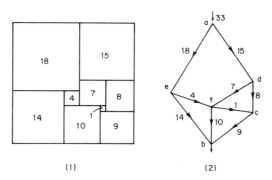

Figure 4 Squared rectangle and p-net.

process can also be thought of as placing a battery in turn in each edge of the c-net and calculating the relative values of the currents in the other edges, assumed to have unit resistance. Not every squared rectangle produced in this manner will be perfect (if the c-net has a symmetry, some rectangles might be imperfect with some of these still simple and others not), but every simple perfect rectangle of order $n - 1$ is produced from the complete set of c-nets of order n. The authors produced all the simple perfect rectangles from orders 9 to 12 (they showed that there are none below 9) but gave a full description of only those of orders 9 and 10. The c-nets to order 13 had to be first constructed, of course.

One reason for constructing simple perfect rectangles was that one with equal sides might be encountered, thus being a perfect square. This did not happen. A theoretical method of constructing perfect squares, based on certain types of 3-pole networks which are to be combined, was developed and perfect squares produced. Five are mentioned (four compound and one simple) but only one, the lowest order obtained, is shown. The square is shown here in Fig. 5. Two others were described before paper [21] appeared, one in [16, 20], shown in Fig. 6, and the other in [20].

The treatment in the paper is graph theoretical and utilizes electrical theory as well. With respect to the latter, a proposition now known as the

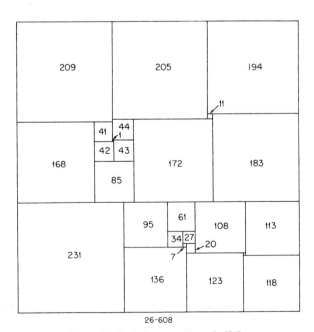

Figure 5 Perfect square shown in [21].

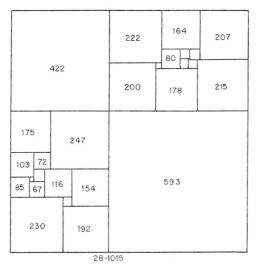

Figure 6 Perfect square [16, 20].

matrix tree theorem appears for the first time (it was subsequently independently discovered by others, at least three times). Though latent in known electrical and matrix theory, it had not previously been made explicit and expressed, perhaps because a use for it had not yet appeared. The theorem expresses a certain square matrix, the first cofactors of which are equal in value and give the number of spanning trees of the connected graph from which the matrix is formed. This number, called the complexity of the graph in [21], plays a role in the theory of squared rectangles. If the incoming current in a p-net is taken as equal to the complexity (number of spanning trees) of the p-net, the values of all the currents in the wires work out as integers; the horizontal side of the resultant squared rectangle is equal to this complexity, and the semi-perimeter is equal to the complexity of the c-net from which the p-net was derived. If the elements as thus calculated have a common factor they and the rectangle are "reduced" by this factor. In the case of a squared square the complexity of its p-net is equal to kn^2, where n is the reduced side of the square.

The concluding section of the paper mentions a number of generalizations including rectangled rectangles and the dissection of polygons (of angles $\pi/2$ and $3\pi/2$) and cylinders. A 3-pole p-net would give a squared hexagon. For rectangulations the wires of the p-net have general resistances, not necessarily equal. Curiously, a professor of electrical engineering some time ago wrote several papers in which some electrical network problems were solved by translating the network into a rectangled rectangle (or polygon) and then treating the latter.

The first perfect square published, Sprague's square, appeared only a few months before the one in [16], and two incidental results in [21] were also anticipated in publication, in [17–19].

Smith and Tutte continued developing the theory of the nets producing squared rectangles, particularly the nets from which squared squares could be constructed [31]. In a companion paper [32] Tutte further developed the theory of constructing perfect squares and described a number of additional examples, including several simple squares.

C. J. Bouwkamp in Holland worked on the problem of squaring rectangles in 1945–1946. His paper, in three parts [25], presented the method of [21] described in a more physical manner and developed various properties of the nets. The c-nets with up to 14 edges were constructed and shown by drawings of each. All the simple squared rectangles of orders 9 to 13 were constructed and Part II of the paper lists them and their elements. A code was developed for describing a squaring by listing the elements (here referring to the lengths of the sides of the component squares) in a certain order; the Bouwkamp code is commonly used. This work reached the practical limit of what could be done by hand. Resumption by computer came later.

In 1948, T. H. Willcocks, an employee of the Bank of England at Bristol, a chess enthusiast, and an amateur mathematician, produced the lowest order

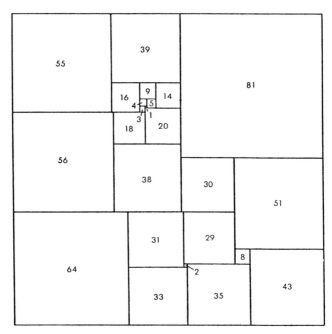

Figure 7 Lowest order compound perfect square known (24-175).

square known until 1978, shown in Fig. 7. He knew of the Brooks, Smith, Stone, and Tutte paper [21] but not the Bouwkamp paper [25]. Without the benefit of a catalogue of squared rectangles, he constructed his own stock and produced a number of compound perfect squares of low order. These he published in *The Fairy Chess Review*, a small periodical (now extinct) devoted to chess and other problems and puzzles [29]. His methods were based on the following. Referring to Fig. 2, if the corner square of R_2 which touches R_1 is removed and R_1 pushed in so that a corner fills the resultant space, another squared square can be produced which would be perfect if the smaller added square is not duplicated. This method is described [21] for a special case and the production of a perfect square of order 39 noted. Willcocks introduced two variations. First, he used a compound perfect rectangle, as shown in Fig. 8. Note that the two basic simple rectangles used are the same as those in Fig. 5 (in fact Fig. 8 can be made by cutting up Figure 5 to show this), and that three more perfect squares can be produced from them. The other variation was the utilization of imperfect squared rectangles. Since a corner element is to be discarded, this element could be the same size as some other element of the rectangle. Willcocks accordingly constructed a stock of squared rectangles with a corner element and an adjacent element

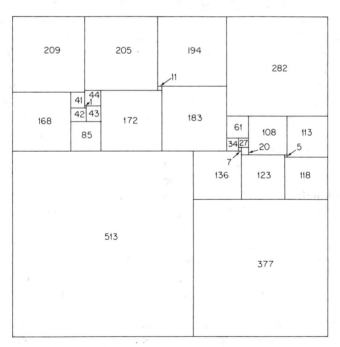

Figure 8 Another Willcocks perfect square.

the same size. This led to the order 24 perfect square [29]. If the squared rectangle and the elements 81 and 64 in Fig. 7 are removed it will be seen that the vacant corner of what is left can be filled with a square of side 30, forming the imperfect squared rectangle with two equal adjacent elements. A paper of 1950 [33] describes the above methods and various other methods utilized by Willcocks, as well as additional perfect squares he had constructed.

The Willcocks perfect square of order 24 was the lowest order perfect squared square known. Despite many attempts, a perfect square of lower order was not found until 1978, not even another one of order 24. Tutte has said "If the merit of a perfect square is measured by the smallness of its order, then the empirical method of cataloging the perfect rectangles had proven superior to our beautiful theoretical method" [36]. It is still the lowest order compound perfect square known.

In 1960 the Dutch group, Bouwkamp, Duijvestijn, and Medema, produced by computer and listed all the c-nets with up to 19 edges and all the simple squared rectangles with up to 15 elements. The former list, a few copies of which had been distributed, subsequently became generally available [39, 62]. Two catalogues of squared rectangles were published [37, 38], the later one evidently superseding the earlier one. The catalogue [38] lists the perfect squared rectangles (3663 in number) in two tables, one according to the c-nets from which they are derived and one according to the nondecreasing ratio of the short to the long side of the rectangle. The simple imperfect rectangles (431 in number) are also listed and the catalogue includes several other tables.

A deterrent to initiating computer work was the absence of a method of constructing all the c-nets with a given number of edges with certainty that there were no omissions. This problem was solved by a graph theory theorem of Tutte which he communicated to Bouwkamp before publication of a paper including it [42]. This theorem as used in the computer work depends upon the operation of forming a c-net with $n + 1$ edges from one with n edges by adding a new edge connecting two nonadjacent vertices of a mesh (face). Given the complete set of c-nets of order n, this operation is performed on each face (not applicable to a triangular face) of each c-net in all possible ways. The result is a collection of $n + 1$ order c-nets which includes every c-net of order $n + 1$ or its dual (with one exception); there will be a considerable amount of duplication. The graph known as a wheel (the description of which is apparent from this name), which has an even number of edges, is not produced by the method and must be added to the collection if $n + 1$ is even.

The method and the program are given in detail in Duijvestijn's thesis of 1962 [43]. This involved the problems of the representation of the graphs for use by the computer, generating new graphs from the old ones, generating dual graphs, and testing the large numbers of graphs obtained to eliminate duplicates and equivalents.

The c-nets having been obtained, a program was also developed for generating and listing the squared rectangles derived from each c-net. Only those through order 15 were listed at this time; later, the perfect ones through order 18 were also listed [47]. Table 1 gives the number of simple perfect rectangles through order 18.

TABLE 1

Number of Simple Perfect Rectangles

Order	Number	Order	Number
9	2	14	744
10	6	15	2609
11	22	16	9016
12	67	17	31427
13	213	18	110384

One reason for systematically generating squared rectangles, as has been stated, is that if the sides of any turn out to be equal, a squared square is obtained. Duijvestijn's thesis was also concerned with this problem. Squared rectangles produced from the c-nets through order 19 were tested for equal sides. The c-nets of order 20 were also generated, those whose complexity showed that a square was not possible eliminated, and the remainder utilized for generating the squared rectangles and testing for equal sides. No perfect square resulted; there were 101 simple imperfect squares of orders from 13 to 19. The computer time was great; a matter of 30 hours is mentioned in connection with the operations dealing with the c-nets of order 20. (See last paragraph.)

One by-product of this construction of complete sets of c-nets up to order 19 was the revival of Euler's problem of enumerating the combinatorially distinct convex polyhedra. As has been stated, the c-nets are 3-connected planar graphs. These are isomorphic with the graphs (vertices and edges) of convex polyhedra and it can be said that simple perfect squared rectangles are produced from convex polyhedra by means of Kirchhoff's laws relating to electric currents. The convex polyhedra with up to 19 edges were now known; previous results, which had not gone this high, could be checked, and further results produced. (An account of the enumeration problem is in [71].)

Tutte's popular article in the *Scientific American* and its reproduction in book form [36] inspired an unknown number of people to work on squaring rectangles and squares. One result was [45], which summarized known methods of constructing compound perfect squares and introduced a new method. This was to first construct squares divided into unequal squares and one rectangle (deficient squared squares). These can be constructed easily; by the theoretical method of [21] simply by letting the resistance of one wire be unknown and imposing the condition that the resultant dissected rectangle have equal sides. This then fixes the unknown resistance, the value of which gives the ratio of the sides of the rectangular element. However, the paper used the patterns of the squared rectangles (perfect and imperfect) given in [38], by assuming one element to be a rectangle, the sides of the dissected rectangle to be equal, and then recalculating the elements, as being simpler for work by hand. Then a perfect rectangle, simple or compound, to fill the rectangular space was sought from the catalogue [38], which has a table arranged by ratio of sides. In this manner a large number of new compound perfect squares of low order (defined as 28 or less) were found, in addition to known ones. The lowest new ones were two of order 25, one of which is shown in Fig. 9; note that in this one a compound perfect rectangle was used. Duijvestijn, who had a prepublication copy of the paper, programmed the method for the computer but did not obtain any perfect square

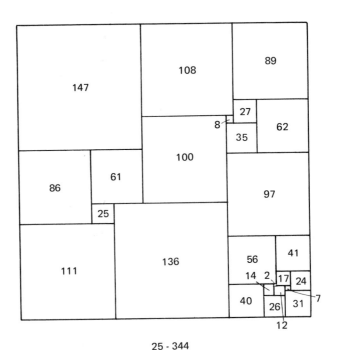

25 - 344

Figure 9 Perfect square from [45].

under order 24 (private communication). The paper mentions deficient squared squares with two (or more) rectangles, but only one such perfect square of low order was found.

Compound perfect squares of medium or high order can be produced in profusion [51], by hand or by computer, and there is little interest now in merely multiplying the number. As has been stated, there have been attempts to lower Willcocks' record of the order 24 square, unsuccessful as to compound perfect squares. The latest attempt in this area is in the two papers of Kazarinoff and Weitzenkamp [67, 68], who show that there is no compound perfect square of order 21 or less. They used a graph theory approach and the treatment is too involved to attempt a brief description. A good deal of the work of the Dutch group, generating c-nets and squared rectangles by computer, was repeated, and additional work was involved in the procedure for searching for compound squared squares. No results of perfect squares actually produced are mentioned. A number of simple perfect rectangles of reduced side ratio p/q with $p + q < 30$ are described; there is one of order 17 with side ratio 3/5 and one with ratio 5/7.

The basic paper of Brooks, Smith, Stone, and Tutte tackled the problem of simple perfect squares. These, as has been indicated, do not include a squared subrectangle in their structure. A highly theoretical method was developed, utilizing 3-pole nets satisfying certain strict conditions, by means of which a simple perfect square of order 55 (side 5468) was produced [21, 26]. Brooks managed to obtain a simple perfect square of order 38 (side 4920), described in [32] and illustrated in [34]. By a clever modification of the manner of connecting the basic nets, Willcocks reduced this square to one of order 37 (side 1947), described in [43] and [45]. Relaxation of the conditions and utilization of some rearrangements enabled the production of simple perfect squares down to order 31 [53a, 54], but this is probably the lowest order possible for this particular theoretical method.

The construction of simple perfect squares is the subject of Wilson's thesis [56]. A novel method is used which is explained by the simplest example. Two p-nets are connected in parallel, that is, the poles of one are made to coincide with the poles of the other, forming a new p-net with the number of wires the sum of the number in each. The complexity of the new p-net is calculated (the unreduced horizontal side of the rectangle from this p-net) and also the potential drop between the poles (the unreduced vertical side of the rectangle); these two numbers are to be coprime. The complexity is written in the form kn^2, where k is an integer free of square factors. Since n is to become the reduced side of any squared square produced, it is to be greater than 53 as it was known that any simple perfect square must have at least 20 elements. The key requirement is that there must be a pair of vertices a and b such that the potential drop between them is divisible by kn. Pairs of p-nets satisfying the above requirements were sought by computer. The basic p-nets used were those for the perfect squared rectangles up to order 15 in Bouwkamp's catalogue [38], excluding those having a reduction factor.

In each of the admissible new p-nets, the vertices a and b are taken as poles and a squared rectangle formed from it with these poles. Its horizontal side (unreduced) will be kn^2 (the number of spanning trees of the graph is independent of which vertices are to be designated as poles). The theory developed by Wilson from graph theory considerations shows that the vertical side of this squared rectangle will be equal to the horizontal side, or some multiple of it (no multiples occurred in the work).

In this manner Wilson obtained 5 simple perfect squares of order 25, and 24 of order 26. It may be noted that each of these can be separated into two squared hexagons, which follows from the method of construction, as is apparent in Fig. 10, reproduced from the thesis.

Other combinations of two basic p-nets, with 1, 2, or 3 added wires, were utilized, and also different orientations of the basic p-nets. Any combinations which would have more than 26 wires were excluded initially. An incidental result was the production of simple squared square cylinders (height equals

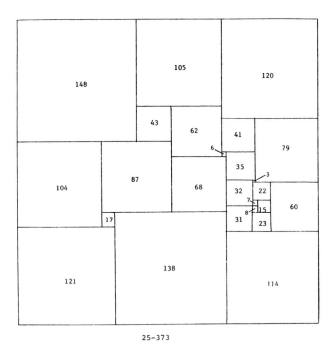

25-373

Figure 10 One of Wilson's five simple perfect squares of order 25.

circumference). If the two new poles a and b are not part of the same mesh, the result is a squared cylinder rather than a squared square, as indicated in [21]. Thirty-one simple perfect squared cylinders were found, of orders 24, 25, and 26.

Table II gives the number of perfect squares of order 31 or less known up to 1978. The 25 and 26 order simple squares included Wilson's 5 and

TABLE II

Number of Known Perfect Squares to Order 31 (1977)

Order	Simple	Cmpd 1	Cmpd 2
24		1	
25	8	2	
26	28	10	1
27	6	19	
28		33	4
29		49	1
30		19	14
31	4	36	1

24, respectively; the others, as well as the 27 order ones, were derived from some of these by a method of partial dissections and recombinations of portions (described in [53a]). The number of compound perfect squares, with one (Cmpd 1) or two (Cmpd 2) subrectangles, are from [51]; the note indicates the sources. The 26 order one with two subrectangles (Fig. 5) is probably the lowest order with two subrectangles possible. No squares with three subrectangles appear in the table as the lowest order one found is of order 38. A simple perfect square of order 21 was produced in 1978 [73].

The present status of the subject of dissecting rectangles and squares into unequal squares is indicated by the review which has been made, in the text along with the Bibliography.

As to general theory, we are very little beyond what was developed by Brooks, Smith, Stone, and Tutte in their series of papers.

As to simple perfect rectangles, it has not yet been shown that *any* given rectangle with commensurate sides can be divided into unequal squares in a simple manner. The general method of producing simple perfect rectangles could be characterized as building up rectangles from unequal squares, as originally put by Moroń. If a simple perfect rectangle with a given ratio of sides does not appear in the systematic building up process to the order that it has been carried, then special methods for the particular case need to be devised, as in the case of simple squares and the 2 × 1 rectangle [63, 64]. The results obtained would necessarily appear in the systematic process if carried further to a sufficiently high order. Proof of the general proposition, which is indicated as plausible in [21], would have to be theoretical.

With respect to compound perfect squares which depend upon availability of perfect rectangles for their construction, it has been positively shown that there are none below order 22. The work of the Dutch group may have bettered this result, but it has not been assembled and described so as to constitute any proof of a minimum. It remains to be proven whether or not there are any compound perfect squares of orders 22 and 23.

Several methods of producing simple perfect squares by computer have been devised. Wilson's method has been described; he introduced some unnecessary limitations to shorten the work and hence it cannot be said that his method is incapable of producing simple perfect squares below order 25. Reference [64] mentions but does not describe a computer program used for producing simple squares; no perfect ones below order 25 were found by this method (and another unpublished program), probably because of failure to include a sufficient range of data. What might be called the brute force method, constructing all the simple perfect squared rectangles up to a given order and testing them for equality of sides, was carried out by Duijvestijn through order 19 in 1962 [43]. Reference [70] indicates the development of faster methods and an intention to carry on the work beyond order 19. This was done, with improvements in the methods, and a simple perfect square of order 21 produced in March 1978 [73]. The square is shown on the cover of this book. There are none below order 21 and only one of order 21. Duijvestijn is now in process of deriving order 22 simple perfect squares but results have not yet been published.

Note This paper was delivered and in press before publication of Duijvestijn's order 21 square. Changes reflecting this fact were made in the proof but had to be kept to a minimum. New references [35a, 58a, 69a, 71, 72, 73] have been added.

ACKNOWLEDGMENTS

I am indebted to Michael Goldberg for making the drawings for Figures 3, 5, 6, and 8. Figures 4 and 9 are reproduced from [71] and [45], respectively, and Figure 7 from Martin Gardner's book [36]; it was gazing at this particular figure that gave the idea for [45].

BIBLIOGRAPHY

The following bibliography is a list of papers and other items concerned with dividing a rectangle or a square into unequal squares, with a few collateral items. There is an attempt at completeness (seldom successful and the author would appreciate being advised of any omissions) with respect to material making some contribution to the subject, substantial or otherwise. Review articles are also included and some other material, either germane to the historical account or having some point of interest. While a few works on mathematical recreations are listed, in general such works and other material which merely show a known squared rectangle or square have been excluded. Nearly all the items have a note, particularly those not mentioned in the text and in some other instances adding to what is said in the text:

1. Max Dehn, Über die Zerlegung von Rechtecken in Rechtecke. *Math. Ann.* **57** (1903) 314–332.
   See text for comment. Cited in [5, 7, 10, 11, 14, 17, 21, 35, 40, 43, 58, 68].
2. Sam Loyd, "Cyclopedia of Puzzles," Privately published, 1914 (not seen); The Patch Quilt Puzzle, reprinted *in* "Mathematical Puzzles of Sam Loyd, Selected and Edited by Martin Gardner," Vol. 1, pp. 73, 147–148. Dover, New York, 1959.
   A square quilt made of 169 square patches of the same size is to be divided into the smallest number of square pieces by cutting along lattice lines. The answer, 11 squares with sides 1, 1, 2, 2, 2, 3, 3, 4, 6, 7, is neither perfect (equal squares) nor simple (adjoining equal squares form a subrectangle). Gardner states that this puzzle first appeared in 1907 in a puzzle magazine edited by Sam Loyd. Referred to and illustrated in [6].
3. Henry E. Dudeney, Puzzle 173, Mrs. Perkin's Quilt, "Amusements in Mathematics," pp. 47, 180. London, 1917; reprinted, Dover, New York, 1958. Puzzle reprinted *in* Dudeney, "536 Puzzles & Curious Problems," (Martin Gardner, ed.), pp. 120, 324–325. Dover, New York, 1967.
   Same problem as in [2]. See comments by the editor and in [53].
4. Henry E. Dudeney, "The Canterbury Puzzles and Other Curious Puzzles," 1st ed. London, 1907 (not seen); 4th ed., 1919; reprinted, Dover, New York, 1958.
   Puzzle 40, pp. 66–67, 191–193, requires the dissection of a given square into 12 unequal squares and one rectangle, the dimensions of the latter being given. This is the puzzle referred to by Tutte in [36] as sparking some of the work there described.
5. Zbigniew Moroń, O rozkladach prostokatow na kwadraty (On the dissection of a rectangle into squares), *Przeglad Mat. Fiz.* **3** (1925) 152–153.
   See text. A manuscript English language translation (by Michael Goldberg) of this Polish paper is available from the author. Cited in [10, 12, 21, 25, 35, 43, 55, 56, 58, 66, 68, 69]. I am indebted to Dr. Stanislaw Dobrzycki of Lublin, Poland, for obtaining information concerning Moroń. The following is quoted from a letter to him by Prof. Wladyslaw Orlicz:

   "Zbigniew Moroń was my younger schoolmate when studying mathematics at the University of Lwów; about 1923–24 we were both junior assistants in the Institute of Mathematics. Professor Stanislaw Ruziewicz (who was then professor of mathematics at the University) communicated to us the problem of the dissection of a rectangle into squares. He had heard of it from the mathematicians of the University Kraków [Cracow] who took interest in it. As young men we enthusiastically engaged ourselves in investigating this problem, but after some time we all came to the conclusion that it was certainly as difficult as many other apparently simple questions in number theory. The examples found by Moroń were to us a great surprise. Before the World War II Moroń was a teacher in secondary schools; after it he was too, and dwelt in Wraclow, where he died some 5 years ago."

   Moroń was born in 1904 and died in 1971. Dr. Dobrzycki has translated the following from a later paper by Moroń (No. 35 or 35A):

   "In the years 1925–28 I found further results in this domain; among others I proved that it is impossible to construct a rectangle with less than 9 different squares; I also knew of the dissection of a square which was later given by Sprague [No. 15]. Nevertheless I did not publish them, but only exposed them at meetings of the mathematical seminar of Professor Ruziewicz."

6. M. Kraitchik, "La Mathématique des Jeux ou Récréations Mathématiques," p. 272. Stevens Frères, Brussels, 1930.
After giving the figure of Sam Loyd's puzzle [2] Kraitchik states, "But it is not possible to decompose a given square into a finite number of squares no two equal. This proposition though not demonstrated appears to be true. It was communicated to us by Monsieur Lusin, professor at Moscow." The next edition, in English (1942), gives an example (from [21]) and states that the problem was "long thought to be impossible." Cited in [14, 20, 21, 25, 40, 58, 68].

7. A. Schoenflies and M. Dehn, Ungelöste Probleme der analytischen Geometrie, in "Einführung in die analytische Geometrie der Ebene und des Raumes," 2nd ed., Appendix VI, pp. 402–411. Springer-Verlag, Berlin, 1931.
Includes parts of [1]. Cited in [10, 14, 21, 25].

8. Michio Abe, On the problem to cover simply and without gap the inside of a square with a finite number of squares which are all different from one another (in Japanese), *Proc. Phys.-Math. Soc. Japan* **4** (1931) 359–366.
This paper, in Japanese, was unknown to workers in the field until just a few years ago and hence is not treated in the text. The results are surprising for the time. Abe attempted to obtain a perfect square by the method of combining two simple squared rectangles of the same size. He constructed over 600 simple perfect rectangles, of which six are described by illustrations. While he found one pair the same size, two 608 × 407 rectangles of order 13, the resultant squared square was imperfect. His total for rectangles of order 13 was not complete and he missed the pair which gives perfect results. He also tried another method, noted under [45], but again with no success. His manner of drawing the squared rectangles suggests that he may have known Moroń's paper [5]. Cited in [9].

9. Michio Abe, On the problem to cover simply and without gap the inside of a square with a finite number of squares which are all different from one another (in English), *Proc. Phys.-Math. Soc. Japan* (3) **14** (1932) 385–387.
See text. Applications of the method are in [20] and [49]. Cited in [14, 17, 21, 35, 58].

10. Jaremkewycz, Mahrenholz, and Sprague, Answer to Problem 1242, *Z. Math. Naturwiss. Unterricht Schulgattungen* **68** (1937) 43; Proposed in **66** (1935) 251.
See text. Cited in [14, 15, 21, 58].

11. H. Toepken, Problem 242, *Jber. Deutsch. Math.-Verein.* **47**, Part 2 (1937), p. 2.
Relates to [1]. Cited in [14, 21].

12. H. Steinhaus, "Mathematical Snapshots," 1st ed., pp. 8, 9, 131. Steckert, New York, Leipzig, 1938.
Gives assembling a rectangle from nine given unequal squares as a problem, citing Moroń for the solution. Cited in [13, 20, 21, 58].

13. S. Chowla, Division of a rectangle into unequal squares, and Problem 1779, *Math. Student* **7** (1939) 69–70, 80.
Solves the trivial problem posed by Steinhaus [12]. A separate item, Problem 1779, asks "It is possible to divide the volume of a rectangular parallelopiped into unequal cubes?" An elegant proof of the impossibility (quoted or paraphrased many times, e.g., in [59]) is given in [21]. Cited in [21, 58, 68].

14. Alfred Stöhr, Über Zerlegungen von Rechtecken in inkogruente Quadrate, *Schr. Math. Inst. Inst. Angew. Math. Univ. Berlin* **4**, Part 5 (1939) 118–140.
Dissertation, University of Berlin, Berlin, 1938. Cited in [15, 17, 21, 35, 40, 58].

15. R. Sprague, Beispiel einer Zerlegung des Quadrats in lauter verschiedene Quadrate, *Math.* **45** (1939) 607–608.
    See text. Cited in [17, 18, 20, 21, 32, 35, 36, 40, 41, 45, 48, 50, 52, 55, 56, 58, 66–69].
16. A. H. Stone, Problem E401, *Amer. Math. Monthly* **47** (Jan. 1940) 48.
    The first part of this problem lists the sides of 28 different squares and asks that they be fitted together to make a single square of side 1015. This is in effect a publication of the perfect squared square (illustrated in Fig 5) since the assembling is trivial. The second part asks if there is any simpler perfect squared square. Cited in [20 (solution); 21, 32, 58].
17. R. Sprague, Über die Zerlegung von Rechtecken in lauter verschiedene Quadrate, *J. Reine angew. Math.* **182** (1940) 60–64.
    See [18]. Cited in [18, 21, 32, 34, 35, 40, 58, 68].
18. R. Sprague, Zur Abschätzung der Mindestzahl inkongruenter Quadrate, die ein gegebenes Rechteck ausfüllen, *Math. Z.* **46** (1940) 460–471.
    Sprague proves the theorem that any given rectangle with commensurate sides can be dissected into unequal squares (in a compound manner). The method requires initial knowledge of two perfect squared rectangles of the same size with no two squares alike, which he gives in [15], and is explained here by this example. The ratio of the sides of each squared rectangle of [15], call them R_1 and R_1', is 13 to 16. Magnify R_1 by 16 and R_1' by 13 and join the two by the now equal short side of one and the long side of the other, forming squared rectangle R_2 with 53 elements and five subrectangles, and sides 61,625 × 30,160. Now magnify R_1 by 13 and R_1' by 16 and form squared rectangle R_2'. R_2 and R_2' are the same size and have no two like elements and form perfect square S_2 which has 108 elements and ten subrectangles. S_1 (the perfect square of [15]) and S_2 do not have any like elements when brought to the same size. In a similar manner square S_3 (with 214 elements and 20 subrectangles) is formed from R_2 and R_2', and so on, forming a series of perfect squares with no like elements when brought to the same size. The rectangle to be dissected, with sides m and n, say, is divided into mn unit squares and these are filled with mn perfect squares from the series. The result would be very highly compounded, a 2 × 3 rectangle, for example, dissected in this manner would appear to have 3351 elements and contain 315 subrectangles. A different and more elegant proof, utilizing graph theory, but still resulting in very high compounding, is in [21]. Cited in [21, 32, 34, 35, 40, 58].
19. H. Reichardt and H. Toepken, Solution to Problem 271, *Jber. Deutsch. Math. Verein.* **50**, Part 2 (1940), pp. 13, 14; (a) Problem posed in **48**, Part 2 (1939) p. 73.
    Shows that there is no simple perfect rectangle less than order 9 and gives the two of order 9. A different proof is in [21]. Cited in [21 (a only); 58].
20. Michael Goldberg and W. T. Tutte, Solution of Problem E401, *Amer. Math. Monthly* **47** (Oct. 1940) 570–572.
    See [16]. Goldberg solves the first part and gives some discussion. Tutte solves the second part by referring to the 26 order perfect square to appear in the forthcoming paper (Fig. 6); he also describes another perfect square of order 28 by listing its elements. There is an added editorial note. Cited in [21, 33, 34, 36, 45, 58].
21. R. L. Brooks, C. A. B. Smith, A. H. Stone, and W. T. Tutte, The dissection of rectangles into squares, *Duke Math. J.* **7** (1940) 312–340.
    See text. Also referred to in the notes to [6, 13, 18–20, 54, 59]. Cited in [25–27, 30–38, 40, 41, 44–46, 48, 50, 52, 55–58, 61, 63, 65–69].
22. Science Service, Science in 1940, Mathematics, *Science* **92**, Suppl. 14 (1940).
    Lists the division of a square into a finite number of smaller squares no two the same size as one of the achievements of 1940. Cited in [34, 45].
23. H. Toepken, Problem 294, *Jber. Deutsch. Math. Verein.* **51**, Part 2 (1941) 2.

24. Michael Goldberg, Squaring the square, *Washington Scientist* **1** (1945) 76.
    Discussion of problem and report of a talk by A. H. Stone before the Philosophical Society of Washington, March 31, 1945.
25. C. J. Bouwkamp, On the dissection of rectangles into squares, Paper I, *Koninkl. Nederl. Akad. Wetensch. Proc. Ser. A* **49** (1946) 1172–1188; Papers II and III, **50** (1947) 58–71, 72–78.
    See text. Cited in [26, 27, 30–34, 36–38, 40, 44–46, 54, 58, 61, 62, 68].
26. R. L. Brooks, C. A. B. Smith, A. H. Stone, and W. T. Tutte, A simple perfect square, *Koninkl. Nederl. Akad. Wetensch. Proc. Ser. A* **50** (1947) 1300–1301.
    Gives the details of the first simple perfect square, of order 55, referred to in [21]. Cited in [32, 36, 45, 55, 58].
27. C. J. Bouwkamp, On the construction of simple perfect squared squares, *Koninkl. Nederl. Akad. Wetensch. Proc. Ser. A* **50** (1947) 1296–1299.
    Revises Section 8 of [25]; gives the details of another simple perfect square, of order 55; corrects a few minor errors in [25]. Cited in [32, 36, 45, 58, 68].
28. T. H. Willcocks, Problem 7523 and solution, *Fairy Chess Rev.* **6** (1947–1948) 114 (Dec.); 123 (Feb.).
    Rediscovery of the perfect square of [16, 20]. Cited in [29].
29. T. H. Willcocks, Problem 7795 and solution, *Fairy Chess Rev.* **7** (1948) 97 (Aug.); 106 (Oct.)
    See text. Cited in [32, 33, 36, 41, 43, 45, 48, 50, 52, 55, 57, 58, 66–68].
30. W. T. Tutte, A note on a paper by C. J. Bouwkamp, *Koninkl. Nederl. Akad. Wetensch. Proc. Ser. A* **51** (1948) 280–228.
    Comment on a point in [25, III].
31. C. A. B. Smith and W. T. Tutte, A class of self-dual maps, *Canad. J. Math.* **2** (1950) 179–196.
    See text. Cited in [32, 34, 36, 56, 58].
32. W. T. Tutte, Squaring the square, *Canad. J. Math.* **2** (1950) 197–209.
    See text. Cited in [33, 34, 36, 40, 45, 46, 52, 54, 56, 58, 61, 63, 67, 68].
33. T. H. Willcocks, A note on some perfect squared squares, *Canad. J. Math.* **3** (1951) 304–308.
    See text. Cited in [41, 45, 46, 48, 49, 52, 54–56, 58, 67–69].
34. Michael Goldberg, The squaring of developable surfaces, *Scripta Math.* **18** (1952) 17–24.
    General discussion with nontrivial examples of squared cylinders and cones, and a Möbius strip. Cited in [40, 56, 58].
35. Zbigniew Moroń, Orozkladach prostokatow na nierowne kwadraty (in Polish), *Wiadom. Mat.* (2) **1** (1955–1956) 75–94.
    Extensive review article (not read). Cited in [50].
35a. Zbigniew Moroń, O prawie doskonalych rozkladach prostokatow (On almost perfect dissections of rectangles). *Wiadom. Mat.* (2) **1** (1955–1956) 175–179.
    Not seen.
36. W. T. Tutte, Squaring the square, *Sci. Amer.* **199** (1958) 136–142, 166. Reprinted with addendum and enlarged bibliography in Martin Gardner, "The 2nd Scientific American Book of Mathematical Puzzles and Diversions," pp. 186–209, 250. Simon & Schuster, New York, 1961; also paperback edition.
    See text. Cited in [45, 52, 54, 57, 60, 61, 64, 67, 68].
37. C. J. Bouwkamp, A. J. W. Duijvestihn, and P. Medema, "Catalogue of Simple Squared Rectangles of Orders Nine through Fourteen and their Elements." Technische Hogeschool, Eindhoven, Netherlands, May 1960, 50 pages.
    See text. Cited in [36, 38, 43, 45].

38. C. J. Bouwkamp, A. J. W. Duijvestijn, and P. Medema, "Tables Relating to Simple Squared Rectangles of Orders Nine through Fifteen." Technische Hogeschool, Eindhoven, Netherlands, Aug. 1960, 360 pages.
    See text. Reviewed in [41]. Cited in [36, 41–45, 48, 49, 52, 55, 56, 62, 65–69].
39. C. J. Bouwkamp, A. J. W. Duijvestijn, and P. Medema, Table of c-nets of Orders 8–19 Inclusive, 2 vols. Philips Research Laboratories, Eindhoven, Netherlands, 1960; unpublished, available in UMT file of Mathematics of Computation.
    See text. Described in [62].
40. Herbert Meschkowski, Die Zerlegung von Rechtecken in inkongruente Quadrate, *in* "Ungelöste und unlösbare Probleme der Geometrie," pp. 92–103. Vieweg and Sohn, Braunschweig, 1960. English translation, The decomposition of rectangles into Incongruent Squares, *in* "Unsolved and Unsolvable Problems in Geometry," pp. 91–102, 164–165. Oliver and Boyd, Edinburgh and London, 1966.
    Cited in [50, 58, 61].
41. Michael Goldberg, Review of [38], *Math. Comp.* **15** (1961) 315.
    Cited in [45].
42. W. T. Tutte, A theory of 3-connected Graphs, *Koninkl. Nederl. Akad. Wetensch. Proc. Ser. A* **64** (1961) 441–455.
    See text. Cited in [43, 48, 52, 56, 68].
43. A. J. W. Duijvestijn, Electronic computation of squared rectangles. Dissertation, Technische Hogeschool, Eindhoven, Netherlands, 1962; also in *Philips Res. Rep.* **17** (1962) 523–612.
    See text. Cited in [45, 48, 52, 54–56, 62, 67, 68, 70].
44. W. T. Tutte, A census of planar maps, *Canad. J. Math.* **15** (1963) 249–271.
    Develops formulas for calculating the number of "rooted" c-nets which can also be used for estimating the number of simple perfect rectangles with a given number of elements, also described in [48].
45. P. J. Federico, Note on some low-order perfect squared squares, *Canad. J. Math.* **15** (1963) 350–362.
    See text. Gives the full description of 35 low order squares, 24 of them new. (The method, described in the text, turns out not to be entirely new as Abe in the unknown Japanese paper [8] distorted some perfect rectangles to form a squared square with one rectangle, but obtained no results as he only applied the method to perfect rectangles with sides differing by one.) Cited in [48, 49, 52, 55, 56, 67–69].
46. S. L. Basin, Generalized Fibonacci sequences and squared rectangles, *Amer. Math. Monthly* **70** (1963) 372–379.
    A series of squares with sides according to the Fibonacci series can be arranged in a rectangle; it is compound and also imperfect as it has two adjoining equal squares of side one. Cited in [49, 58].
47. C. J. Bouwkamp, A. J. W. Duijvestijn, and J. Haubich, Catalogue of Simple Perfect Squared Rectangles of Orders 9 through 18, Philips Research Laboratories, Eindhoven, Netherlands, 1964 (unpublished, 12 vol., 3090 pp. listing 154,490 simple squared rectangles).
    See text. It would take over 6000 volumes the same size to list the simple perfect rectangles of orders 19 to 23. Cited in [62, 65].
48. W. T. Tutte, Squared rectangles, *Proc. IBM Sci. Symp. Combinatorial Problems*, Thomas J. Watson Research Center, Yorktown Heights, N.Y. (1964) 3–9.
    Review article with discussion, announces and illustrates Wilson's first 25 order simple perfect square (see [56]). Cited in [68].

49. P. J. Federico, A Fibonacci perfect squared square, *Amer. Math. Monthly* **71** (1964) 404–406.

   A Fibonacci rectangle (see [46]) is found with which a compound perfect square is formed by removing the corner unit square and combining with a simple perfect rectangle. Cited in [56].

50. L'Udovit Vittek, A perfect decomposition of the square into 25 squares (in Slovak), *Mat. Fyz. Casopis* **14** (1964) 234–235.

   Rediscovery of a perfect square published in [45].

51. P. J. Federico, List of Perfect Squared Squares, 119 pp., May 1964; Supplement, 40 pp., Aug. 1971.

   Manuscript catalogue of over 800 compound perfect squares; includes published squares, and unpublished ones derived by Bouwkamp, Federico, Willcocks, and E. Lainez. Listed here, as copies were distributed. (Lainez is a Spanish metallurgical engineer who read the French translation of [36] and, without the aid of a catalogue of rectangles, constructed his own stock and derived a number of new perfect squares of low order, including the first one in which the subrectangle is necessarily on a side of the square instead of in a corner.)

52. W. T. Tutte, The Quest of the Perfect Square. *Amer. Math. Monthly* **72**, Part II (1965) 29–35.

   Review article with discussion. Cited in [54, 60, 61, 67–69].

53. Martin Gardner, Mrs. Perkins' Quilt and other square-packing problems, reprinted with additions from *Sci. Amer.* **215** (1966) 264–272, **216** (Jan. 1967) 118–121; in Martin Gardner, "Mathematical Carnival," pp. 139–149, 272. Knopf, New York, 1975.

   Includes the problem: Can 24 squares with sides 1 to 24 (the sum of whose squares equals 70^2) be assembled into a square of side 70?

53a. P. J. Federico, Note on Some Simple Perfect Squares, 22 pp., March 1967.

   Manuscript paper; listed here as copies were distributed and results are referred to in [54] and [69].

54. T. H. Willcocks, Some squared squares and rectangles, *J. Combinatorial Theory* **3** (1967) 54–56.

   Tutte says in [69], "The two last-mentioned authors (Federico and Willcocks) have modified the theoretical method (of [21]) to obtain simple perfect squares of orders ranging down to 31." Cited in [55].

55. W. T. Tutte, Topics in graph theory, *in* "Graph Theory and Theoretical Physics," (F. Harary, ed.), pp. 301–312. Academic Press, New York, 1967.

56. John C. Wilson, A method for finding simple perfect squared squarings, Ph.D. Thesis, University of Waterloo, Waterloo, Ontario, 1967 (80 pp. plus 72 pp. computer output). See text.

57. G. H. Morley, Networks and squared squares, *Eureka* (*J. Archimedeans, Cambridge Univ. Math. Soc.*) No. 30, Oct. (1967) 14–16.

   A highly ingenious method of constructing simple perfect squares of high order, 60 or above, from specially related pairs of squared rectangles, discovered by the author while in high school.

58. I. M. Yaglom, "How to Divide a Square" (in Russian). Nauka, Moscow, 1968.

   The first and only book on the subject; contains some new material (not read). The bibliography includes three Russian items (not seen) which appear to merely show a known perfect square. Cited in [68].

58a. P. J. Van Albada, La dissection du carré en carrés, *Bull. Soc. Math. Belg.* **20** (1968) 161–170.

59. Mark Kac and Stanislaw M. Ulam, "Mathematics and Logic," p. 32. Mentor paperback, New York, 1969.
    The electrical analogy (of [21]) cited as an "illustration of the remarkable and wholly unexpected connections of which mathematics is full."
60. Sherman K. Stein, "Mathematics, the Man-Made Universe," 2nd ed. Freeman, San Francisco, California, 1969; 3rd ed., 1975.
    Chapters 7 and 8 give an interesting elementary exposition of "tiling" a rectangle with unequal squares. Cited in [61].
61. Ross Honsberger, Squaring the square, *in* "Ingenuity in Mathematics," New Mathematical Library, No. 23, pp. 46–60. Random House, New York; now by Math. Assoc. Amer., Washington, D.C., 1970.
62. C. J. Bouwkamp, Review of [39], *Math. Comp.* **24** (1970) 995–997.
63. R. L. Brooks, A procedure for dissecting a rectangle into squares, and an example for the rectangle whose sides are in the ratio 2 : 1, *J. Combinatorial Theory* **8** (1970) 232–243.
    First solution of the problem "to find a simple perfect rectangle whose horizontal side is twice the vertical side" [36]. The solution has 1323 elements. Cited in [64, 68, 69].
64. P. J. Federico, Some simple perfect 2×1 rectangles, *J. Combinatorial Theory* **8** (1970) 244–246.
    Second and further solutions of the problem of [63]; 23, 24, and 25 order 2×1 simple perfect rectangles are shown. Cited in [67–69].
64a. R. D. Hollands, Developing a problem, *Math. in Teaching (Assoc. Teaching Acids in Math.)* No. 50, Spring (1970) 64–66.
    Elementary introduction from standpoint of problem solving and teaching.
65. C. J. Bouwkamp, On some special squared rectangles, *J. Combinational Theory Ser. B* **10** (1971) 206–211.
    Discusses simple perfect rectangles having the property that the set of elements can be arranged in two different ways forming two dissections; lists 59 examples. Cited in [68].
66. Howard Eves, "A Survey of Geometry," Rev. ed., Vol. 1, pp. 229–231. Allyn & Bacon, Boston, Massachusetts, 1972; 1st ed., 1963.
    Listed to show the inclusion of the subject in a textbook on geometry.
67. N. D. Kazarinoff and R. Weitzenkamp, On existence of compound perfect squared squares of low order, *J. Combinatorial Theory Ser. B* **14** (1973) 163–179.
    See text. Cited in [68].
68. N. D. Kazarinoff and R. Weitzenkamp, Squaring rectangles and squares, *Amer. Math. Monthly* **80** (1973) 877–888.
    See text.
69. W. W. Rouse Ball and H. S. M. Coxeter, "Mathematical Recreations and Essays," 12th ed. Univ. of Toronto Press, Toronto, 1974; also in paperback.
    Section on squared rectangles and squares contributed by W. T. Tutte.
69a. Lionel March and Philip Steadman, Electrical networks and mosaics of rectangles, *in* "The Geometry of Environment," pp. 263–283. MIT Press, Cambridge, Massachusetts, 1974.
    Application to architectural design.
70. A. J. W. Duijvestijn, Fast calculation of inverse matrices occurring in squared-rectangle calculation, *Philips Res. Rep.* **30** (1975) 329–339.
71. P. J. Federico, The number of polyhedra. *Philips Res. Rep.* **30** (1975) 220*–231*.
    Application to the enumeration of polyhedra. Figure 4 is reproduced from this article.

72. Albert A. Mullin, on arithmetic aspects of geometric problems. *Am. Math. Soc. Notices*, **25** (Feb. 1978) A-227.
    States several "Lemmas" on the nature of the integers forming the side-lengths of the elements of a perfect squared rectangle. One of these is that the sides cannot all be in arithmetic progression. This settles the problem mentioned under [53].
73. A. J. W. Duijvestijn, Simple perfect squared square of lowest order. *J. Combinatorial Theory*, *B* **25** (1978) 260–263; See also *Sci. Amer.* (June 1978) 86–87.
    See last paragraph of text.

AMS 05-01

WASHINGTON, D.C.

Subsemigroups and Supergraphs

S. FOLDES

and

G. SABIDUSSI

We make no restriction to finite cardinalities. The axiom of choice is assumed.

A *graph* G is an ordered pair $G = (V, E)$, where $V = V(G)$ is any set (vertices), and $E = E(G)$ (edges) is a set of 2-subsets of V. If $S \subseteq V$, then the *subgraph* $G[S]$ of G *induced by* S is defined by

$$V(G[S]) = S, \quad E(G[S]) = \{A \in E(G): A \subseteq S\}.$$

If G and H are graphs, then a *homomorphism* from H to G is a mapping $h: V(H) \to V(G)$ such that for every $A \in E(H)$, $h(A) \in E(G)$. If $H = G$, then h is called an *endomorphism*. The set of homomorphisms from H to G is denoted Hom(H, G), and we write Hom(G, G) = End G. If $h \in$ Hom(H, G) and $g \in$ Hom(G, K), then the product $gh \in$ Hom(H, K) is defined by

$$gh(x) = g(h(x))$$

for every $x \in V(H)$. A homomorphism $h: H \to G$ is an *isomorphism* if there is a $g \in$ Hom(G, H) with $hg = id_{V(G)}$, $gh = id_{V(H)}$. The set of isomorphisms from G to G, called *automorphisms* of G, is denoted Aut G. Obviously

Aut $G \subseteq$ End G. A subset $S \subseteq V(G)$ is a *faithful constituent* of End G if for every $h \in$ End G, $h(S) \subseteq S$, and h is determined by its restriction $h|S$ to S.

A *semigroup* is a set A together with an associative binary operation called product, possessing a unit element. Any subset B of A closed under product and containing the unit element is a *subsemigroup*. For every graph G, End G is a semigroup, $id_{V(G)}$ being the unit. Aut G is a subsemigroup of End G and it is in fact a group. If S is a faithful constituent of End G, then every subsemigroup A of End G is isomorphic to its restriction $A|S$.

Proposition 1 (Frucht [5, 6]; Sabidussi [9]) *Every group is isomorphic to the automorphism group of some graph.*

This result was generalized to semigroups as follows.

Proposition 2 (Hedrlin and Pultr [7]) *Every semigroup is isomorphic to the endomorphism semigroup of some graph.*

Proposition 3 (Bouwer [2]; Babai [1]) *Given a finite graph H and a subgroup B of Aut H, there exists a graph G such that*

(i) *H is a subgraph of G induced by a faithful constituent of Aut G,*
(ii) *the restriction Aut $G|V(H)$ is B.*

Again, in the case of finite groups, Proposition 1 can be viewed as a consequence. We have obtained the following extension of Proposition 3.

Proposition 4 *Given a graph H and a subsemigroup B of End H, there exists a graph G such that*

(i) *H is a subgraph of G induced by a faithful constituent of End G,*
(ii) *the restriction End $G|V(H)$ is B.*

Proposition 2 can be deduced as a corollary. Thus, in the case of finite graphs, we have the following diagram of generalizations:

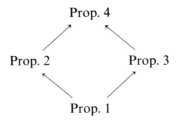

The proof of Proposition 4 relies on a combination of results of Erdös and Hajnal [3, 4] on the existence of graphs with arbitrarily large chromatic number and containing no "small" odd circuits, and a result of Hell [8] on mutually rigid graphs. The latter result asserts the existence of arbitrarily large families of graphs ($G_i: i \in I$) with the property that if $f \in \text{Hom}(G_i, G_j)$, then $i = j$ and f is the identity transformation of $V(G_i) = V(G_j)$. In other terms, the full subcategory induced by the G_i's in the category of graphs has only identities. Such a family of graphs is called mutually rigid. It is then possible to show the following.

Lemma *Let n and k be cardinal numbers. Then there exists a family ($G_i: i \in I$) of n mutually rigid connected graphs such that for every $i \in I$*
  (i) *the chromatic number of G_i is (strictly) larger than k,*
  (ii) *G_i does not have odd circuits of girth less than 9,*
  (iii) *every vertex of G_i lies on some circuit of length 9.*

With the aid of the lemma, the "supergraph" G of Proposition 4 can be constructed using a technique similar to the one used in [1] by Babai to prove Proposition 2. Actually, the number 9 appearing in the Lemma could be replaced by any larger odd integer, and the only reason we take 9 is because it is the smallest value that allows the proof of Proposition 4 to be carried out smoothly.

REFERENCES

1. L. Babai, Representation of permutation groups by graphs, *Colloq. Math. Soc. Janos Bolyai*, 4, *Combinatorial Theory and Its Applications, Balatonfüred, 1969* (1970) 55–80.
2. I. Z. Bouwer, Section graphs for finite permutation groups, *J. Combinatorial Theory* **6** (1969) 378–386.
3. P. Erdös, Graph theory and probability I, *Canad. J. Math.* **11** (1959) 34–38.
4. P. Erdös and A. Hajnal, Some remarks on set theory IX, *Michigan Math. J.* **11** (1964) 107–127.
5. R. Frucht, Herstellung von Graphen mit vorgegebener abstrakter Gruppe, *Compositio Math.* **6** (1939) 239–250.
6. R. Frucht, Sobre la construccion de sistemas parcialmente ordenados con grupo de automorfismos dado, *Rev. Un. Mat. Argentina* **13** (1948) 12–18.
7. Z. Hedrlin and A. Pultr, Symmetric relations (undirected graphs) with given semigroup, *Monatsh. Math.* **68** (1965) 318–322.
8. P. Hell, On some strongly rigid families of graphs and the full embeddings they induce, *Algebra Universalis* **4** (1974) 108–126.
9. G. Sabidussi, Graphs with given infinite group, *Monatsh. Math.* **64** (1960) 64–67.

AMS 05C99

DÉPARTEMENT DE MATHÉMATIQUES
UNIVERSITÉ DE MONTRÉAL
MONTRÉAL, QUÉBEC

GRAPH THEORY AND RELATED TOPICS

There Are Finitely Many Kuratowski Graphs for the Projective Plane

HENRY H. GLOVER

and

J. P. HUNEKE

In 1930 Kuratowski [5] proved that a graph embeds in the 2-sphere if and only if it does not contain a subgraph homeomorphic to one of two graphs, K_5 or $K_{3,3}$. No similar list of graphs for any other surface has been proved to be complete; prior to the result described in this paper no proof that a similar list of graphs for any other surface is even finite is known to exist. This paper summarizes the proof of this result for the real projective plane. Namely, that a graph embeds in the projective plane if and only if it does not contain a subgraph homeomorphic to one of a finite list of graphs. The proof is independent of any detailed list of graphs.

For a 2-dimensional manifold M let $I(M)$ denote a smallest set of graphs such that every graph which does not embed in M contains a subgraph homeomorphic to one of the graphs in $I(M)$. Let $\{I_n(M) | n \in N\}$ be a partition of $I(M)$ such that $K \in I_n(M)$ if and only if $K \in I(M)$ and the maximal valence of a vertex in K is n. There has been limited success in describing $I(M)$ or even $I_n(M)$. The known results for S^2, the two sphere, and P, the real projective plane, are:

Theorem 1 (Kuratowski [5]) $I(S^2)$ *contains exactly two graphs, K_5 (the complete graph on five vertices) and $K_{3,3}$ (the complete 3, 3 bipartite graph).*

The first generalization of Theorem 1 appeared 42 years later as Theorem 2. This was improved in Theorem 3 and then generalized in the Main Theorem of this paper.

Theorem 2 (Miligram [6]) $I_3(P)$ *is finite.*

Theorem 3 (Milgram [7]; Glover and Huneke [1]) $I_3(P)$ *contains exactly six graphs.*

Theorem 4 (Glover et al. [4]) $I(P)$ *contains at least* 103 *graphs.*

Other brief lists of graphs in $I(M)$ for other surfaces have been published. The results outlined in this paper are

Theorem 5 (Glover and Huneke [2]) $I_n(P)$ *is finite for each n.*

Theorem 6 (Glover and Huneke [3]) $I_n(P)$ *is empty if $n > 26$.*

As an immediate corollary of these two Theorems [2, 3] we have the

Main Theorem $I(P)$ *is finite.*

Notation If v is a vertex of a graph K, let $st(v)$ denote the subgraph of K of all edges containing v. For a subgraph $L \subset K$ define $st(L) = \bigcup_{v \in L} \{st(v)\}$. Let K_n denote the complete n-graph and $K_{n,m}$ the complete n, m-bipartite graph for any natural numbers n, m. A *θ-graph* is any graph homeomorphic to $K_{2,3}$.

In order to identify nonbounding cycles of a graph in a surface the following definition is useful. Let L be a subgraph of a graph K. L is a *k-graph of K* provided there exists a subgraph L' of K, L a subgraph of L', such that either (a) L is homeomorphic to $K_{2,3}$, L' is homeomorphic to $K_{3,3}$, and one of the vertices of L' contained in three edges of L' is not in L, or (b) L is homeomorphic to K_4, L' is homeomorphic to K_5, and one of the vertices in four edges of L' is not in L, or (c) L is homeomorphic to K_4, L' contains an arc A in a cycle in L' with the quotient graph L'/A homeomorphic to K_5, and A is disjoint from L. See Fig. 1, in which L has dashed edges, and L' has both dashed edges and dotted edges.

For the sketch of the proof of the Main Theorem we need Lemma 1, a method to find Kuratowski graphs within a graph in $I(M)$, or in particular in $I(P)$. A proof of this result is included. From Lemma 1, all graphs in $I(P)$ which contain a θ-graph disjoint from a k-graph are explicitly characterized; this class of graphs includes all those graphs in $I(P)$ which are not 3-con-

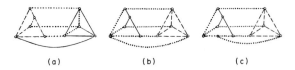

Figure 1

nected. For each graph K in $I(P)$ which does not contain a θ-subgraph disjoint from a k-graph of K, Lemma 1 also gives a bound on the complement of a Kuratowski subgraph of K in terms of the largest valence of a vertex of K. Then bounds on this largest valence are found, first if K does not contain a pair of disjoint θ-subgraphs, and second (again using Lemma 1) if K contains a pair of disjoint θ-subgraphs. For complete details of each lemma, the reader is referred to [2] and [3].

Lemma 1 *Let K be a graph without valency 2 vertices, let e be an edge of K, let L_1 be a 2-connected subgraph of K not containing e, let L_2 be a connected component of $K - \text{st}(L_1)$ which does not intersect e, and let M be a compact 2-manifold. If K does not embed in M but $\phi: K - e \to M$ is an embedding such that $\phi(L_1)$ is contractible in M, then*

(i) *the quotient graph K/L_2 is nonplanar, and*
(ii) *there is a k-graph of K disjoint from L_2.*

Proof (i) implies (ii) since a k-graph of K/L_2 disjoint from $\{L_2\}$ is a k-graph of K disjoint from L_2. Hence, it is sufficient to prove (i). Suppose $\psi: K/L_2 \to S^2$ is an embedding; we now show K embeds in M, a contradiction. Let C be the subgraph of K such that $\psi(C)$ bounds the component D of $S^2 - \psi(K/L_2 - \text{st}\{L_2\})$ which contains $\psi(\{L_2\})$. Let C_1 be a simple cycle in L_1 such that $\psi(C_1)$ bounds the component D_1 of $S^2 - \psi(L_1)$ which contains $\psi(\{L_2\})$. There exists an embedding $f: (S^2 - D) \to M$ such that $f\psi|_{C_1} = \phi|_{C_1}$ and $f(S^2 - D) \cap \phi(C_1 \cup \text{st}(L_2)) = \phi(C_1)$ by the following comments. First observe that $\psi(C_1)$ bounds $S^2 - D_1$, and $\phi(C_1)$ is contractible in M so bounds D_2 which is homeomorphic to $\mathbb{R}^2$. Hence there exists a homeomorphism $h: S^2 - D_1 \to D_2 \cup \phi(C_1)$ such that $h\psi|_{C_1} = \phi|_{C_1}$. Next observe, $\phi(C_1 \cup \text{st}(L_2)) \cap (D_2 \cup \phi(C_1)) = \phi(C_1)$, because otherwise $\phi(C_1 \cup \text{st}(L_2)) \supseteq D_2 \cup \phi(C_1)$ since L_2 is connected. Thus we have an embedding $h_0: D \cup \psi(C \cap C_1) \to D_2 \cup \phi(C \cap C_1)$ such that $h_0^{-1}\phi|_{C \cap C_1} = \psi|_{C \cap C_1}$, and hence an embedding of K into S^2 is given by ψ on $K/L_2 - \text{st}\{L_2\}$ and by $h_0^{-1}\phi$ on $\text{st}(L_2)$. This contradicts the fact that K does not embed in M. Hence if $D = D_1$ (equivalently, $C = C_1$) then h is a corestriction of the desired function f and so we have f. If $D \neq D_1$ then $D_1 \supset D$, so $S^2 - D_1 \subset S^2 - D$ and we need an extension of h to $S^2 - D$. Recall $S^2 - \psi(\{L_2\})$ is homeomorphic to $\mathbb{R}^2$ and by the Jordan curve theorem there is a neighborhood N of $D_2 \cup \phi(C_1)$ with N homoemorphic to $\mathbb{R}^2$ and $N \cap \phi(L_2) = \phi$. Hence there is a homeomorphism $\bar{h}: S^2 - \psi(\{L_2\}) \to N$ such that $\bar{h}|_{S^2 - D_1} = h$ and $\bar{h}(\psi \text{st}\{L_2\} - \psi\{L_2\}) \subset \phi(\text{st}(L_2) - L_2)$. Finally note $\bar{h}|_{S^2 - D}$ is a co-restriction of the desired function f, so f is established. Define $\bar{\phi}: K \to M$ by ϕ on $C_1 \cup \text{st}(L_2)$ and $f\psi$ on $K/L_2 - \text{st}(L_2)$. Since (i) ϕ and $f\psi$ are embeddings whose images intersect at $\phi(C_1)$, (ii) $C_1 \cup \text{st}(L_2)$ and $K/L_2 - \text{st}\{L_2\}$ are closed in K, cover K, and intersect in C_1, and (iii) $\phi|_{C_1} = f\psi|_{C_1}$ it follows that $\bar{\phi}$ is a well-defined embedding. Hence the result follows. ∎

The following lemmas establish the Main Theorem of this chapter. They are proved in detail in [2] and [3].

Lemma 2 *If $K \in I(P)$ is not 3-connected, or if $K \in I(P)$ is 3-connected and contains a θ-subgraph disjoint from a k-graph of K, then K contains a pair of disjoint k-graphs of K, K is the union of the two corresponding Kuratowski graphs, and (hence) K has no more than 16 vertices.*

The proof of Lemma 2 uses Lemma 1 to show that a θ-graph disjoint from a k-graph implies the existence of disjoint k-graphs. K is the union described because the union of these two Kuratowski graphs does not embed in P. The upper bound 16 is established in [2], though in fact the largest such example has 14 vertices. A combinatorial check shows that all graphs of this form appear in the list of 103 graphs in [4].

In light of Lemma 2, to study $I(P)$ it is sufficient to make the following:

Standing Hypothesis In each of the remaining lemmas, assume $K \in I_n(P)$ ($n \geq 3$) is a 3-connected graph which does not contain a θ-graph disjoint from a k-graph.

Lemma 3 *Let $L \in I(S^2)$ and let $f: L \to K$ be an embedding. Let e_1, e_2 be distinct edges of L and assume $\mathcal{A}$ is a set of m pairwise disjoint arcs in K from $f(e_1)$ to $f(e_2)$ which are disjoint from $f(L - (e_1 \cup e_2))$. Then $m \leq 10$.*

A proof of Lemma 3 uses Lemma 1 to contradict the Standing Hypothesis if $m \geq 11$. Also basic to this lemma is the unique property of the projective plane, that a neighborhood of a noncontractible cycle in P cannot be an annulus (because it contains a Möbius band).

Lemma 4 *There is a Kuratowski graph $L \subset K$ containing not more than $880n$ vertices, and at most $(880n)(2n - 4)$ vertices of K are not in L.*

The bound on L is derived from Lemma 3. The bound on $K - L$ uses a generalization (to graphs of girth at least three without θ-subgraphs) of the fact that each tree has more valence one vertices than vertices with valence greater than two; observe that the Standing Hypothesis guarantees there are no θ-graphs in K disjoint from L.

Hence it remains to find a bound on n, the maximal valence of a vertex of K.

Lemma 5 *If v is a vertex of K with valence n, then there is a cycle in K not containing v but containing at least $(n - 1)/2$ vertices of $\mathrm{st}(v)$.*

The proof of Lemma 5 follows from properties of P using embeddings of $K - e$ in P for e an edge of K.

Lemma 6 *If K does not contain a pair of disjoint θ-graphs, then $n \leq 26$.*

To prove the contrapositive of Lemma 6, assume $v \in K$ is a vertex with valence greater than 26. By Lemma 5 there is a cycle $C \subset K$ containing at least 13 vertices of $\mathrm{st}(v)$ but not v. Since K does not embed in P and is 3-connected, the disjoint θ-graphs can be found in K.

Lemma 7 *If v is a vertex of K such that $K - \mathrm{st}(v)$ is nonplanar, then the valence of v is not more than 12.*

To prove Lemma 7, assume v has valence more than 12 and find a θ-graph in K which is disjoint from a k-graph in a Kuratowski graph in the nonplanar graph $K - \mathrm{st}(v)$, thereby contradicting the Standing Hypothesis.

Finally, if $K \in I(P)$ does not satisfy the hypotheses of Lemma 2 (i.e., does satisfy the Standing Hypothesis) and does not satisfy the hypothesis of Lemma 6, and if v is a vertex of K which does not satisfy the hypothesis of Lemma 7, then enough of the graph K can be reconstructed so that by using Lemma 1, a bound on the valence of v can be established by

Lemma 8 *If K contains a pair of disjoint θ-graphs and if v is a vertex of K such that $K - \mathrm{st}(v)$ is planar then subgraphs L_1, L_2, C_1, C_2 and E of K can be found such that:*

(1) L_1 and L_2 are each 2-connected,
(2) L_1 and L_2 each contain a θ-subgraph,
(3) $L_1 \cap L_2$ is empty,
(4) $L_1 \cup L_2$ spans K (i.e., each vertex of K is in $L_1 \cup L_2$),
(5) $K = E \cup L_1 \cup L_2$,
(6) each edge of E has an endpoint in L_1 and an endpoint in L_2,
(7) $C_1 \subset L_1$ and $C_2 \subset L_2$,
(8) C_1 and C_2 are each a simple cycle,
(9) $E \cap L_1 \subset C_1$ and $E \cap L_2 \subset C_2$,
(10) $v \in C_1$,
(11) $E \cap \mathrm{st}(v)$ has at most eight edges, and
(12) $L_1 \cap \mathrm{st}(v)$ has at most six edges.

Observe that Lemma 8 implies the valence of the vertex v in its hypothesis has valence at most 14. Hence the previous lemmas establish the Main Theorem, that $I(P)$ is finite.

REFERENCES

1. H. Glover and J. P. Huneke, Cubic irreducible graphs for the projective plane, *Discrete Math.* **13** (1975) 341–355.
2. H. Glover and J. P. Huneke, Graphs with bounded valency that do not embed in the projective plane, *Discrete Math.* **18** (1977) 155–165.
3. H. Glover and J. P. Huneke, The set of irreducible graphs for the projective plane is finite, *Discrete Math.* **22** (1978) 243–256.
4. H. Glover, J. P. Huneke, and C.-S. Wang, 103 graphs which are irreducible for the projective plane, *J. Combinatorial Theory Ser. B*, to appear.
5. K. Kuratowski, Sur le problème des courbes gauches en topologie, *Fund. Math.* **15** (1930) 271–283.
6. M. Milgram, Irreducible graphs, *J. Combinatorial Theory Ser. B* **12** (1972)
7. M. Milgram, Irreducible graphs—Part 2, *J. Combinatorial Theory Ser. B* **14** (1973) 7–45.

AMS 05C10

THE OHIO STATE UNIVERSITY
COLUMBUS, OHIO

On a Kuratowski Theorem for the Projective Plane

HENRY H. GLOVER, J. PHILIP HUNEKE,
and
CHIN SAN WANG

Introduction

In an earlier paper [5] the writers exhibited a list of 103 graphs which are irreducible for the real projective plane P. In this paper, we give an account of how this list of 103 graphs can be systematically constructed from five particular irreducible graphs for P, and how these 103 are the only graphs that can be so constructed. Our method gives a partial ordering to the set of all irreducible graphs of P, $I(P)$. If we call these graphs generated by (or below) the five graphs mentioned above J, then $J \subset I(P)$ is a poset. If we call the set of 103 graphs in the appendix L then the main result of this paper is

Theorem 2.1 *As sets $J = L$.*

A longer account of the results in this paper will appear elsewhere.

A brief account of the sections in this paper follows. In Section 1 we describe the poset $I(M)$ for any surface M. In Section 2 we describe the poset $J \subset I(P)$ that we construct in this paper and state the main result just mentioned. In Section 3 we give the part of $I(P)$ (or equivalently J) that lies below $K_{3,5}$ in our ordering of $I(P)$. In Section 4 we give an account of the proof of Theorem 2.1. An appendix with the 103 graphs is at the end of the paper just before the references.

1. The Poset $I(M)$

Given a (finite) graph X and a surface M we say that X is *irreducible* for M if X does not embed in M ($X \not\subset M$) but every proper subgraph Y of X ($Y \subset X$) does embed in M ($Y \subset M$). We denote by $I(M)$ the set of all distinct (nonhomeomorphic) irreducible graphs for M.

As an example we note that Kuratowski's theorem says that

$$I(\mathbb{R}^2) = \{K_5, K_{3,3}\},$$

the two Kuratowski graphs [6].

We next introduce a partial ordering on the set $I(M)$. As motivation for this process observe that in $I(\mathbb{R}^2)$, $K_{3,3}$ may be obtained from K_5 by splitting any vertex and deleting two edges as shown in the figure.

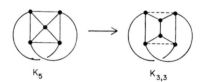

In the general case, given a graph X we form a new graph $S_v X$ constructed by replacing a vertex $v \in X$ by two vertices v' and v'' connected by an edge e and then connecting some of the edges incident to $v \in X$ so that they become incident to v' in $S_v X$ and the remaining edges in X incident to v so that they become incident to v'' in $S_v X$. Note that X is obtained from $S_v X$ by collapsing e. We say briefly that $S_v X$ is obtained from X by *a splitting of the vertex v in X*. In general, for given graph X with given vertex v, a number of distinct graphs $S_v X$ can be obtained from X by different splittings of v.

As an example illustrating $S_v X$ notice that any vertex v of any graph X can be split to give $S_v X$ which is homeomorphic to X. Similarly any splitting of any cubic (valency 3) vertex of X gives a graph $S_v X$ homeomorphic to X. As another example notice that we can obtain exactly two nonhomeomorphic graphs X by splitting a vertex of $X = K_5$, namely $S_v X$ homeomorphic to K_5 or the graph just depicted (including the dotted edges).

The following result is basic in our construction of the poset $J \subset I(P)$.

Lemma 1.1 *Let $S_v X$ be obtained from X by a splitting of a vertex $v \in X$. If $X \not\subset M$ then $S_v X \not\subset M$.*

Proof Since X is homeomorphic to the quotient space $S_v X/e$ and M is homeomorphic to the quotient space M/e, an embedding $S_v X \subset M$ induces an embedding $X \subset M$ establishing the contrapositive of the desired result. ∎

Let $X \in I(M)$ and let $S_v X$ be obtained from X by a splitting of a vertex $v \in X$. Then since $S_v X \not\subset M$, there exists at least one subgraph $Y \subseteq S_v X$ such that $Y \in I(M)$. In general one will obtain several distinct Y's from a given $S_v X$. Let us say that Y is obtained from X by an *SD operation* on a vertex v.

Consider now the reflexive transitive relation $\geq$ generated by successive SD operations.

Lemma 1.2 *The relation $\geq$ on $I(M)$ is a partial ordering.*

Proof We need only check that $\geq$ is antisymmetric. Observe that the lexicographical ordering of the valency sequences (each written as a non-increasing sequence) of graphs $X \in I(M)$ is a linear ordering of these valency sequences and that the function which assigns to each graph its valency sequence is ordering preserving. That is, $Y > X$ implies $\sigma(Y) > \sigma(X)$, where $\sigma(X)$ denotes the valency sequence of X, and also $X > Y$ implies $\sigma(X) > \sigma(Y)$. Hence $\geq$ is antisymmetric as desired. ∎

2. The Main Result

We now turn our attention to the projective plane P. We are interested in the set $I(P)$. Consider the five graphs from the appendix (Fig. 2.1). In [5] it is shown that each of these graphs belongs to $I(P)$. By applying successive SD operations to these five graphs we obtain a poset $J \subset I(P)$. Call the set of graphs in the appendix L. The main result of this paper is

Theorem 2.1 $J = L$ *as sets.*

In a longer account of the results about the poset J we will give all of its order relations. Space does not permit us to do so here.

Of course the maximal elements of the poset J are given in Fig. 2.1. The minimal elements of J are the cubic graphs E_{42}, F_{11}, F_{12}, F_{13}, F_{14}, and G.

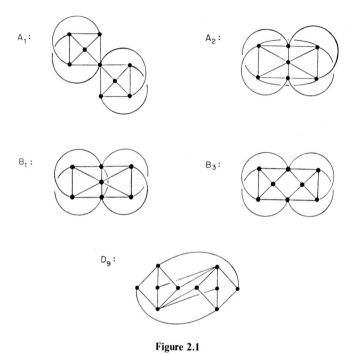

Figure 2.1

As an example of the relations in J consider the four graphs

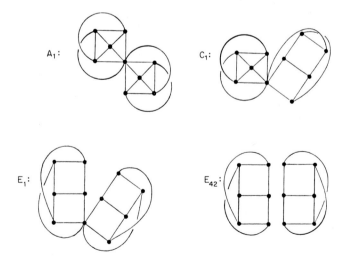

and the "obvious" path in J

$$\underset{A_1}{\bullet}\text{———}\underset{C_1}{\bullet}\text{———}\underset{E_1}{\bullet}\text{———}\underset{E_{42}}{\bullet}$$

The question left unanswered by this paper is whether $J = I(P)$ as posets. We believe the answer is yes, and offer as evidence that (i) $I(P)$ is finite [2, 3], (ii) min $I(P)$ = min J [1, 7] and (iii) for X with at most eight vertices $I(P)$ and J agree [4].

3. Graphs in $I(P)$ below $K_{3,5}$

In [5] it is shown that $K_{3,5} \in I(P)$. There follows a directed graph (Fig. 3.1) giving those graphs in $I(P)$ (or equivalently in J) below $K_{3,5}$. To check that E_4 is the only graph immediately below $K_{3,5}$ it suffices to notice that up

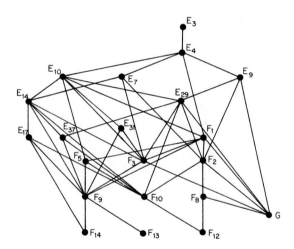

Figure 3.1

to isomorphism there is only one splitting of $K_{3,5}$ yielding a graph not homeomorphic to $K_{3,5}$. To see that E_7, E_9, E_{10}, and F_1 are the only graphs immediately below E_4 it again suffices to check that up to isomorphism there are only four splittings possible. The first three give E_7, E_9, E_{10}, the last, by deleting an edge, F_1. It is more difficult to check other parts of this poset. This will be done in our extended account.

For completeness we note that $K_{3,5}$ can be obtained from two of the five graphs we started with, A_2 and B_1, by SD operations as shown in Fig. 3.2.

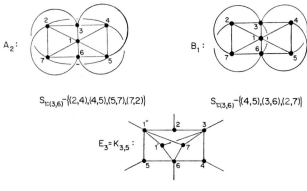

Figure 3.2

For an explanation of the notation see [5], e.g., $S_{1:(36)} - \{(4, 5), (3, 6), (2, 7)\}$ means split 1 into 1' and 1" so that 3 and 6 become incident to 1', then delete (4, 5), (3, 6) and (2, 7).

4. The Proof of Theorem 2.1

The proof of Theorem 2.1 consists in showing the 103 graphs in the appendix are the only graphs obtained by successive SD operations applied to the five graphs of Fig. 2.1. Call the set of graphs in the appendix L. The proof consists in systematically finding all the distinct splittings of a given graph $X \in L$ then checking whether each different splitting $S_v X \in L$. If $S_v X \in L$ then we continue. If not then we must find all proper subgraphs $Y \subset X$ such that $Y \in L$. There must be one such or we have made a mistake. We find all of these by systematically deleting individual edges and if necessary pairs, triples, etc., of edges. We continue in this way until we obtain the cubic graphs of [1] and [7].

Appendix: 103 Graphs in $I(P)$

The following 103 graphs are given by the order A_i, $1 \leq i \leq 5$; B_i, $1 \leq i \leq 11$; C_i, $1 \leq i \leq 11$; D_i, $1 \leq i \leq 19$; E_i, $1 \leq i \leq 42$; F_i, $1 \leq i \leq 14$, and G. A_i's are graphs with Betti number $\beta = 12$. Similarly, B_i's C_i's, D_i's, E_i's, F_i's, G are graphs with $\beta = 11, 10, 9, 8, 7, 6$, respectively.

The valency sequence of each graph is given directly below the picture. The general form of the valency sequence is $n(s_1, s_2, \ldots, s_n)$ where n is the number of vertices of the graph and s_i's are valencies of vertices arranged in nonincreasing order.

KURATOWSKI THEOREM FOR THE PROJECTIVE PLANE

For most of the 103 graphs, each contains a subgraph homeomorphic to $K_{3,3}$. To study the (non)embedding of these graphs we often embed this $K_{3,3}$ into P. As an example the following graph represents a graph X in which $(4, 5), (6, 7), (8, 9)$ are edges in X (through points of identification in P).

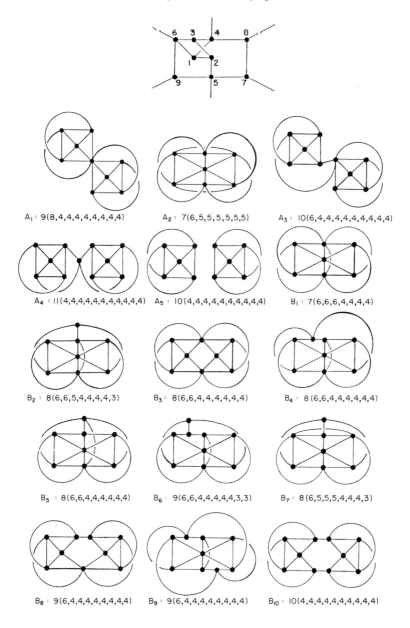

A_1: 9(8,4,4,4,4,4,4,4,4) A_2: 7(6,5,5,5,5,5) A_3: 10(6,4,4,4,4,4,4,4,4,4)

A_4: 11(4,4,4,4,4,4,4,4,4,4,4) A_5: 10(4,4,4,4,4,4,4,4,4,4) B_1: 7(6,6,6,4,4,4,4)

B_2: 8(6,6,5,4,4,4,4,3) B_3: 8(6,6,4,4,4,4,4,4) B_4: 8(6,6,4,4,4,4,4,4)

B_5: 8(6,6,4,4,4,4,4,4) B_6: 9(6,6,4,4,4,4,4,3,3) B_7: 8(6,5,5,5,4,4,4,3)

B_8: 9(6,4,4,4,4,4,4,4,4) B_9: 9(6,4,4,4,4,4,4,4,4) B_{10}: 10(4,4,4,4,4,4,4,4,4,4)

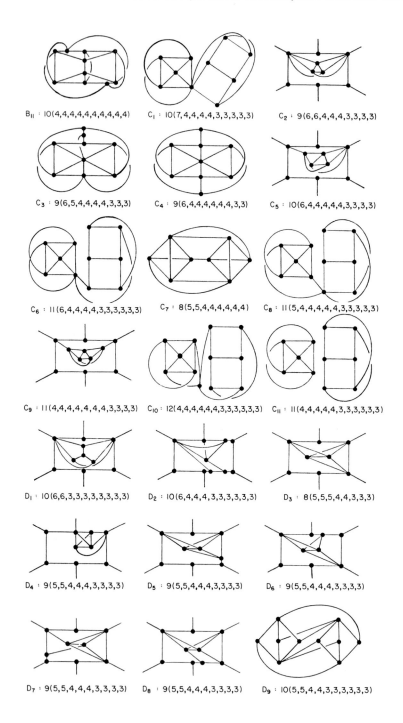

KURATOWSKI THEOREM FOR THE PROJECTIVE PLANE

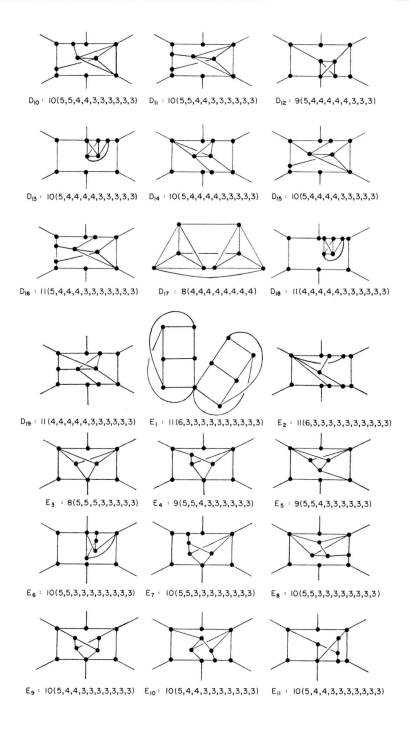

D_{10} : 10(5,5,4,4,3,3,3,3,3,3) D_{11} : 10(5,5,4,4,3,3,3,3,3,3) D_{12} : 9(5,4,4,4,4,4,3,3,3)

D_{13} : 10(5,4,4,4,4,3,3,3,3,3) D_{14} : 10(5,4,4,4,4,3,3,3,3,3) D_{15} : 10(5,4,4,4,4,3,3,3,3,3)

D_{16} : 11(5,4,4,4,3,3,3,3,3,3,3) D_{17} : 8(4,4,4,4,4,4,4,4) D_{18} : 11(4,4,4,4,4,3,3,3,3,3,3)

D_{19} : 11(4,4,4,4,4,3,3,3,3,3,3) E_1 : 11(6,3,3,3,3,3,3,3,3,3,3) E_2 : 11(6,3,3,3,3,3,3,3,3,3,3)

E_3 : 8(5,5,5,3,3,3,3,3) E_4 : 9(5,5,4,3,3,3,3,3,3) E_5 : 9(5,5,4,3,3,3,3,3,3)

E_6 : 10(5,5,3,3,3,3,3,3,3,3) E_7 : 10(5,5,3,3,3,3,3,3,3,3) E_8 : 10(5,5,3,3,3,3,3,3,3,3)

E_9 : 10(5,4,4,3,3,3,3,3,3,3) E_{10} : 10(5,4,4,3,3,3,3,3,3,3) E_{11} : 10(5,4,4,3,3,3,3,3,3,3)

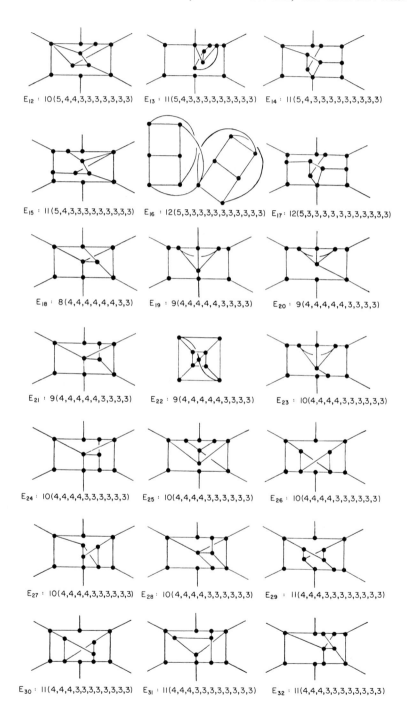

E_{12} : 10(5,4,4,3,3,3,3,3,3) E_{13} : 11(5,4,3,3,3,3,3,3,3,3) E_{14} : 11(5,4,3,3,3,3,3,3,3,3)

E_{15} : 11(5,4,3,3,3,3,3,3,3,3) E_{16} : 12(5,3,3,3,3,3,3,3,3,3,3) E_{17} : 12(5,3,3,3,3,3,3,3,3,3,3)

E_{18} : 8(4,4,4,4,4,4,3,3) E_{19} : 9(4,4,4,4,4,3,3,3,3) E_{20} : 9(4,4,4,4,4,3,3,3,3)

E_{21} : 9(4,4,4,4,4,3,3,3,3) E_{22} : 9(4,4,4,4,4,3,3,3,3) E_{23} : 10(4,4,4,4,3,3,3,3,3,3)

E_{24} : 10(4,4,4,4,3,3,3,3,3,3) E_{25} : 10(4,4,4,4,3,3,3,3,3,3) E_{26} : 10(4,4,4,4,3,3,3,3,3,3)

E_{27} : 10(4,4,4,4,3,3,3,3,3,3) E_{28} : 10(4,4,4,4,3,3,3,3,3,3) E_{29} : 11(4,4,4,3,3,3,3,3,3,3,3)

E_{30} : 11(4,4,4,3,3,3,3,3,3,3,3) E_{31} : 11(4,4,4,3,3,3,3,3,3,3,3) E_{32} : 11(4,4,4,3,3,3,3,3,3,3,3)

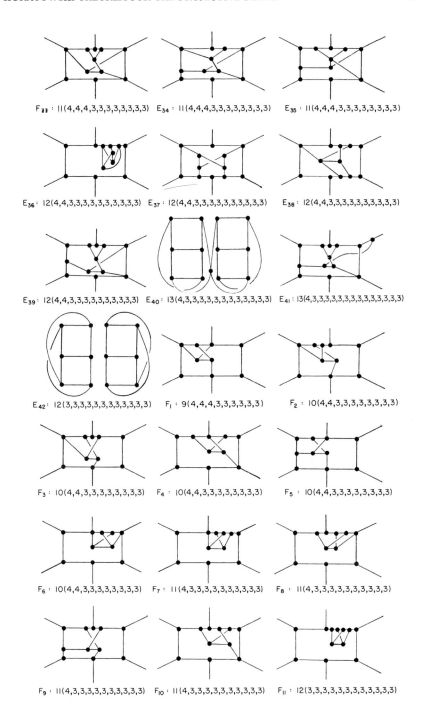

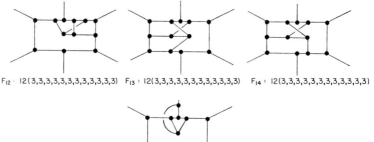

F_{12} : 12(3,3,3,3,3,3,3,3,3,3,3) F_{13} : 12(3,3,3,3,3,3,3,3,3,3,3) F_{14} : 12(3,3,3,3,3,3,3,3,3,3,3)

G : 10(3,3,3,3,3,3,3,3,3,3)

REFERENCES

1. H. H. Glover and J. P. Huneke, Cubic irreducible graphs for the projective plane, *Discrete Math.* **13** (1975) 341–355.
2. H. H. Glover and J. P. Huneke, Graphs with bounded valency that do not embed in the projective plane, *Discrete Math.* **18** (1977) 155–165.
3. H. H. Glover and J. P. Huneke, The set of irreducible graphs for the projective plane is finite, *Discrete Math.* **22** (1978) 243–256.
4. H. H. Glover and J. P. Huneke, Embedding graphs with at most 8 vertices, in preparation.
5. H. H. Glover, J. P. Huneke, and C. S. Wang, 103 graphs which are irreducible for the projective plane, *J. Combinatorial Theory Ser. B*, to appear.
6. K. Kuratowski, Sur le probleme des courbes gauches en topologie, *Fund. Math.* **15** (1930) 271–283.
7. M. Milgram, Irreducible graphs—Part 2, *J. Combinatorial Theory Ser. B* **14** (1973) 7–45.

AMS 05C10

Henry H. Glover and
J. Philip Huneke
THE OHIO STATE UNIVERSITY
COLUMBUS, OHIO

Chin San Wang
CHRISTOPHER NEWPORT COLLEGE
NEWPORT NEWS, VIRGINIA

On F-Hamiltonian Graphs

ROLAND HÄGGKVIST

Introduction

In this paper we shall study questions of the type: "Suppose that we are given a graph G and a set F of independent paths in G. What conditions should be imposed on G and F in order for F to be contained in a hamiltonian cycle or path in G?"

This question can of course never be fully answered, but one might look for partial characterizations. It is the purpose of this note to present some results of this flavour, with emphasis on the case where F is a 1-factor in G.

In order to summarize the results obtained we make some definitions. (It will be implicitly understood that we only consider graphs of order at least four.) We say that G is F-*hamiltonian* (F-*semihamiltonian*) if

 (i) F is a set of independent paths in G;
 (ii) F is contained in a hamiltonian cycle (path) in G. Such a cycle (path) is an F-*hamiltonian cycle* (*path*).

By a HAC-graph (abbreviation for *hamiltonian alternating cycle* graph) we mean a graph with a 1-factor and where every 1-factor is contained in a hamiltonian cycle in G. A HAP-graph (*hamiltonian alternating path* graph) is similarly defined for hamiltonian paths.

Call G a c_k-*graph* if

$$d(x) + d(y) \geq n + k$$

for every pair of nonadjacent vertices x, y in G. Here as in the rest of the paper, n is the order of G.

It is well known (Ore [5]) that every c_0-graph is hamiltonian, and every c_{-1}-graph is semihamiltonian (i.e., contains a hamiltonian path). A result of Kronk [2] contains the statement that every c_k-graph is k-line-hamiltonian (for $0 \leq k$). In particular, every $c_{n/2}$-graph G of even order is a HAC-graph. This is far from best possible, however.

In this paper it is shown that every c_1-graph of even order is a HAC-graph, and every c_{-1}-graph of even order is a HAP-graph. Both these statements are sharp. These results are generalized to Ore-type conditions for G to be F-hamiltonian for general F.

We also study another class of conditions on F and G ensuring G to be F-hamiltonian.

We specialize to the case where F is a 1-factor in G and assume that for every edge $e = (x, y)$ in F we have

Case 1 $d(x) + d(y) \geq n - 1$.

Case 2 $d(x) + d(y) \geq 4n/3 - 1$.

In case 1 we conclude that G is F-semihamiltonian, in case 2 that G is F-hamiltonian.

Definitions

The terminology is fairly standard; see Harary [1] for concepts not defined here or in the introduction.

We denote the vertex set of a graph G by $V(G)$ and the edge set by $E(G)$. The *order* of G is the cardinality of $V(G)$. When it is implicitly understood that G is a graph we put $|G| = |V(G)|$. If E is a set of edges then $V(E)$ denotes the set of vertices incident with E.

$d_T(x)$ denotes the degree of x in T. If T is a subgraph of G, and x belongs to $G - T$, then $d_T(x)$ is the number of edges between x and T in G. Here x can be a vertex or an edge (x_1, x_2), say, in which case $d_T(x) = d_T(x_1) + d_T(x_2)$. $E(A, B)$ denotes the set of edges between A and B in G.

A graph G is k-*line-hamiltonian* if any set F of independent paths is contained in a hamiltonian cycle whenever $|E(F)| \leq k$. An F-*alternating cycle* (*path*) is a cycle (path) using an edge of F whenever it uses one of its end vertices. Whenever a cycle $S: x_1 - x_2 \cdots x_m - x_1$ is given the indices are counted modulo m (i.e., modulo the length of the cycle).

Meyniel's Theorem

The results in this paper heavily depend on Meyniel's theorem, as now stated. The original proof of Meyniel has been shortened by Overbeck-Larisch [6] and Thomassen and Bondy [7].

Theorem A (Meyniel [4]) *Let D be a strongly connected digraph of order n fulfilling*
$$d(x) + d(y) \geq 2n - 1$$
for every pair of nonadjacent vertices x, y in D. Then D is hamiltonian.

Theorem B *Let D be a digraph of order n fulfilling*
$$d(x) + d(y) \geq 2n - 3$$
for every pair of nonadjacent vertices x, y in D. Then D contains a hamiltonian path.

Theorem B is an immediate corollary to Theorem A (join a new vertex z to every vertex in D by double edges and apply Theorem A). We note that D in Theorem B need not be strongly connected.

Main Theorems and Proofs

In this section we consider the case where F is a 1-factor.

Theorem 1 *Let F be a 1-factor in a c_1-graph G. Then G is F-hamiltonian.*

Remark Hence every c_1-graph of even order is a HAC-graph.

Proof of Theorem 1 Let B be a bipartite graph with bipartition (A_1, A_2) fulfilling

(i) $F \subset B \subset G$ and
(ii) $|E(B)|$ is maximal among graphs fulfilling (i).

By (i) $|A_1| = |A_2| = n/2$ and by (ii) $E(B) = E(A_1, A_2)$.
Put $G_1 = G - E(B)$ and $B_1 = B - E(F)$. We note some properties of B. Let e be an edge in F. Then
$$d_{B_1}(e) \geq d_{G_1}(e). \tag{1}$$

Proof of (1) Let (x, y) be the edge e with $x \in A_1$. We define another bipartite graph, B', with bipartition (A_1', A_2') as follows. Put $A_1' = (A_1 - x) \cup \{y\}$, $A_2' = (A_2 - y) \cup \{x\}$ and $E(B') = E(A_1', A_2')$. We note that
$$|E(B')| = |E(B)| + d_{G_1}(e) - d_{B_1}(e).$$
By (ii) $|E(B')| \leq |E(B)|$, and hence (1) holds.

Let $e = (x, x')$, $f = (y, y')$, $x, y \in A_1$ be two edges in F fulfilling $(x, y') \notin E(B)$ and $(y, x') \notin E(B)$. Then

$$d_{B_1}(e) + d_{B_1}(f) \geq n - 1. \tag{2}$$

Proof of (2) By (1)

$$d_{B_1}(e) + d_{B_1}(f) \geq d_{G_1}(e) + d_{G_1}(f).$$

Hence

$$2(d_{B_1}(e) + d_{B_1}(f)) \geq d_{G_1}(e) + d_{G_1}(f) + d_{B_1}(e) + d_{B_1}(f)$$
$$= d(e) + d(f) - 4 \geq 2n - 2$$

since $(x, y') \notin D(G)$ and $(y, x') \notin E(G)$, by the maximality of $|E(B)|$. (2) follows immediately.

Direct the edges of B from A_2 to A_1 and contract along the edges of F. Let the resulting digraph be D. For $e \in E(F)$ we denote the corresponding vertex in D by e. We note that for every vertex e in D we have

$$d_D(e) = d_{B_1}(e). \tag{3}$$

Since the vertices e and f in D are nonadjacent in D if and only if the edges e and f fulfill the assumptions of (2), we immediately deduce from (2) that D has the property

$$d_D(e) + d_D(f) \geq 2|V(D)| - 1 \tag{4}$$

for every pair of nonadjacent vertices e, f in D. Using Theorem B we conclude that D contains a hamiltonian path:

$$P: x_1 - x_2 \cdots x_m \quad \text{(where} \quad m = n/2).$$

Let the edge x_i in F be (v_{2i-1}, v_{2i}) for $i = 1, 2, \ldots, m$, with $v_{2_i - 1} \in A_1$. Then P corresponds to the F-hamiltonian path

$$P': v_1 - v_2 - \cdots - v_{2m} \quad \text{in} \quad B.$$

If $(v_1, v_{2m}) \in E(G)$, then G is F-hamiltonian; hence without loss of generality, $(v_1, v_{2m}) \notin E(G)$ which gives

$$d(v_1) + d(v_{2m}) \geq n + 1$$

by hypothesis.
Consider the sum

$$\sum_{i=2}^{2m} (|E(v_1, v_i)| + |E(v_{2m}, v_{i-1})|) = d(v_1) + d(v_{2m}) \geq n + 1.$$

Since there are fewer than n summands, at least one, the jth, say, is greater than one, and hence

$$(v_1, v_j) \in E(G) \quad \text{and} \quad (v_{2m}, v_{j-1}) \in E(G).$$

If j is odd, $j = 2k + 1$, then the edge (v_{j-1}, v_j) does not belong to F and hence

$$S: v_1 - v_2 - \cdots - v_{2k} - v_{2m} - v_{2m-1} - \cdots - v_{2k+1} - v_1$$

is an F-hamiltonian cycle in G.

If j is even, $j = 2k$, then the edges (v_1, v_{2k}) and (v_{2k-1}, v_{2m}) belong to B, and hence the edges $e_1 = (x_k, x_1)$ and $e_2 = (x_m, x_k)$ belong to D. Consequently D is strongly connected. This, together with (4), implies that D fulfills the assumptions in Theorem A and hence is Hamiltonian. Thus B, and hence G, is F-hamiltonian. ∎

Theorem 2 *Let F be a 1-factor in a graph G. Suppose that*

$$d(e) + d(f) \geq 2n - 3$$

for every pair of edges $e = (x, x')$, $f = (y, y')$ for which either

$$(x, y') \notin E(G) \text{ and } (y, x') \notin E(G) \text{ or } (x, y) \notin E(G) \text{ and } (x', y') \notin E(G).$$

Then G is F-semihamiltonian.

Proof of Theorem 2 The proof of Theorem 1 carries over with some minor changes. In (2) we replace $n - 1$ by $n - 3$ and in (4) $2|V(D)| - 1$ by $2|V(D)| - 3$. The proof of (2) now runs as follows:

$$2(d_{B_1}(e) + d_{B_1}(f)) \geq 2n - 7$$

which implies that

$$d_{B_1}(e) + d_{B_1}(f) \geq n - 3$$

since the left side is an integer. P' is the F-hamiltonian path which we are looking for. ∎

Theorem 3 *Let F be a 1-factor in a c_{-1}-graph G. Then G is F-semihamiltonian.*

Theorem 4 *Let G be a graph with a 1-factor F such that*

$$d(f) \geq n - 1$$

for every edge f in F. Then G is F-semihamiltonian.

Remark Theorems 3 and 4 are special cases of Theorem 2.

Theorem 5 *Let G be a graph with a 1-factor F such that $d(f) \geq 4n/3 - 1$ for every edge f in F. Then G is F-hamiltonian.*

Proof of Theorem 5 By contradiction. Assume that the theorem is false and that G is a graph with a minimal number of vertices n for which the theorem fails for the 1-factor F. The theorem is clearly true for $n = 4$. Let e be an edge in F. Since e is incident with at least $4n/3 - 3 > 2(n/2 - 1)$ other edges in G, there exists an edge f in $F - e$ such that $V(e)$ is joined to $V(f)$ by at least three edges; consequently G contains an F-alternating cycle. Let $S: x_1 - x_2 \cdots x_{2m} - x_1$ be a longest F-alternating cycle in G, where the edges (x_{2i}, x_{2i+1}) belong to F for $i = 1, 2, \ldots, m$. By assumption $|S| < n$.

Put $T = G - S$, $F_1 = T \cap F$, and $F_2 = S \cap F$. We list some properties of G.

Let e be an edge in F_1. Then

$$d_S(e) \leq 2m \quad (=|S|). \tag{5}$$

Proof of (5) Let e be the edge (z_1, z_2). Since S is maximal we have

$$|E(x_{2i-1}, z_1)| + |E(x_{2i}, z_2)| \leq 1$$

and

$$|E(x_{2i-1}, z_2)| + |E(x_{2i}, z_1)| \leq 1$$

for $i = 1, 2, \ldots, m$.

This means that

$$d_s(e) = \sum_{i=1}^{m} (|E(x_{2i-1}, z_1)| + |E(x_{2i}, z_2)|)$$

$$+ \sum_{i=1}^{m} (|(E(x_{2i-1}, z_2)| + |E(x_{2i}, z_1)|) \leq 2m.$$

This proves (5).

Every edge e in F_1 fulfills

$$d_T(e) \geq \frac{4|T|}{3} - 1. \tag{6}$$

Proof of (6)

$$d_T(e) \geq \frac{4n}{3} - 1 - d_S(e) \geq \frac{4n}{3} - 1 - |S| \geq \frac{4|T|}{3} - 1$$

since $n - |T| = |S|$.

$$|S| \geq \frac{n}{2}. \tag{7}$$

Proof of (7) By the minimality of G, T is F_1-hamiltonian [by (6)]. Hence $|T| \leq |S|$ since $|S|$ was maximal. This proves (7).

$$|T| \geq \frac{n}{3} + 1. \qquad (8)$$

Proof of (8) By (5), $d_S(e) \leq |S|$ for $e \in E(F_1)$. Hence, for such an e

$$d_T(e) \geq \frac{4n}{3} - 1 - |S|,$$

i.e.,

$$2(|T| - 1) \geq d_T(e) \geq \frac{4n}{3} - 1 - |S|.$$

This gives, since $|S| = n - |T|$, that

$$2|T| + n - |T| \geq \frac{4n}{3} + 1$$

which proves (8).

Let z_1 and z_2 be two vertices in T. Then

$$\begin{array}{c} T \text{ contains an } F_1\text{-hamiltonian cycle } S_1 \text{ using an} \\ F_1\text{-alternating path } P_1 : z_1 - \cdots - z_2 \end{array} \qquad (9)$$

Proof of (9) The statement is trivial, by (6), if $(z_1, z_2) \in E(F_1)$. Hence, without loss of generality, there exist independent edges

$$e_1 = (z_1, y_1) \in E(F_1) \quad \text{and} \quad e_2 = (z_2, y_2) \in E(F_1).$$

Put $G_1 = T - z_1 - y_1$. Let the edges in $F_1' = F_1 - e_1$ be denoted by $e_2, e_3, \ldots, e_k$ where $2k = |T|$. F_1' is a 1-factor of G_1. We have $d_{G_1}(e_i) \geq d_T(e_i) - 4$ for $i = 2, 3, \ldots, k$. Since $|S| \geq n/2$, $|T| \leq n/2$ follows. Moreover $n \geq 4$, $|T|$ even, and (8) imply $|T| \geq 4$; hence by (7) $n \geq 8$ holds. Therefore,

$$d_{G_1}(e_i) \geq \frac{4n}{3} - 1 - |S| - 4 \geq \frac{n}{3} - 5 + |T| \geq \frac{4|T| + (n/2)}{3} - 5$$

$$\geq \frac{4(|T| - 2)}{3} - 1$$

We conclude that G_1 contains an F_1'-hamiltonian cycle

$$S_1' : v_1 - v_2 \cdots v_{2k-2} - v_1.$$

Without loss of generality, $z_2 = v_{2k-2}$ and $y_2 = v_1$. This means that the edges $(v_{2i}, v_{2i+1}) \in E(F_1')$ for $i = 1, \ldots, k-1$. We consider the inequalities

$$\frac{4n}{3} - 1 - |S| - 2 \le d_{G_1}(e_1)$$

$$= d_T(e_1) - 2$$

$$= \sum_{i=1}^{k-1}(|E(v_{2i-1}, y_1)| + |E(v_{2i}, z_1)|)$$

$$+ \sum_{i=1}^{k-1}(|E(v_{2i-1}, z_1)| + |E(v_{2i}, y_1)|)$$

$$\le \sum_{i=1}^{k-1}(|E(v_{2i-1}, y_1)| + |E(v_{2i}, z_1)|) + 2k - 2.$$

Rearranging the inequality we obtain

$$\frac{4n}{3} - |S| - 1 - 2k \le \sum_{i=1}^{k-1}(|E(v_{2i-1}, y_1)| + |E(v_{2i}, z_1)|).$$

Since $2k + |S| = |T| + |S| = n$ and $|T| \le n/2$, i.e., $|T|/2 < n/3$ we obtain

$$k - 1 < \frac{n}{3} - 1 \le \sum_{i=1}^{k-1}(|E(v_{2i-1}, y_1)| + |E(v_{2i}, z_1)|).$$

Hence one of the summands, the jth, say, equals two which means that

$$(v_{2j-1}, y_1) \in E(G) \quad \text{and} \quad (v_{2j}, z_1) \in E(G).$$

Then T contains the F_1-hamiltonian cycle

$$S_1': v_{2k-2} - v_1 - v_2 \cdots v_{2j-1} - y_1 - z_1 - v_{2j} - v_{2j+1} \cdots v_{2k-2}$$

which contains the F_1-alternating path

$$P_1: z_1 - y_1 - v_{2j-1} - v_{2j-2} \cdots v_2 - y_2 - z_2$$

as claimed. Hence (9) holds.

There exists an F_2-hamiltonian cycle S_2 in $G_2 = G - T$ containing two consecutive vertices x, y such that

(i) $(x, y) \notin E(F_2)$,
(ii) $(x, z_1) \in E(G)$ for some $z_1 \in V(T)$ and $(y, z_2) \in E(G)$ (10)
for some $z_2 \in V(T) - z_1$.

Proof of (10) Because of (9) and the maximality of S, S does not contain any such pair of vertices. Let A denote the set of vertices in G which have degree $\ge [2n/3]$. By assumption every edge in F is incident with a vertex of A.

Since every vertex in $A \cap V(S)$ is joined to T by at least two edges [this follows from the fact that $|S| \leq [2n/3] - 1$ by (8)], we may assume that

$$A \cap V(S) = \{x_{2i} : i = 1, 2, \ldots, m\}, \tag{11}$$

$$d((x_{2j-1}, x_{2j})) \geq (4n/3) - 1 \quad \text{for some} \quad j \in \{1, 2, \ldots, m\}. \tag{12}$$

Proof of (12) We have

$$m\left(\frac{4n}{3} - 1\right) \leq \sum_{e \in E(F_2)} d(e) = \sum_{i=1}^{m} d((x_{2i}, x_{2i+1})) = \sum_{i=1}^{m} d((x_{2i-1}, x_{2i})).$$

Since we have m summands at least one, the jth say, fulfills (12). This proves (12).

Without loss of generality, $j = 1$. We consider the following string of inequalities

$$\sum_{i=2}^{m} (|E(x_1, x_{2i-1})| + |E(x_2, x_{2i})|)$$

$$= d_{G_2}((x_1, x_2)) - \sum_{i=1}^{m} (|E(x_1, x_{2i})| + |E(x_2, x_{2i-1})|)$$

$$\geq (4n/3) - 1 - |T| - 2m \quad \text{by (12) and since } S \text{ does not fulfill (10)}$$
$$\geq (n/3) - 1 \quad \text{since } 2m = |S| = n - |T|$$
$$> m - 1 \quad \text{since } |S| \leq [2n/3] - 1.$$

The sum has $m - 1$ terms, each an integer a_i, $0 \leq a_i \leq 2$. The sum is greater than $m - 1$, i.e., at least m. Therefore, there exists a $q \in \{2, 3, \ldots, m\}$ such that

$$|E(x_1, (x_{2q-1}))| + |E(x_2, x_{2q})| = 2,$$

i.e., $(x_1, x_{2q-1}) \in E(G)$ and $(x_2, x_{2q}) \in E(G)$. We know that

$(x_{2q-1}, x_{2q}) \notin E(F_2)$ and $(x_1, x_2) \notin E(F_2)$ and $x_2 \in A$, $x_{2q} \in A$.

Hence the cycle

$$S_2 : x_1 - x_{2q-1} - x_{2q-2} - \cdots x_2 - x_{2q} - x_{2q+1} \cdots - x_1$$

fulfills (10) with $x = x_2$ and $y = x_{2q}$.
This proves (10).

Combining (10) and (9) we find an F-alternating cycle S_3 in G [consisting of $S_2 - (x, y)$, P_1, and the edges (x, z_1) and (y, z_2)] which is longer than S. This contradicts the choice of S, and proves the theorem. ∎

Generalisations

In this section we deal with the case where F is an arbitrary set of independent paths. Recall that n is the order of G. We state without proof the following.

Lemma (Kronk [2]) *Let E be a set of k independent edges in a c_k-graph G. Then E is contained in a hamiltonian cycle in G.*

Put

$$f(n, k) = \min\{k, n - 2k + 1\}, \qquad g(n, k) = \min\{k - 1, n - 2k - 1\}.$$

Theorem 6 *Let E be a set of k independent edges in a graph G.*

(a) *If G is a $c_{f(n,k)}$-graph then G is E-hamiltonian.*
(b) *If G is a $c_{g(n,k)}$-graph then G is E-semihamiltonian.*

Proof of Theorem 6 We only prove (a). [The proof of (b) is similar.] Induction is used on the number of vertices in G. The theorem is true if $n = 4$. Assume that the theorem is true for all graphs of order $< n$ and let G be a graph of order n, and let E be a set of k independent edges in G. Without loss of generality we may assume that $n > 2k$, since otherwise Theorem 1 applies. Moreover, by Lemma 1, we may assume that $f(n, k) < k$. This means that

$$f(n, k) - 1 = f(n - 1, k). \tag{13}$$

We know that G is connected, by Ore's theorem, and $n < 2k$ by assumption. Hence, in G, a path $P: x_1 - x_2 - x$ exists for which $(x_1, x_2) \in E$ while x is a vertex which is not incident with E. Let G' denote the graph obtained from $G - x_2$ by adding the edge (x_1, x). Put

$$E' = (E - (x_1, x_2)) \cup \{(x_1, x)\}.$$

Then G' is a $c_{f(n-1,k)}$-graph of order $n - 1$, and E' is a set of k independent edges in G'. Hence E' is contained in a hamiltonian cycle in G', which implies that E is contained in a hamiltonian cycle in G. This proves the theorem by induction. ∎

Corollary 1 *Any set E of independent edges in a $c_{n/3}$-graph G is contained in a hamiltonian cycle.*

Corollary 2 *Let F be a set of k nontrivial independent paths $P_1, P_2, \ldots, P_k$ each of length at least one and having altogether exactly q interior vertices in a graph G.*

(a) *If G is a $c_{f(n,k)+q}$-graph then G is F-hamiltonian.*
(b) *If G is a $c_{g(n,k)+q}$-graph then G is F-semihamiltonian.*

Proof We apply Theorem 6 on the graph obtained from G by replacing each path P_i by an edge while deleting all the interior vertices of F. ∎

Constructions

In this section we present constructions of non-F-hamiltonian graphs for the case where F does not contain any interior vertices. These constructions show that Theorems 1–6 are fairly sharp.

We now list some examples.

1. There exist c_0-graphs of even order which are not HAC-graphs. One such example is the graph G of order $4m + 2$ for $m = 1, 2, \ldots$ defined by $G - E(F) = K_{2m, 2m+2}$ where F is a 1-factor in G. G is not F-hamiltonian.

2. If we define G to fulfill $G - E(F) = K_{2m, 2m+4}$ for a 1-factor F in G, then we have a c_{-2}-graph which is not F-semihamiltonian. This example also shows that Theorem 4 cannot be sharpened in general.

3. For every $m > 0$ there exists a nonhamiltonian graph G of order $n = 6m$, containing a 1-factor F such that

$$d(f) \geq \frac{4n}{3} - 2 \quad \text{for every} \quad f \in E(F).$$

One example, where every 1-factor has this property, is the graph G obtained from $G_1 = K_{2m} \cup 2mK_1$ and $G_2 = K_{2m}$ by joining every vertex in G_1 to every vertex in G_2. This graph is nonhamiltonian since $G - V(G_2)$ has $> |V(G_2)|$ components.

4. For $k = m + s$ there exists a $c_{f(n,k)-2}$-graph (a $c_{g(n,k)-1}$-graph) of order $n = 3m + 2s + 1$ ($n = 3m + 2s + 2$) with the property that some set of $m + s$ independent edges fail to be contained in a hamiltonian cycle (path).

Take G to be a graph with vertex set

$$V(G) = \{v_i : 1 \leq i \leq n\}$$

and edge set

$$E(G) = \{(v_i, v_j): 1 \leq i \leq 2m + s \wedge 1 \leq i < j \leq n\}.$$

Let E be the set $\{(v_{2i-1}, v_{2i}): 1 \leq i \leq m + s\}$. Then G is a $c_{f(n,k)-2}$-graph (a $c_{g(n,k)-1}$-graph). E is not contained in any hamiltonian cycle S (hamiltonian path S) since else

$$S_1 = S - \{(v_{2i-1}, v_{2i}): 1 \leq i \leq m\} - \{(v_j : 2m \leq j \leq 2m + s\}$$

would have at most $m + s$ components ($m + s + 1$ components) contradicting the fact that $V(S_1)$ is a set of $m + s + 1$ independent vertices ($m + s + 2$ independent vertices) in G.

The Bipartite Case

The proof of Theorems 1 and 2 used the fact that conditions for a bipartite graph B, with a 1-factor F, ensuring B to be F-hamiltonian coincide with conditions for a digraph to be hamiltonian. This fact has also been noted by other authors, e.g., Las Vergnas [3] who, using this, proved

Theorem 7 (Las Vergnas [3]) *Let B be a bipartite graph with bipartition (A_1, A_2), such that $|A_1| = |A_2| = n$. If, for each pair x, y of nonadjacent vertices with $x \in A_1$, $y \in A_2$, we have*

Case 1 $d_B(x) + d_B(y) \geq n + 2$.
Case 2 $d_B(x) + d_B(y) \geq n + 1$.

Then, in case 1, B is a HAC-graph, in case 2, B is a HAP-graph.

Remark It should be noted that this theorem amounts to saying that the Ore-type conditions for a bipartite graph B to be 1-line-hamiltonian coincide with the Ore-type conditions for B to be a HAC-graph; the same can be said of semihamiltonian versus HAP-graphs.

It seems likely that this situation continues, i.e., that degree-conditions on a graph G of even order to be 1-line-hamiltonian imply that G is a HAC graph, etc., but so far no further progress has been made in this direction.

A Conjecture

It is tempting to flood the market with conjectures related to the concept of F-hamiltonian graphs and its generalisations. I have been able to restrict myself to the following:

Conjecture *Every set of independent edges in a c_1-graph is contained in a cycle.*

However, I hope to return to this subject.

REFERENCES

1. F. Harary, "Graph Theory," Addison-Wesley, Reading, Massachusetts, 1969
2. H. V. Kronk, Variations on a theorem of Posa, *in* "The Many Facets of Graph Theory" (G. Chartrand and S. F. Kapoor, eds.), pp. 193–197. Springer-Verlag, Berlin and New York, 1969.
3. M. Las Vergnas, Thesis, University of Paris, Paris, 1972.
4. M. Meyniel, Une condition sufficante d'existence d'un circuit Hamiltonien dans un graph orienté, *J. Combinatorial Theory Ser. B* **14** (1973) 137–147.

5. O. Ore, Note on Hamilton circuits, *Amer. Math. Monthly* **67** (1960) 55.
6. M. Overbeck-Larisch, Hamiltonian paths in oriented graphs, *J. Combinatorial Theory Ser. B* **21** (1976) 76–80.
7. C. Thomassen and J. A. Bondy, A short proof of Meyniel's theorem, *Discrete Math.* **19** (1977) 195–197.

AMS 05C35

DEPARTMENT OF MATHEMATICS
UNIVERSITY OF UMEÅ
UMEÅ, SWEDEN

Rochromials and the Colourings of Circuits

DICK WICK HALL

1. An Example

We consider a 7-circuit in the plane, in which the vertices are labeled with the positive integers from 1 to 7. Nineteen additional nonisomorphic graphs can be obtained by adding single edges to this circuit, subject to the conditions: (i) all additional edges are outside the circuit, (ii) each additional edge joins two vertices of the circuit, (iii) the graph consisting of the circuit and the additional edges is planar.

We use the symbol E_i, $1 \leq i \leq 20$, to indicate the edges which have been added to our circuit with the stipulation that $E_1 = (00)$ means that no edges have been added. Table 1 gives the definitions of the E_i which we shall use.

TABLE 1

i	E_i	i	E_i	i	E_i	i	E_i
1	(00)	6	(13, 16)	11	(13, 14, 16)	16	(13, 16, 35)
2	(13)	7	(13, 46)	12	(13, 14, 46)	17	(13, 14, 15, 16)
3	(14)	8	(13, 47)	13	(13, 14, 47)	18	(13, 14, 15, 57)
4	(13, 14)	9	(14, 15)	14	(13, 14, 57)	19	(13, 14, 16, 46)
5	(13, 15)	10	(13, 14, 15)	15	(13, 15, 35)	20	(13, 14, 47, 57)

We now choose three colours, denoted by A, B, and C, and colour the vertices of our 7-circuit in such a way that going around the circuit (in the direction agreeing with the labeling) the colours appear in the order $ABABCAC$. It is clear that we may do this in seven ways, depending upon which vertex is attached to our starting colour A. Once this starting point has been decided upon, the addition of outside edges will destroy the colouring if two vertices with the same colour are joined by such an edge. Table 2 lists our seven colourings together with a number of the E_i. This table lists a 0 or 1 after the colouring according as E_i does or does not destroy the colouring. The extremely simple task of completing Table 2 is left to the reader.

TABLE 2

1	2	3	4	5	6	7	E_1	E_2	E_3	E_5	E_7	E_8
A	B	A	B	C	A	C	1	0	1	0	0	0
C	A	B	A	B	C	A	1	1	1	1	1	0
A	C	A	B	A	B	C	1	0	1	0	0	0
C	A	C	A	B	A	B	1	0	1	0	0	0
B	C	A	C	A	B	A	1	1	1	1	1	1
A	B	C	A	C	A	B	1	1	0	1	0	1
B	A	B	C	A	C	A	1	0	1	0	0	0

For each i, $1 \leq i \leq 20$, we define the nonnegative integer t_i as the sum of the column E_i in the completed Table 2. This gives us a vector

$$T = \langle t_i \rangle$$

which we call the vector determined by the colouring of the circuit we have chosen. For the particular colouring used in our example

$$T = \langle 7, 3, 6, 2, 3, 0, 2, 2, 5, 2, 0, 2, 1, 1, 0, 0, 0, 1, 0, 0 \rangle.$$

When we colour the 7-circuit we insist that no two adjacent vertices have the same colour. Subject to this stipulation, it will be clear from the results in the remainder of this paper that the following theorem holds.

Theorem 1 *For any colouring of the 7-circuit, the vector T determined by this colouring has rank 6. A basis for this vector consists of t_1, t_2, t_3, t_5, t_7, and t_8, and the following equations hold*

$$t_4 = -t_1 + t_2 + t_3$$
$$t_9 = -t_1 + 2t_3$$
$$t_{11} = -t_3 + 2t_5$$
$$t_{13} = -t_1 + t_3 + t_8$$
$$t_{15} = -t_3 + 2t_5$$
$$t_{17} = -t_2 - t_3 + 3t_5$$
$$t_{19} = t_1 - 2t_2 - 2t_3 + 3t_5 + t_7$$

$$t_6 = t_2 - t_3 + t_5$$
$$t_{10} = -t_1 + t_3 + t_5$$
$$t_{12} = -t_2 + t_5 + t_7$$
$$t_{14} = -t_2 + t_7 + t_8$$
$$t_{16} = t_1 - t_2 - 2t_3 + 2t_5 + t_7$$
$$t_{18} = -2t_2 + t_5 + t_7 + t_8$$
$$t_{20} = -2t_2 + t_7 + 2t_8$$

Two different colourings of the 7-circuit can determine the same vector. For example, each of the colourings

$$BACBDEA \quad \text{and} \quad BACDBEA$$

determines the vector T in which $t_1, t_2, t_3, t_5, t_7, t_8$ are, respectively, 7, 6, 6, 5, 5, 5.

2. Constrained Chromials

We suppose that we are given a planar graph containing a 7-circuit, and that there are no edges or vertices of the graph inside this circuit. We assume that the vertices of the circuit have been coloured, perhaps with the colouring used in our example in Section 1. The number of ways of colouring the vertices of the entire graph, using some or all of λ colours, is a polynomial in λ, known as the *constrained chromial* of the graph. By a *free chromial* we mean the same thing without the constraint placed upon the circuit of having the assigned colouring.

The fundamental problem associated with constrained chromials is that of expressing the constrained chromial of a planar graph in terms of free chromials. This problem was solved for graphs containing a 4-circuit or a 5-circuit, and partially solved for graphs containing a 6-circuit, by Birkhoff and Lewis [1], using an extremely laborious method. The solution for the 6-circuit was completed by Hall and Lewis [3], using brute force.

There are 162 constrainted chromials for the 7-circuit. One of these has all seven colours different (in which case the vector T introduced in Section 1 consists of all 7's), and the remaining 161 divide themselves into 23 groups of seven each. The sum of the constrained chromials in each of these groups is called a *constrained rochromial*. The sum of the corresponding seven free chromials in each group is called a *free rochromial*. Hall [2] has solved the problem of expressing constrained rochromials in terms of free rochromials. In this paper we present a greatly simplified method of solving the same problem. This simplified method is interesting because it seems likely to lead to solutions for the 8-circuit and 9-circuit.

Extensive tables of free chromials and free rochromials for graphs containing n-circuits, $3 \leq n \leq 9$, have been prepared for the author by Robert King. Those for free rochromials for $n = 6$ and $n = 7$ are given in the Appendix to this paper.

The chromial of a connected graph having p vertices is a polynomial of degree p in λ, where λ is the number of colours. All of our graphs will be connected and have more than two vertices. Thus all chromials and rochromials which we consider will be divisible by $\lambda(\lambda - 1)$. We divide out this common factor and express all chromials and rochromials in terms of

$$y = \lambda - 1.$$

This amounts precisely to expressing our chromials in the Tutte form with the obvious common factor removed.

From Theorem 1, we see that given any colouring of a 7-circuit, the vector T determined by this colouring has a basis consisting of $t_1, t_2, t_3, t_5, t_7, t_8$. From Table 1 we see that this corresponds to adding the following edges to our 7-circuit

$$E_1 = (00), \quad E_2 = (13), \quad E_3 = (14),$$
$$E_5 = (13, 15), \quad E_7 = (13, 46), \quad E_8 = (13, 47).$$

From these six graphs we obtain the six basic contracted graphs we shall use for the 7-circuit. These are given in the following tabulation.

Contraction number	Vertices to be identified
1	None
2	1 and 3
3	1 and 4
4	1 and 3; also 1 and 5
5	1 and 3; also 4 and 6
6	1 and 3; also 4 and 7

As an illustration of how we shall use these ideas, we consider the graph G formed by the 7-circuit and the four additional outside edges 13, 14, 47, 57. We wish to determine the free rochromial corresponding to contraction number 3, that is, to the graph obtained by adding these four additional edges after identifying the vertices 1 and 4. We must, of course, find the free chromials of the graphs we get by leaving G unrotated, and by rotating it 1, 2, 3, 4, 5, or 6 steps. This amounts to leaving our circuit unrotated and changing the added edges appropriately. We now list the free chromials we get, noting that the obvious common factor has been removed, and, as in our tables at the end of this paper, writing our polynomials in increasing powers of y.

Added outside edges		Constant	y	y^2	y^3	y^4
13, 14, 47, 57		0	0	0	0	0
24, 25, 51, 61		1	−4	6	−4	1
35, 36, 62, 72		7	−17	15	−6	1
46, 47, 73, 13		1	−4	6	−4	1
57, 51, 14, 24		0	0	0	0	0
16, 26, 25, 35		4	−12	13	−6	1
27, 37, 36, 46		4	−12	13	−6	1
	Sum	17	−49	53	−26	5

The sum we get is the free rochromial for the uncontracted graph 13, 14, 47, 57 with contraction number 3. This particular graph 13, 14, 47, 57 is number 20 in our tables, and its free rochromial is listed there.

For a given planar graph G containing our 7-circuit, but having no edges or vertices inside this circuit, we denote by [1], [2], [3], [4], [5], [6], the free rochromials obtained for the six basic contracted graphs obtained from G.

It needs to be pointed out that these contracted maps are not the same as those used in earlier papers by Birkhoff, Lewis, and Hall, since we are not adding edges to triangulate the interiors of the circuits which we obtain.

The method used in proving Theorem 1, and similar theorems for other circuits, is quite general. It can be best understood by examining in detail what happens for the 5-circuit. To this end we give a listing of all the free rochromials for the 5-circuit. Inasmuch as these are the rochromials for graphs of five vertices, and the appropriate quadratic factor has been divided out, none of the polynomials listed will have degree exceeding three.

	Contracted					
	No.	Edges	Constant	y	y^2	y^3
No. 1	Uncontracted graph:		00			
	1	00	-5	5	-5	5
	2	13	0	-5	5	
No. 2	Uncontracted graph:		13			
	1	00	-5	10	-10	5
	2	13	2	-6	4	
No. 3	Uncontracted graph:		13, 14			
	1	00	-5	15	-15	5
	2	13	4	-7	3	

Using the notation of Section 1, we let $E_1 = (00)$, $E_2 = (13)$, $E_3 = (13, 14)$. Colouring our 5-circuit with three colours A, B, C we have the following tabulation.

1	2	3	4	5	(00)	(13)	(13, 14)
A	B	A	B	C	1	0	0
C	A	B	A	B	1	1	1
B	C	A	B	A	1	1	0
A	B	C	A	B	1	1	0
B	A	B	C	A	1	0	0
					5	3	1

It follows that
$$t_1 = 5, \quad t_2 = 3, \quad t_3 = 1.$$

We now consider the three graphs which we have numbered 1, 2, 3, and let $[1]_i$, $[2]_i$ be, for $i = 1, 2, 3$, the free rochromials for the two basic contracted graphs obtained from G, namely, those obtained from the contractions (00) and (13). We denote by $R_i(y)$ the constrained rochromial corresponding to the colouring $ABABC$ of the 5-circuit, where we agree, as usual, that the factor $y(y + 1)$ has been removed from $R_i(y)$. Our theory tells us that there must be polynomials $f(y)$, $A(y)$, $B(y)$, each independent of the graph G, such that

$$f(y)R_i(y) = A(y)[1]_i + B(y)[2]_i \tag{1}$$

is an identity for $i = 1, 2, 3$. It is our task to determine the polynomials $f(y)$, $A(y)$, $B(y)$, and we see quite easily that this may be done by comparing coefficients.

It is evident that
$$R_i(y) = t_i(y - 1) \quad \text{for} \quad i = 1, 2, 3$$
and we try to determine the polynomials
$$f(y), A(y), B(y)$$
in such a way that each has the lowest possible degree consistent with our other requirements. We may thus assume that

$$(y - 1)f(y) = a_3 y^3 + a_2 y^2 + a_1 y + a_0,$$
$$A(y) = A_2 y^2 + A_1 y + A_0,$$
$$B(y) = B_2 y^2 + B_1 y + B_0,$$

and that
$$a_0 \neq 0.$$

Substituting in (1), making use of our table of free rochromials just given, and comparing coefficients of the constant terms, we obtain equations which may be indicated as a matrix in the following form

A_0	B_0	$a_0 t_1$	$a_0 t_2$	$a_0 t_3$
-5	0	1	0	0
-5	2	0	1	0
-5	4	0	0	1

Row reduction enables us to put this matrix in the form

A_0	B_0	$a_0 t_1$	$a_0 t_2$	$a_0 t_3$
-5	0	1	0	0
0	2	-1	1	0
0	0	1	-2	1

Thus we have the three equations

$$-5A_0 = a_0 t_1,$$
$$2B_0 = a_0(-t_1 + t_2),$$
$$0 = a_0(t_1 - 2t_2 + t_3).$$

From the last of these equations, we see that

$$t_3 = -t_1 + 2t_2. \qquad (2)$$

Now (2) is important, since it enables us to express t_3 in terms of the t_i for our two basic contracted maps. *Precisely the same type of argument, using the much larger matrices for the free rochromials for the 7-circuit, gives us the proof for Theorem 1.*

Returning to our specific graph, we have seen that

$$t_1 = 5, \qquad t_2 = 3, \qquad t_3 = 1.$$

We observe that these values satisfy (2).

Thus we have

$$A_0 = -a_0, \qquad B_0 = -a_0.$$

Now the coefficients of the linear terms we get by comparing coefficients must be $5a_1$, $3a_1$, $1a_1$. It is thus a simple matter to find that

$$A_1 = -a_1, \qquad B_1 = -a_1 + 2a_0.$$

We now make the trial assumption that $A(y)$ and $B(y)$ are linear functions and substitite in (1) to see what we get.

Graph number 1 gives

$$-5a_1 y^4 + 5a_0 y^3 - 10a_0 y^2 + 5a_1 y + 5a_0$$
$$-5a_1 y^4 + (6a_1 + 3a_0)y^3 + (-4a_1 - 6a_0)y^2 + 3a_1 y + 3a_0$$
$$-5a_1 y^4 + (12a_1 + a_0)y^3 + (-8a_1 - 2a_0)y^2 + a_1 y + a_0.$$

If we choose $a_1 = 0$, $a_0 = 1$, these expressions are divisible by 5, 3, and 1, respectively. The common quotient is

$$y^3 - 2y^2 + 0y + 1 = (y^2 - y - 1)(y - 1) = (\lambda^2 - 3\lambda + 1)(\lambda - 2)$$

which is exactly what we want. See Tutte [4].

Using the standard reduction for computing chromials and rochromials by deletion and contraction (with the understanding that the edge involved is not an edge of the coloured circuit), it follows that since (1) is true for graphs number 1, 2, 3, it is true for all planar graphs by induction on the total number of edges and vertices outside the circuit. Letting $G(\lambda)$ be the constrained rochromial for our colouring, [1] the free rochromial for G, and [2] the free rochromial for G when the added edge (13) is contracted, (1) becomes

$$(\lambda^2 - 3\lambda + 1)G(\lambda) = -[1] + (2\lambda - 3)[2].$$

3. Concluding Remarks

As the size of the circuit increases, the number of uncontracted graphs increases more rapidly than the factorial function. Considering only free rochromials, there are 20 uncontracted graphs for the 7-circuit each with six contractions. Theorem 1 enables us to solve the 7-circuit problem for rochromials, using far fewer than the 120 rochromials given by these graphs.

For the 8-circuit, yet to be conquered, there are 75 uncontracted graphs, each with 14 contractions.

The 9-circuit gives us 262 uncontracted graphs, each with 22 contractions, a total of 5764 free rochromials to be considered if we do not make use of results similar to those in Theorem 1.

Appendix

Free Rochromials in Tutte Form, Each Divided by $y(y + 1)$

The six circuit

	Contracted						
No.	Edges	Constant	y	y^2	y^3	y^4
No. 1	Uncontracted graph:	00					
1	00	6	−6	6	−6	6
2	13	0	6	−6	6	
3	14	0	6	−12	6	
4	13, 15	0	0	6		
5	13, 46	0	0	6		
No. 2	Uncontracted graph:	13					
1	00	6	−12	12	−12	6
2	13	−2	7	−10	5	
3	14	−2	10	−14	6	
4	13, 15	0	−3	3		
5	13, 46	0	−4	4		

Free Rochromials in Tutte Form, Each Divided by $y(y + 1)$ (continued)

			The six circuit				
	Contracted						
	No.	Edges	Constant	y	y^2	y^3	y^4
No. 3	Uncontracted graph:		14				
	1	00	6	−12	18	−12	6
	2	13	−2	10	−8	6	
	3	14	−4	12	−12	4	
	4	13, 15	0	0	6		
	5	13, 46	2	−2	6		
No. 4	Uncontracted graph:		13, 14				
	1	00	6	−18	24	−18	6
	2	13	−4	11	−12	5	
	3	14	−6	16	−14	4	
	4	13, 15	0	−3	3		
	5	13, 46	2	−6	4		
No. 5	Uncontracted graph:		13, 15				
	1	00	6	−18	24	−18	6
	2	13	−4	11	−11	4	
	3	14	−4	14	−16	6	
	4	13, 15	3	−6	3		
	5	13, 46	0	−2	2		
No. 6	Uncontracted graph:		13, 46				
	1	00	6	−18	24	−18	6
	2	13	−4	12	−12	4	
	3	14	−6	16	−16	6	
	4	13, 15	0	0	0		
	5	13, 46	4	−8	4		
No. 7	Uncontracted graph:		13, 14, 15				
	1	00	6	−24	36	−24	6
	2	13	−6	15	−13	4	
	3	14	−8	20	−16	4	
	4	13, 15	3	−6	3		
	5	13, 46	2	−4	2		
No. 8	Uncontracted graph:		13, 14, 46				
	1	00	6	−24	36	−24	6
	2	13	−6	16	−14	4	
	3	14	−10	22	−16	4	
	4	13, 15	0	0	0		
	5	13, 46	6	−10	4		
No. 9	Uncontracted graph:		13, 15, 35				
	1	00	6	−24	36	−24	6
	2	13	−6	15	−12	3	
	3	14	−6	18	−18	6	
	4	13, 15	6	−9	3		
	5.	13, 46	0	0	0		

Free Rochromials in Tutte Form, Each Divided by $y(y + 1)$

			The seven circuit						
		Contracted							
	No.	Edges	Constant	y	y^2	y^3	y^4	y^5
No. 1	Uncontracted graph:		00						
	1	00	−7	7	−7	7	−7	7
	2	13	0	−7	7	−7	7	
	3	14	0	−7	14	−14	7	
	4	13, 15	0	0	−7	7		
	5	13, 46	0	0	−7	7		
	6	13, 47	0	0	−7	7		
No. 2	Uncontracted graph:		13						
	1	00	−7	14	−14	14	−14	7
	2	13	2	−8	12	−12	6	
	3	14	2	−11	21	−19	7	
	4	13, 15	0	3	−8	5		
	5	13, 46	0	4	−9	5		
	6	13, 47	0	4	−10	6		
No. 3	Uncontracted graph:		14						
	1	00	−7	14	−21	21	−14	7
	2	13	2	−11	14	−12	7	
	3	14	4	−16	22	−16	6	
	4	13, 15	0	2	−8	6		
	5	13, 46	−1	4	−10	7		
	6	13, 47	−2	6	−10	6		
No. 4	Uncontracted graph:		13, 14						
	1	00	−7	21	−28	28	−21	7
	2	13	4	−12	19	−17	6	
	3	14	6	−20	29	−21	6	
	4	13, 15	0	5	−9	4		
	5	13, 46	−1	8	−12	5		
	6	13, 47	−2	10	−13	5		
No. 5	Uncontracted graph:		13, 15						
	1	00	−7	21	−35	35	−21	7
	2	13	4	−15	20	−15	6	
	3	14	6	−22	30	−20	6	
	4	13, 15	−2	7	−10	5		
	5	13, 46	−2	7	−10	5		
	6	13, 47	−2	7	−10	5		
No. 6	Uncontracted graph:		13, 16						
	1	00	−7	21	−28	28	−21	7
	2	13	4	−12	18	−15	5	
	3	14	−4	−17	29	−23	7	
	4	13, 15	−2	8	−10	4		
	5	13, 46	−1	7	−9	3		
	6	13, 47	0	5	−10	5		

Free Rochromials in Tutte Form, Each Divided by $y(y + 1)$ (continued)

			The seven circuit					
	Contracted							
	No.	Edges	Constant	y	y^2	y^3	y^4	y^5
No. 7	Uncontracted graph:		13, 46					
	1	00	−7	21	−28	28	−21	7
	2	13	4	−13	19	−15	5	
	3	14	5	−19	31	−24	7	
	4	13, 15	−1	5	−7	3		
	5	13, 46	−2	8	−10	4		
	6	13, 47	−2	10	−13	5		
No. 8	Uncontracted graph:		13, 47					
	1	00	−7	21	−35	35	−21	7
	2	13	4	−16	22	−16	6	
	3	14	8	−26	32	−20	6	
	4	13, 15	0	2	−6	4		
	5	13, 46	−3	10	−12	5		
	6	13, 47	−6	14	−14	6		
No. 9	Uncontracted graph:		14, 15					
	1	00	−7	21	−35	35	−21	7
	2	13	4	−15	21	−17	7	
	3	14	8	−25	30	−18	5	
	4	13, 15	0	4	−9	5		
	5	13, 46	−2	8	−13	7		
	6	13, 47	−4	12	−13	5		
No. 10	Uncontracted graph:		13, 14, 15					
	1	00	−7	28	−49	49	−28	7
	2	13	6	−19	27	−20	6	
	3	14	10	−31	38	−22	5	
	4	13, 15	−2	9	−11	4		
	5	13, 46	−3	11	−13	5		
	6	13, 47	−4	13	−13	4		
No. 11	Uncontracted graph:		13, 14, 16					
	1	00	−7	28	−49	49	−28	7
	2	13	6	−19	26	−18	5	
	3	14	8	−28	38	−24	6	
	4	13, 15	−4	12	−12	4		
	5	13, 46	−3	10	−10	3		
	6	13, 47	−2	8	−10	4		
No. 12	Uncontracted graph:		13, 14, 46					
	1	00	−7	28	−49	49	−28	7
	2	13	6	−20	27	−18	5	
	3	14	9	−30	40	−25	6	
	4	13, 15	−3	9	−9	3		
	5	13, 46	−4	11	−11	4		
	6	13, 47	−4	13	−13	4		

Free Rochromials in Tutte Form, Each Divided by $y(y + 1)$ (continued)

			The seven circuit					
	Contracted							
	No.	Edges	Constant	y	y^2	y^3	y^4	y^5
No. 13	Uncontracted graph:		13, 14, 47					
	1	00	−7	28	−49	49	−28	7
	2	13	6	−20	29	−21	6	
	3	14	12	−35	40	−22	5	
	4	13, 15	0	4	−7	3		
	5	13, 46	−4	14	−15	5		
	6	13, 47	−8	20	−17	5		
No. 14	Uncontracted graph:		13, 14, 57					
	1	00	−7	28	−49	49	−28	7
	2	13	6	−21	29	−19	5	
	3	14	11	−34	42	−25	6	
	4	13, 15	−1	4	−5	2		
	5	13, 46	−5	14	−13	4		
	6	13, 47	−8	20	−17	5		
No. 15	Uncontracted graph:		13, 15, 35					
	1	00	−7	28	−49	49	−28	7
	2	13	6	−19	26	−18	5	
	3	14	8	−28	38	−24	6	
	4	13, 15	−4	12	−12	4		
	5	13, 46	−3	10	−10	3		
	6	13, 47	−2	8	−10	4		
No. 16	Uncontracted graph:		13, 16, 35					
	1	00	−7	28	−49	49	−28	7
	2	13	6	−20	26	−16	4	
	3	14	7	−27	40	−27	7	
	4	13, 15	−5	12	−10	3		
	5	13, 46	−4	10	−8	2		
	6	13, 47	−2	8	−10	4		
No. 17	Uncontracted graph:		13, 14, 15, 16					
	1	00	−7	35	−70	70	−35	7
	2	13	8	−26	34	−21	5	
	3	14	12	−39	47	−25	5	
	4	13, 15	−6	16	−14	4		
	5	13, 46	−5	13	−11	3		
	6	13, 47	−4	11	−10	3		
No. 18	Uncontracted graph:		13, 14, 15, 57					
	1	00	−7	35	−70	70	−35	7
	2	13	8	−28	37	−22	5	
	3	14	15	−45	51	−26	5	
	4	13, 15	−3	8	−7	2		
	5	13, 46	−7	17	−14	4		
	6	13, 47	−10	23	−17	4		

Free Rochromials in Tutte Form, Each Divided by $y(y+1)$ (continued)

			The seven circuit					
	Contracted							
	No.	Edges	Constant	y	y^2	y^3	y^4	y^5
No. 19	Uncontracted graph:		13, 14, 16, 46					
	1	00	−7	35	−70	70	−35	7
	2	13	8	−27	34	−19	4	
	3	14	11	−38	49	−28	6	
	4	13, 15	−7	16	−12	3		
	5	13, 46	−6	13	−9	2		
	6	13, 47	−4	11	−10	3		
No. 20	Uncontracted graph:		13, 14, 47, 57					
	1	00	−7	35	−70	70	−35	7
	2	13	8	−29	39	−23	5	
	3	14	17	−49	53	−26	5	
	4	13, 15	−1	3	−3	1		
	5	13, 46	−8	20	−16	4		
	6	13, 47	−14	30	−21	5		

REFERENCES

1. G. D. Birkhoff and D. C. Lewis, Chromatic polynomials, *Trans. Amer. Math. Soc.* **60** (1946) 355–451.
2. D. W. Hall, Colouring seven-circuits, "Graphs and Combinatorics," *in* Lecture Notes in Mathematics (R. A. Bari and F. Harary, eds.), No. 406, pp. 273–290. Springer-Verlag, Berlin and New York, 1974.
3. D. W. Hall, and D. C. Lewis, Colouring six-rings, *Trans. Amer. Math. Soc.* **64** (1948) 184–191.
4. W. T. Tutte, Chromials, *MAA Studies in Math.* **12**, Part II (1975) 361–377.

AMS 05C15

STATE UNIVERSITY OF NEW YORK AT BINGHAMTON
BINGHAMTON, NEW YORK

A General Construction for Equidistant Permutation Arrays

KATHERINE HEINRICH,† G. H. J. VAN REES

and

W. D. WALLIS†

1. Introduction

An *equidistant permutation array* (EPA) is a $v \times r$ array in which every row is a permutation of the integers $1, 2, \ldots, r$ and every pair of distinct rows has precisely λ columns in common. As in [10] this array will be denoted by $A(r, \lambda; v)$ and λ is said to be the *index* of the array.

Bolton [1] defines $R(r, \lambda)$ to be the maximum v for which there exists an $A(r, \lambda; v)$. Deza [3] has shown that $R(r, \lambda) \leq \max\{\lambda + 2, n^2 + n + 1\}$, where $n = r - \lambda$. Mullin and Nemeth [7] have established $R(r, 1) \leq r^2 - 4r - 1$ for $r \geq 6$.

Bolton [1] has also established that $R(r, k) \geq 2 + [k/\{\frac{1}{3}(r - k)\}]$, $r > k + 2$, where $[x]$ denotes "the greatest integer $\leq x$" function and $\{x\}$ denotes "the least integer $\geq x$" function. Woodall [11] has proved that

---

† The first and third authors gratefully acknowledge the hospitality of the University of Waterloo during the winter of 1977.

$R(4q, q) \geq q(q - 1)$ for q a prime power. Vanstone [9] has generalized this to
$$R\left(\frac{3q(q^{n-1} - 1)}{q - 1} + q, \frac{3q(q^{n-2} - 1)}{q - 1} + q\right) \geq (q - 1)q^{n-1}$$
for q a prime and $n \geq 2$ a positive integer.

Heinrich and van Rees [6] have established that $R(r, 1) \geq 2r - 4$ for $r > 6$ which implies $R(r + \lambda, 1 + \lambda) \geq 2r - 4$ for $r > 6$.

The constructions in this paper will be described in the language of generalized Room squares which we now define. A *generalized Room square* (GRS) is an $r \times r$ array defined on a symbol set V of cardinality v such that every element of V is contained in each row and column of the array precisely once, every cell contains a (possibly empty) subset of V and every pair of distinct elements of V is contained in λ of the cells. Such an array will be denoted by $S(r, \lambda; v)$. Deza et al. [4] have proved the following theorem.

Theorem 1.1 *An $A(r, \lambda; v)$ exists if and only if an $S(r, \lambda; v)$ exists.*

2. The Construction

The basic definitions concerning Latin squares needed in our construction can be found in Dénes and Keedwell [2]. The following definitions will also be useful.

A *transversal* in a Latin square of order n is a set of n cells, one in each row and column, such that no two cells contain the same element. A set of Latin squares is said to have a *common transversal* if the n cells of a transversal in one square are also the cells of a transversal in each of the other Latin squares.

The main theorem will now be proven by constructing a generalized Room square from a set of mutually orthogonal Latin squares (MOLS).

This construction is a direct generalization of the construction in [6] which uses a pair of orthogonal Latin squares with a common transversal.

Theorem 2.1 *If there exists a set of $\lambda + 1$ mutually orthogonal Latin squares of order n with two disjoint common transversals then $R(n + \lambda + 2, \lambda) \geq (\lambda + 1)n + 1$.*

Proof Without loss of generality we may assume that the ith Latin square is based on the set $S_i = \{1_i, 2_i, \ldots, n_i\}$. Superimpose the $(\lambda + 1)$ mutually orthogonal Latin squares in the first n rows and n columns of an $(n + \lambda + 2) \times (n + \lambda + 2)$ array.

Let a typical cell of the first common transversal be (i, j) with entry $\{a_1, a_2, \ldots, a_{\lambda+1}\}$ where $a_i \in S_i$. Remove $\{a_1, a_2, \ldots, a_{\lambda+1}\}$ from cell (i, j) and place it in cell $(i, n + 1)$. Place the element a_k in cell $(n + 1 + k, j)$. Repeat the procedure for the other common transversal but with "row" and "column" interchanged. A new element, ∞, is placed in the now empty cells corresponding to the cells of the second transversal and in cell $(n + 1, n + 1)$. In cells $(n + 1 + i, n + 1 + i)$, for $1 \leq i \leq \lambda + 1$, place every element except those with subscript i.

It is easily seen that each element occurs exactly once in each row and column of the array. Since we superimposed a set of mutually orthogonal Latin squares every pair $\{a_i, a_j\}$, $i \neq j$, occurs exactly once in the first $n + 1$ rows and columns. In the cells $(n + 1 + i, n + 1 + i)$ every pair $\{a_i, a_i\}$ and $\{\infty, a_i\}$ occurs λ times and every pair $\{a_i, a_j\}$, $i \neq j$, occurs $\lambda - 1$ times. Thus we have constructed an $S(n + \lambda + 2, \lambda; (\lambda + 1)n + 1)$ and so by Theorem 1.1 $R(n + \lambda + 2, \lambda) \geq (\lambda + 1)n + 1$. ∎

Corollary 2.2 *If n is a prime power then $R(2n - 1, n - 3) \geq (n - 1)^2$.*

Proof It is well known (see [2, Chap. 5]) that if n is a prime power then there exists a set of $n - 1$ mutually orthogonal Latin squares of order n. Let a and b be two elements in one of the Latin squares. Then the positions of the entries a and b in this square will be the positions of two disjoint common transversals in the remaining squares. Hence there exists a set of $n - 2$ mutually orthogonal Latin squares of order n with two disjoint common transversals and so by Theorem 2.1, $R(2n - 1, n - 3) \geq (n - 1)^2$. ∎

It is quite interesting to note that using this result and the Deza bound [3], the following inequality holds for n a prime power;

$$n^2 - 2n + 1 \leq R(2n - 1, n - 3) \leq n^2 + 5n + 7.$$

3. Specific Results for $\lambda = 1$

When $\lambda = 1$, the construction of Theorem 2.1 gives $R(n, 1) \geq 2n - 5$ which is not as good a bound as that given in [6]; specifically $R(n, 1) \geq 2n - 4, n > 6$.

The following theorem, under certain conditions, can improve this bound to $R(n, 1) \geq 2n - 3$. The remainder of the section will then be concerned with finding cases in which the required conditions hold.

Theorem 3.1 *If there exist a pair of orthogonal Latin squares of order $n - 3$ which contain four disjoint common transversals t_1, t_2, t_3, and t_4 and one other common transversal t_5 which is disjoint from t_1, t_2, and t_3 but intersects t_4 in exactly one element, then $R(n, 1) \geq 2n - 3$.*

Proof Using t_1 and t_2 as the two common transversals in the construction of Theorem 2.1 we can construct an $S(n, 1; 2n - 5)$. Then delete the ∞ element and add the three elements ∞_1, ∞_2, and ∞_3 to the GRS in the following way. Let ∞_1, ∞_2, and ∞_3 be adjoined to those cells of the GRS which correspond to the transversals t_3, t_4 and t_5, respectively. Furthermore, ∞_1 must be added to cells $(n, n - 1)$, $(n - 2, n - 2)$, and $(n - 1, n)$. We now add ∞_2 to cells $(n, n - 2)$, $(n - 2, n - 1)$, and $(n - 1, n)$. Similarly, ∞_3 is added to cells $(n - 1, n - 2)$, $(n, n - 1)$, and $(n - 2, n)$. It is straightforward to check that we have now constructed an $S(n, 1; 2n - 3)$. ■

In order to make use of this theorem appropriate Latin squares must be constructed. The next two theorems give constructions for such squares.

Theorem 3.2 *If $n - 3$ is a prime number congruent to 1 or 3 modulo 8 and greater than 6, then $R(n, 1) \geq 2n - 3$.*

Proof Let $p = n - 3$. Construct the following Latin square with row and column labels $0, 1, 2, \ldots, p - 1$. Let the entry in cell (x, y) of the Latin square be $(2r + 3)x + (-2r - 2)y$ where $2r + 3 \not\equiv 0, 1$ or $\frac{1}{2} \pmod{p}$. It is well known (see [2, Chap. 12]) that such a Latin square is orthogonal to its own transpose. We will show that this pair of orthogonal Latin squares has common transversals t_1, t_2, t_3, t_4 and t_5 as described in Theorem 3.1, and so we can use that theorem to prove $R(n, 1) \geq 2n - 3$.

Let t_5 consist of the entries on the main diagonal except that elements $r + 1$, $r + 2$, $-r - 1$, $-r - 2$, $3r + 4$, and $-3r - 4$ are replaced by the elements $r + 1$ in cell $(-r - 1, -r - 2)$, $-r - 1$ in cell $(r + 1, r + 2)$, $3r + 4$ in cell $(r + 2, r + 1)$, and $-3r - 4$ in cell $(-r - 2, -r - 1)$. Since t_5 is a transversal then the entry $-r - 2$ must be in cell $(3r + 4, -3r - 4)$ and entry $r + 2$ in cell $(-3r - 4, 3r + 4)$.

However the cell $(3r + 4, -3r - 4)$ must contain $((2r + 3)(3r + 4) + (2r + 2)(3r + 4)) \pmod{p}$. Therefore,

$$(4r + 5)(3r + 4) \equiv -r - 2 \pmod{p}.$$

Hence

$$6r^2 + 16r + 11 \equiv 0 \pmod{p}$$

and so

$$r \equiv \frac{-8 \pm \sqrt{-2}}{6} \pmod{p}.$$

By the reciprocity theorem in Hardy and Wright [5], if $p \equiv 1$ or $3 \pmod{8}$ then -2 is a quadratic residue and hence two values for r are valid. Since $p > 6$, $2r + 3$ will never equal 0, 1, or $\frac{1}{2}$. Hence t_5 is a transversal of the Latin square.

We define a *back diagonal* as the cells (i, j) where $i + j \equiv k \pmod{p}$ and a *front diagonal* as the cells (i, j) where $i - j \equiv k \pmod{p}$. Note that each back diagonal and each front diagonal is a transversal in this construction.

A GENERAL CONSTRUCTION FOR EQUIDISTANT PERMUTATION ARRAYS   251

Since $p > 6$ there must be a back diagonal which intersects t_5 in one position. Let this be the transversal t_4. By the way t_5 was chosen there are three disjoint back diagonals which are also disjoint from t_4 to t_5. Since these transversals are symmetric about the main diagonal they are also transversals in the transposed Latin square. We can now apply Theorem 3.1 to obtain $R(n, 1) \geq 2n - 3$ for prime $n \equiv 1$ or $3 \pmod 8$. ∎

Lemma 3.3 *If a pair of orthogonal Latin squares, L_1 and L_2 of order a, contain a set of three common disjoint transversals r_1, r_2, and r_3 and a pair of common transversals r_4 and r_5† which intersect in exactly one element and if a pair of orthogonal Latin squares, M_1 and M_2 of order b, contain two disjoint common transversals s_1 and s_2 then $R(ab + 3, 1) \geq 2ab + 3$.*

Proof The direct product of L and M, denoted by $L \times M$, is formed by replacing each element α of M by the Latin square L of order a in which each symbol of L is subscripted with α. This $a \times a$ subsquare shall be called a block of $L \times M$. It is well known (see [2, Chap. 11]) that if L_1 is orthogonal to L_2 and M_1 is orthogonal to M_2 then $L_1 \times M_1$ and $L_2 \times M_2$ are a pair of orthogonal Latin squares of order ab.

Consider the blocks of $L_1 \times M_1$ which correspond to the transversal s_1 of M_1. These blocks are isomorphic to L_1 and thus contain the transversals r_1, r_2, and r_3 with the elements of the transversals having the appropriate subscript. Since no block has the same subscript, the union of the r_1 transversals in the blocks corresponding to s_1 form a transversal t_1 in $L_1 \times M_1$. Similarly the union of the r_2 transversals and the union of the r_3 transversals form transversals t_2 and t_3, respectively, in $L_1 \times M_1$.

Thus t_1, t_2, and t_3 are a set of three disjoint transversals of $L_1 \times M_1$.

Consider the blocks of $L_1 \times M_1$ which correspond to the transversal s_2. Let A be the first of these blocks. The block A has two transversals r_4 and r_5 which intersect in exactly one element. Form the transversal t_4 of $L_1 \times M_1$ by taking the union of the r_4 transversal in block A and the r_1 transversals in the other blocks corresponding to s_2. Form the transversal t_5 of $L_1 \times M_1$ by taking the union of the r_5 transversal in block A and the r_2 transversals in the other blocks corresponding to s_2.

Clearly t_4 and t_5 are transversals of $L_1 \times M_1$ which intersect each other exactly once and are disjoint from t_1, t_2, and t_3.

Since r_1, r_2, r_3, r_4, and r_5 were common transversals of L_1 and L_2, and s_1 and s_2 were common transversals of M_1 and M_2 then t_1, t_2, t_3, t_4, and t_5 are common transversals of $L_1 \times M_1$ and $L_2 \times M_2$. Thus these transversals satisfy the conditions of Theorem 3.1 and so $R(ab + 3, 1) \geq 2ab + 3$. ∎

† Note that r_4 and r_5 are not necessarily distinct from r_1, r_2, and r_3.

Theorem 3.4 *If* $n - 3 = ab$ *where both* a *and* b *are either relatively prime to six or greater than* 53, *then* $R(n, 1) \geq 2n - 3$.

This is easily proved using the fact that if $a \geq 53$ there exists a set of four mutually orthogonal Latin squares of order a (see [2, Chap. 11]). If $(a, 6) = 1$ there is a self-orthogonal Latin square of order a with rows and columns labeled $0, 1, 2, \ldots, a - 1$ and with the elements $2x - y \pmod{a}$ in cell (x, y). In both cases the transversals required in Lemma 3.3 can be found.

For large n it is probably true that one can always find the appropriate pair of orthogonal squares so that $R(n, 1) \geq 2n - 3$. But other methods seem likely to increase $R(n, 1)$ beyond $2n - 3$.

Since writing the paper, the second author and Vanstone have succeeded in proving that if

$$\lambda > \frac{n + 6}{2} + \frac{7}{n}$$

then the inequality of Theorem 2.1 can be replaced by an equality.

REFERENCES

1. D. W. Bolton, unpublished manuscript.
2. J. Dénes and A. D. Keedwell, "Latin Squares and Their Applications," Akadémiai Kiadó, Budapest, 1974.
3. M. Deza, Matrices dont deux lignes quelconques coincident dans un nombre donné de positions communes, *J. Combinatorial Theory Ser. A* **20** (1976) 306–318.
4. M. Deza, R. C. Mullin, and S. A. Vanstone, Room squares and equidistant permutation arrays, *Ars Combinatoria* **2** (1976) 235–244.
5. G. H. Hardy and E. M. Wright, "Introduction to the Theory of Numbers," 4th ed. Oxford Univ. Press (Clarendon), London and New York, 1962.
6. K. Heinrich and G. H. J. van Rees, Some constructions for equidistant permutation arrays of index one, *Utilitas Math.* **13** (1978) 193–200.
7. R. C. Mullin and E. Nemeth, An improved bound for equidistant permutation arrays of index one, *Utilitas Math.* **13** (1978) 77–85.
8. H. J. Ryser, "Combinatorial Mathematics," Carus Mathematical Monographs, No. 14. Math. Assoc. Amer., New York, 1963.
9. S. A. Vanstone, Pairwise orthogonal generalized room squares and equidistant permutation arrays, *J. Combinatorial Theory Ser. A*, **25** (1978) 84–89.
10. S. A. Vanstone and P. J. Schellenberg, A construction for equidistant permutation arrays of index one, *J. Combinatorial Theory Ser. A*, **23** (1977) 180–186.
11. D. R. Woodall, unpublished manuscript.

AMS 05B30

Katherine Heinrich and W. D. Wallis
DEPARTMENT OF MATHEMATICS
UNIVERSITY OF NEWCASTLE
NEW SOUTH WALES, AUSTRALIA

G. H. J. van Rees
DEPARTMENT OF COMBINATORICS AND OPTIMIZATION
UNIVERSITY OF WATERLOO
WATERLOO, ONTARIO

J-Components, Bridges, and J-Fragments

ARTHUR M. HOBBS

In this paper, we use the terminology of [1]. A *link graph* is a graph with two vertices and exactly one edge joining them, and a *loop graph* is a graph with one vertex and exactly one loop on it. We denote the number of elements of a set A by $|A|$. $V_k(G)$ is the set of vertices of degree k in graph G. The *interior* $I(P)$ of a path P is the set of vertices of P other than the end vertices of P. $I(P)$ may be empty. Two paths P and Q are *internally disjoint* iff $I(P) \cap I(Q) = \emptyset$.

Given a subgraph J of a graph G, the *J-fragments* of G are the subgraphs of G of the following types:

(1) a link graph with both ends in J whose edge is not in J;
(2) a loop graph with its end in J whose edge is not in J; and
(3) a subgraph F formed from a component H of $G - V(J)$ by adding to H the link graphs in G which join vertices in H to vertices in J.

The preceding definition is essentially that of Robertson [14]; following Tutte [18], the J-fragments of types (1) and (2) are called *inner* J-fragments and the J-fragments of type (3) are called *outer* J-fragments. The word *fragment* may be used alone to refer to a J-fragment for some subgraph J of G.

J-fragments were introduced by Tutte under the names "J-component" [18] and "bridge" [16]. I have chosen to call these subgraphs of a graph by the name "J-fragment" in order to avoid the prior uses of the other two names. The term "fragment" has been used before in graph theory (e.g., see [15, 9, 10]). However, I do not believe the term has been previously widely adopted with any one meaning.

This paper is intended to advocate the general introduction in elementary texts of the J-fragment. In furtherance of this goal, many theorems in connectivity are here shown to have short and natural proofs using fragments. Further, several other aspects of graph theory in which fragments have been used (though not by that name) are mentioned.

The following lemmas are basic. Given a J-fragment F, we call the elements of $V(F) \cap V(J)$ the *vertices of attachment* of F. The first lemma is trivial.

Lemma 1 *If G is k-connected, if J is a subgraph of G, and if F is an outer J-fragment of G, then $|V(F) \cap V(J)| \geq \min(k, |V(J)|)$. Further, if $|V(J)| < k$, then there is just one outer J-fragment.*

Lemma 2 *Let J be a subgraph of G and let F be a J-fragment with at least two vertices of attachment. Then there are trees T and T' in F such that T' spans F with $V(F) \cap V(J) \subseteq V_1(T')$ and $V_1(T) = V(F) \cap V(J)$.*

Proof If F is an inner fragment, $T = T' = F$ will do. Otherwise, suppose F was formed from the component H of $G - V(J)$, and let T'' be a spanning tree of H. Form T' by adding to T'' each element of $V(F) \cap V(J)$ and one edge from that element to $V(T'')$. Then T' has the required form. Further, starting with tree T', we may remove one by one the vertices of degree one which are not in $V(F) \cap V(J)$, continuing until a tree T with $V_1(T) = V(F) \cap V(J)$ is reached. ∎

A subgraph H of a graph G *avoids* a subgraph J of G iff $V(H) \cap V(J) \subseteq V_1(H)$ and $E(H) \cap E(J) = \emptyset$.

Corollary 2a *Let J be a subgraph of G, and let F be a J-fragment of G. Let v and w be distinct vertices of F. Then there is a path in F joining v and w which avoids J. Further, if F is an outer J-fragment with at least three vertices of attachment, then there is a tree T in F such that $|V_1(T)| = 3$, T avoids J, and $V_1(T) \subseteq V(J)$.*

In a brilliant piece of work which generalizes a theorem of Whitney [23], Tutte [16, 20] showed:

Theorem 1 (Tutte) *If G is a 4-connected planar graph, then G is Hamiltonian.*

The proof of this theorem relies extensively on the theory of fragments of a cycle of a planar graph. The theory of fragments of a cycle, begun in [16], is repeated with modifications in [12], and is carried further in [1, 21], where it is used in new proofs of Kuratowski's theorem characterizing planar graphs. These new proofs are uncommonly clear because only the construction of copies of K_5 and $K_{3,3}$ in the various cases are left for the proof itself; the properties of a minimal counterexample to Kuratowski's theorem, developed in the first few pages of each of the earlier published proofs, are here immediate consequences of theorems about fragments. Fragments have also been used in fast algorithms for determining the planarity or nonplanarity of a graph [2, 17]. Further, they have been generalized in matroid theory and are used there (under the name "bridge") in a deep way in many theorems (see [19, beginning with Chap. 4]).

We use the term *block* to mean a vertex, two vertices with one or more edges joining them, a loop graph, or a 2-connected graph. If H is a subgraph of a graph G, then $G - H = (G - E(H)) - V_0(G - E(H))$. A path P is *suspended* in a graph G if $I(P) \subseteq V_2(G)$ and $V(P) - I(P) \subseteq V(G) - V_2(G)$. A path P of a block G is *removable* if P is a suspended path of G and $G - P$ is a block. In [18], Tutte gives a beautiful proof (using fragments) of the following extension of a theorem of Whitney [22]:

Theorem 2 (Tutte–Whitney theorem) *Let G be a 2-connected graph. Let H be a block with at least one edge such that H is a proper subgraph of G. Then there is a block K and a path L which avoids K such that $H \subseteq K$ and $K \cup L = G$ (i.e., G has a removable path L which avoids H).*

The proof in [18] does not directly give a good algorithm for finding L (see [5] for the definition of "good" in this context). I will remedy that shortly, by giving a new proof of Theorem 2. In [8], Mitchem and I showed the following:

Lemma 3 *Let G be a loopless 2-connected graph and let $v \in V(G)$. Let C be a cycle of G such that $G \neq C$. Then there is a removable path P of G such that $P \subset C$ and $v \notin I(P)$.*

Proof Let Λ_F be the list of all paths in C which join successive vertices of attachment around C of a C-fragment F of G such that the paths in Λ_F do not contain vertices of attachment of F in their interiors. Note that Λ_F has at least two members, for each C-fragment F, by Lemma 1. Let $\Lambda = \bigcup_F \Lambda_F$. Delete from Λ all paths whose interiors contain v, to get a list Λ'. Since v cannot be in the interior of more than one member of Λ_F for any C-fragment F, and since $G \neq C$, Λ' is not empty. Let Q be a shortest member of Λ' and let P be a suspended path of G in Q. Unless $P = Q$, the ends x and y of P must meet distinct C-fragments whose other vertices of attachment are outside of Q, because Q is a shortest member of Λ'. The cases that arise are shown in Fig. 1, where C is shown as a circle, Q is the dashed portion of C, P

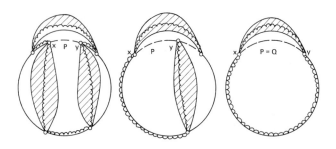

Figure 1

is the portion of Q between x and y, C-fragments are shown by shaded half-moons, and paths are shown by wavy lines. The existence of the indicated paths in the C-fragments is guaranteed by Corollary 2a. Thus the two internally disjoint paths joining x and y in $G - P$ shown in each case in Fig. 1 verify that P is removable. ∎

The proof of Lemma 3 immediately yields a good algorithm for finding a suitable path P. Forming the list Λ requires first a list Φ of C-fragments. Φ can be formed in at most $c_1 n^2$ steps, where $n = |V(G)|$. Λ then takes at most $c_2 n^2$ further steps to be formed, and selecting Λ', choosing Q, and choosing P in Q takes at most $c_3 n^2$ steps. Thus P can be found in at most cn^2 steps, for some constant c.

Corollary 3a *Every cycle C of a 2-connected graph G, G not a cycle, contains two distinct removable paths.*

Proof Use Lemma 3 to select a removable path P_1 in C. If P_1 includes an internal vertex, let it be v. If not, subdivide the edge of P_1 and let the subdividing vertex be v. In the latter case, the resulting graph formed from G is 2-connected, since G is 2-connected. Now use Lemma 3 again, picking a removable path P_2 in C which avoids v. Since P_2 avoids v and P_1 and P_2 are suspended paths in G, P_2 is internally disjoint from P_1. ∎

The following corollaries and theorems were first proved by the named authors, but the proofs provided here are generally much shorter than the previously published proofs and are new, I believe. A graph G is *minimally k-connected* if G is k-connected but $G - e$ is not k-connected for each edge e of G.

Corollary 3b (Dirac [3]; Plummer [13]) *Suppose G is a minimally 2-connected graph and let C be a cycle of G. If $G \neq C$, then C includes two vertices of degree two in G separated on C by two vertices whose degrees in G are higher than two.*

Proof Select the removable paths P_1 and P_2 guaranteed by Corollary 3a. Since G is minimally 2-connected, neither path can have just one edge. Thus each contains a vertex whose degree in G is two. Since P_1 is suspended in G, the end vertices of P_1 have degree greater than two in G. Since these end vertices also separate on C the two previously noted vertices of degree two, the corollary is proved. ∎

Now we come to the proof of the Tutte–Whitney theorem:

Proof Let F be an H-fragment of G. If F includes a cycle, let L be a removable path of G in a cycle of F. If F is a path, let $L = F$. Otherwise, F is a tree; select a longest path L in F such that all of the internal vertices of L have degree two in G and one of the ends of L is of degree one in F. Then $G - L$ is 2-connected. In each case, let $K = G - L$. The theorem follows. ∎

Note that the operations described in the preceding proof can be carried out as described in at most cn^2 steps, where $n = |V(G)|$ and c is a constant. Thus the promised good algorithm for finding L has been produced.

Corollary 2A (Corollary of the proof) *Let G be a 2-connected graph, C a cycle in G, $G \neq C$, and let G have k C-fragments with cycles and q_i C-fragments which are trees with i vertices of attachment, $i = 2, 3, \ldots$. Then G has at least*

$$2 + 2k + q_2 + \sum_{i=3}^{\infty} i q_i$$

distinct removable paths.

The next theorem is an immediate consequence of Corollary 2A, but an independent proof is given here, since it is short.

Theorem 3 (Fleischner [6]) *Let G be 2-connected and suppose G is not a cycle. Then there are three distinct removable paths in G.*

Proof Let C be a cycle of G, $G \neq C$. Then by Lemma 3, C includes two removable internally disjoint paths P_1 and P_2. By the Tutte–Whitney theorem, G includes a removable path P_3 which avoids C. These three paths are the ones required. ∎

The following is an extension of Menger's theorem for 2-connected graphs:

Corollary 3A (Fleischner [6]) *Let G be a 2-connected graph, G not a cycle. Let v and w be vertices of G. Then there are internally disjoint paths P_1, P_2, and P_3 in G such that $\{v, w\} \cap I(P_3) = \emptyset$, P_3 is removable from G, and P_1 and P_2 are paths joining v to w.*

Proof The corollary is clearly correct if G is the union of a cycle with a path which avoids the cycle (i.e., G is a theta-graph). Suppose the corollary is true for all 2-connected graphs which are not cycles and which have fewer edges than G. By Theorem 3, let Q_1, Q_2, and Q_3 be internally disjoint removable paths of G. Then one, say Q_j, does not have either v or w as an internal vertex. Let $P_3 = Q_j$. By induction, in $G - P_3$ we can find paths P_1 and P_2 which are internally disjoint and which join v to w. Thus the paths found are those required. ∎

Theorem 4 *Let G_2 be a cycle. For $i = 2, 3, 4, \ldots,$ form G_{i+1} from G_i by adding a suspended path of arbitrary positive length to two distinct vertices of G_i. Then for any $k \geq 3$, G_k is 2-connected and contains at least k internally disjoint removable paths. Further, there is a graph formed in this way which has exactly k such paths.*

Proof G_k is proved to be 2-connected in [18, Theorem 9.16]. For $i \in \{4, 5, \ldots, k\}$, let P be the path added to G_{i-1} to form G_i. Then P is clearly removable from G_i. If Q is removable from G_{i-1} and P does not end on an internal vertex of Q, then Q is also removable from G_i. If P meets an internal vertex of path R which is removable in G_{i-1}, we have one of the situations shown in Fig. 2. In each of these cases, a path which is removable in G_i

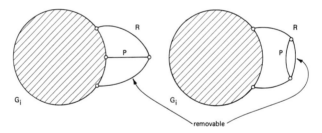

Figure 2

and which is a subpath of R is marked. Thus G_i has i removable internally disjoint paths if G_{i-1} has $i - 1$ of them. Since the theorem clearly holds for G_3, the first part of the theorem thus follows by induction. The graph consisting exactly of two vertices v and w joined by k internally disjoint suspended paths can be formed by the procedure described and has exactly k internally disjoint removable paths. The theorem follows. ∎

Further work on connectivity, using fragments, has been carried out both here and in Europe. Tutte [18] used fragments for a deep study of 2- and 3-connected graphs. Robertson [15] carried Tutte's study on to cyclic-4-connected graphs. Robertson [14] also used fragments for studying other properties of 2-connected graphs. Dirac [4] used fragments (although not any of the names for fragments) to prove Menger's theorem for connectivity $k \geq 3$. Mader, Jung, and Halin have made a substantial contribution to the study of k-connected graphs in a large group of papers (see [7] for an English language summary of some of their results and a partial listing of further papers). One of the more important results, due to Mader [11], states:

Theorem 5 *Let G be minimally k-connected, $k > 1$. Then every cycle of G includes a vertex of degree k.*

In addition, Mader [11] showed that, in any minimally k-connected graph G, $|V_k(G)| \geq \max\{k + 1, \Delta(G)\}$. Mader also proved two conjectures of Halin [7] using Theorem 5: First, for every minimally k-connected graph G, $|V_k(G)| > ((k - 1)/(2k - 1))|V(G)|$; and second, if S is a subgraph of a minimally k-connected graph G such that there are at least two outer S-fragments and if $|V(S)| = k$, then each outer S-fragment contains a vertex of degree k.

REFERENCES

1. J. A. Bondy and U. S. R. Murty, "Graph Theory with Applications." Amer. Elsevier, New York, 1976.
2. G. Demoucron, Y. Malgrange, and R. Pertuiset, Graphes planaires: Reconnaissance et construction de representations planaires topologiques, *Rev. Informat. Française Recherche Opérationalle* **8** (1964) 33–41.
3. G. A. Dirac, Minimally 2-connected graphs, *J. Reine Angew. Math.* **228** (1967) 204–216.
4. G. A. Dirac, Short proof of Menger's graph theorem, *Mathematika* **13** (1966) 42–44.
5. J. Edmonds, Paths, trees, and flowers, *Canad. J. Math.* **17** (1965) 449–467.
6. H. Fleischner, In the square of graphs, Hamiltonicity and pancyclicity, Hamiltonian connectedness and panconnectedness are equivalent concepts, *Monatsh. Math.* **82** (1976), 125–149.
7. R. Halin, On the structure of n-connected graphs, *in* "Recent Progress in Combinatorics" (W. T. Tutte, ed.), pp. 91–102. Academic Press, New York, 1969.
8. A. M. Hobbs and J. Mitchem, The entire graph of a bridgeless connected plane graph is Hamiltonian, *Discrete Math.* **16** (1976), 233–239.
9. H. A. Jung, Über den Zusammenhang von Graphen, mit Anwendungen auf symmetrische Graphen, *Math. Ann.* **202** (1973) 307–320.
10. H. A. Jung, Die Zusammenhangsstruktur symmetrischer Graphen, *J. Reine Angew. Math.* **283/284** (1976) 202–221.
11. W. Mader, Ecken vom Grad n in minimalen n-fach zusammenhängenden Graphen, *Arch. Math. (Basel)* **23** (1972) 219–224.
12. O. Ore, "The Four Color Problem," Academic Press, New York, 1967.
13. M. D. Plummer, On minimal blocks, *Trans. Amer. Math. Soc.* **134** (1968) 85–94.
14. N. Robertson, Traversing Edge Triples by Simple Circular Paths, Preprint, The Ohio State University, Columbus, Ohio.
15. N. Robertson, Graphs minimal under girth and connectivity constraints, Ph.D. Thesis, University of Waterloo, Waterloo, Ontario, 1969.

16. W. T. Tutte, A theorem on planar graphs, *Trans. Amer. Math. Soc.* **82** (1956) 99–116.
17. W. T. Tutte, How to draw a graph, *Proc. London Math. Soc.* **13** (1963) 743–768.
18. W. T. Tutte, "Connectivity in Graphs," Univ. of Toronto Press, Toronto, 1966.
19. W. T. Tutte, "Introduction to the Theory of Matroids," Amer. Elsevier, New York, 1971.
20. W. T. Tutte, Bridges and Hamiltonian circuits in planar graphs, *Aequationes Math.* **15** (1977) 1–33.
21. W. T. Tutte, Separation of vertices by a circuit, *Discrete Math.* **12** (1975) 173–184.
22. H. Whitney, Non-separable and planar graphs, *Trans. Amer. Math. Soc.* **34** (1932) 339–362.
23. H. Whitney, A theorem on graphs, *Ann. of Math.* **32** (1931) 378–390.

AMS 05C99

DEPARTMENT OF MATHEMATICS
TEXAS A & M UNIVERSITY
COLLEGE STATION, TEXAS

Hamilton Cycles in Regular Two-Connected Graphs

BILL JACKSON[†]

All graphs considered are simple. We shall give an outline of the proof of the following result [7].

Theorem A *Let G be a 2-connected, k-regular graph on n vertices, where $3k \geq n$. Then G is hamiltonian.*

This result is best possible for $k = 3$, since the Petersen graph is a nonhamiltonian, 2-connected, 3-regular graph on ten vertices. It is essentially best possible for $k \geq 4$ since there exist nonhamiltonian, 2-connected, k-regular graphs on $3k + 4$ vertices for k even, and $3k + 5$ vertices for all k. For example, consider the following graph consisting of three disjoint copies of K_{k+1}, A, B, and C, together with two extra vertices u and v. One first deletes a set $S = \{u_1 v_1, u_2 v_2, \ldots, u_k v_k\}$ of k independent edges where S contains at least one edge from each of A, B, and C. The vertex u is then joined to the vertices u_i, and the vertex v to the vertices v_i, for all i, $1 \leq i \leq k$, to form a 2-connected, k-regular graph G on $3k + 5$ vertices. Moreover, G is easily seen to be nonhamiltonian since deleting the vertices u and v from G leaves a graph with three components.

[†] The research reported here has been sponsored by the Canadian Commonwealth Association.

One may construct a nonhamiltonian graph on $3k + 4$ vertices for k even, say, $k = 2t$, by taking one copy A' of K_{2t}, two disjoint copies B' and C' of K_{2t+1}, and two extra vertices u' and v'. One deletes a set $S = \{u_1'v_1', u_2'v_2', \ldots, u_t'v_t'\}$ of t independent edges of B' and C', where S' contains at least one edge from each of B' and C'. The vertex u' is then joined to t vertices of A' together with the vertices u_i', and the vertex v' to the remaining t vertices of A' together with the vertices v_i', for all i, $1 \leq i \leq t$. In fact, k-regular graphs on $3k + 4$ vertices exist only when k is even.

There are many degree conditions for a graph to be hamiltonian. The first was the following result due to Dirac.

Theorem 1 [2, Theorem 3] *Let G be a graph on n vertices, where $n \geq 3$. If the degree of each vertex is greater than or equal to $n/2$ then G is hamiltonian.*

Again this result is, in a sense, best possible since the complete bipartite graph $K_{k+1,k}$ is a nonhamiltonian graph on $2k + 1$ vertices and every vertex has degree greater than or equal to k. One can, however, reduce the bound on the degrees by adding a regularity condition, as can be seen from the following result due to Nash-Williams.

Theorem 2 [8, Theorem 3] *Let G be a k-regular graph on $2k + 1$ vertices. Then G is hamiltonian.*

The bound was reduced still further by Erdös and Hobbs by adding the condition that the graph be 2-connected.

Theorem 3 [3, Theorem 3] *Let G be a 2-connected, k-regular graph on $2k + 4$ vertices, where $k \geq 4$. Then G is hamiltonian.*

Theorem 4 [4] *Let G be a 2-connected, k-regular graph on n vertices, where $k \geq \frac{1}{2}(n - c\sqrt{n})$ and c is equal to $\sqrt{2}$ if n is even and to one if n is odd. Then G is hamiltonian.*

Bollobás and Hobbs later obtained the following, much stronger, result.

Theorem 5 [1] *Let G be a 2-connected, k-regular graph on n vertices, where $9k/4 \geq n$. Then G is hamiltonian.*

Before outlining the proof of Theorem A, we first need some definitions. Let G be any graph and $V(G)$ denote the set of vertices of G. For $v \in V(G)$ let $N(v)$ denote the set of neighbours of v and for $A \subseteq V(G)$ put

$$N(A) = \bigcup_{v \in A} N(v).$$

Let $E(A)$ denote the set, and $\varepsilon(A)$ the number of edges in G between the vertices of A. Further, for $A, B \subseteq V(G)$, let $E(A, B)$ denote the set, and $\varepsilon(A, B)$ the number of edges in G joining vertices of A to vertices of B. For $v \in V(G)$, let $d(v)$ denote the degree of v in G.

The proof of Theorem A is by contradiction. Hence assume that G is a nonhamiltonian, 2-connected, k-regular graph on n vertices and that $n \le 3k$. Choose a cycle C of maximum length in G so that the number of components in $R = G - C$ is minimal. Let $c_1, c_2, \ldots, c_{n-r}, c_1$ be the vertices in order round C; the subscripts of the c_i will be reduced modulo $n - r$ throughout. For $A \subseteq V(C)$ put

$$A^+ = \{c_{i+1} | c_i \in A\} \quad \text{and} \quad A^- = \{c_{i-1} | c_i \in A\}.$$

The proof is in two parts. In the first part we consider the case when R contains an isolated vertex v and use a modification of the following very nice result due to Woodall.

The Hopping Lemma [10, Lemma 12.3]   *Let $c_1, c_2, \ldots, c_{n-r}, c_1$ be the vertices in order round a longest cycle C in a graph G, chosen so that the number of components of $G - C$ is minimal. Suppose v is an isolated vertex of $G - C$. Let $Y_0 = \emptyset$ and, for $j \ge 1$ put*

$$X_j = N(Y_{j-1} \cup \{v\}) \quad \text{and} \quad Y_j = \{c_i \in C | c_{i-1}, c_{i+1} \in X_j\}.$$

Put $X = \cup X_j$. Then $X \subseteq C$ and X does not contain two consecutive vertices of C.

We slightly generalise a result obtained in the proof of the hopping lemma to prove the following.

Lemma   *For all $j \ge 1$ there does not exist a path Q_j in $G - v$, $Q_j = q_1 q_2 \cdots q_g$ such that*

(1) $V(C) \subseteq V(Q_j)$,
(2) $q_1, q_g \in X_j$, and
(3) *if $q_i \in Y_k$ and $k \le j - 1$, then $q_{i-1}, q_{i+1} \in X_k$.*

Corollary   *Let $Y = \cup Y_j$, $Z^+ = X^+ \backslash Y$, and $Z^- = X^- \backslash Y$. Then*

(a) Z^+ *and* Z^- *are independent sets of vertices.*
(b) *Given $c_i \in Z^+$ and $c_j \in Z^-$, there do not exist neighbours b_i of c_i and b_j of c_j which are adjacent on C and lie in the segment $\{c_{i-2}, c_{i-3}, \ldots, c_{j+2}\}$.*
(c) *Given $c_i, c_j \in Z^+$ or $c_i, c_j \in Z^-$, then for both $l = i$ and $m = j$ or $l = j$ and $m = i$ there does not exist $c_k \in \{c_{l+2}, c_{l+3}, \ldots, c_{m-1}\}$ such that c_l is joined to c_k and c_m to c_{k-1}.*
(d) *No vertex of $R - v$ is joined to two vertices of Z^+ or two vertices of Z^-.*

Roughly speaking, the above corollary means there cannot be very many edges between the vertices of $V(G)\backslash X$. Using the corollary we are able to obtain an upper bound for $\varepsilon(V(G)\backslash X)$. Since each vertex of $V(G)\backslash X$ has degree k,

$$\varepsilon(V(G)\backslash X, X) = (n - |X|)k - 2\varepsilon(V(G)\backslash X)$$

and hence we obtain a lower bound for $\varepsilon(V(G)\backslash X, X)$. However, since each vertex of X has degree k, $\varepsilon(X, V(G)\backslash X) \leq |X|k$. This turns out to be less than the lower bound for $\varepsilon(V(G)\backslash X, X)$ and hence gives a contradiction.

In the second part of the proof we assume that R contains no isolated vertices. In fact, Nash-Williams [9, Lemma 3] has shown that R consists entirely of isolated vertices if $n \leq 3k - 2$, so we need only consider graphs on $3k - 1$ or $3k$ vertices. We choose a longest path P in R and investigate $E(V(C), V(P))$. We are able to conclude that there are only three possible structures and then show, by a similar type of edge counting as in the first part of the proof, that none of these structures can exist.

One way to generalise Theorem A would be to show that all 2-connected, k-regular graphs on at most $3k + 3$ vertices are hamiltonian. It would be possible to show that all such graphs on $3k + 1$ vertices, with the exception of the Petersen graph, are hamiltonian by showing that equality can occur in the first part of the proof of Theorem A, only for the Petersen graph and then extending the second part of the proof to exclude all other graphs on $3k + 1$ vertices. Such a proof would, however, be rather tedious.

Another possible generalisation is given in the following conjecture due to Häggkvist.

Conjecture 1 [5] *If G is an m-connected, k-regular graph on at most $(m + 1)k$ vertices, then G is hamiltonian.*

This conjecture is, of course, false when k and m are both equal to three because of the Petersen graph. It would be almost best possible, however, for values of k and m such that $k \geq m + 1$, for m even, and $k \geq m + 2$ for m odd, since one can construct nonhamiltonian, m-connected, k-regular graphs on roughly $(m + 1)k$ vertices by generalising the construction for 2-connected graphs described in the first part of the paper.

A second conjecture due to Häggkvist concerns regular bipartite graphs.

Conjecture 2 [6] *If G is a 2-connected, k-regular, bipartite graph on at most $6k$ vertices, then G is hamiltonian.*

One may also consider "almost regular" graphs. A possible conjecture is the following:

Conjecture 3 *If G is a 2-connected graph on at most $3k + 2$ vertices with degree sequence $(k, k, \ldots k, k + 1, k + 1)$ then G is hamiltonian.*

We have constructions to show that Conjectures 2 and 3 would also be best possible.

ACKNOWLEDGMENT

I should like to thank Professor J. A. Bondy for many helpful discussions on this problem.

REFERENCES

1. B. Bollobás and A. M. Hobbs, Hamiltonian cycles in regular graphs, *in* "Advances in Graph Theory" (B. Bollobás, ed.), pp. 43–48. North-Holland Publ., Amsterdam, 1978.
2. G. A. Dirac, Some theorems on abstract graphs, *Proc. London Math. Soc.* **2** (1952) 69–81.
3. P. Erdös and A. M. Hobbs, A class of hamiltonian regular graphs, *J. Graph Theory* **2** (1978) 129–135.
4. P. Erdös and A. M. Hobbs, Hamilton cycles in regular graphs of moderate degree, *J. Combinatorial Theory Ser. B* **23** (1977) 139-142.
5. R. Häggkvist, personal communication.
6. R. Häggkvist, Unsolved problem, *Proc. Hungar. Colloq. Combinatorics, 5th, Keszthély* (1976) 1203–1204.
7. Bill Jackson, Hamilton cycles in regular 2-connected graphs, *J. Combinatorial Theory Ser. B*, to appear.
8. C. St. J. A. Nash-Williams, Valency Sequences which Force Graphs to have Hamiltonian Circuits, Interim Rep., University of Waterloo Res. Rep., Waterloo, Ontario, 1969.
9. C. St. J. A. Nash-Williams, Edge-disjoint hamiltonian circuits in graphs with vertices of large valency, *in* "Studies in Pure Mathematics" (L. Mirsky, ed.), pp. 157–183. Academic Press, New York, 1971.
10. D. R. Woodall, The binding number of a graph and its Anderson number, *J. Combinatorial Theory Ser. B* **15** (1973) 225–255.

AMS 05B30

UNIVERSITY OF READING
WHITEKNIGHTS, READING
BERKSHIRE, ENGLAND

Representations of Matroids

CHRISTOPHER LANDAUER

Notation If X is a set, then $|X|$ denotes the cardinality of X. If X, Y are sets, then $Y\backslash X$ denotes the difference set. If

$$\{e(x) | x \in I\}$$

is a set of expressions, then the sum of the values will be denoted by

$$\sum (x \in I) e(x).$$

We will quite often be taking elements out of sets and putting others in, so it will be convenient to write $A\backslash x$ instead of $A\backslash\{x\}$, and $A \cup x$ instead of $A \cup \{x\}$.

Convention By a matroid M on a set E, we mean several collections of subsets of E (see [9]): the collection $\mathscr{C}$ of circuits, the collection $\mathscr{E}$ of independent subsets of E, and the collection $\mathscr{B}$ of bases of E. We do not assume that E is a finite set, or that the rank of M (the size of an element of $\mathscr{B}$) is finite, but we do insist that each circuit be finite (so M has "finite character"). A *zero* of M is an element of E which is contained in no independent set, i.e., a one-point circuit. The *rank* of a set $A \subseteq E$ is the size of a maximal independent subset of A. Throughout this paper, M will be a fixed matroid on a set E.

We first define regular sets in a matroid.

Definition If B is a basis of M and $x \in E$, then the *B-support* of x is the subset of B given by:

$$S_B(x) = \begin{cases} \emptyset & \text{if } x \text{ is a zero of } M, \\ \{x\} & \text{if } x \in B, \\ C \setminus x & \text{if } x \notin B, \text{ and } C \subseteq B \cup x \text{ is a circuit (the} \\ & \text{circuit is known to be unique).} \end{cases}$$

The *B-support* of a subset A of E is $S_B(A) = \cup S_B(a)$, over all $a \in A$. A subset A of E is *regular* if $|S_B A| = |A|$. When there is only one basis B under consideration, we write $S(a)$ for $S_B(a)$, and $S(A)$ or even SA for $S_B(A)$, and call these simply *supports* rather than *B-supports*. Henceforth, B will be a fixed basis for the matroid M.

Note For $x \in E$, $b \in B$, we have $b \in S(x)$ iff $B \setminus b \cup x$ is a basis for M.

Lemma 1 (Brualdi [2]) *If A, B are bases of M, then there is a bijection $\psi: A \to B$ with $\psi(a) = a$ for $a \in A \cap B$, such that for all $a \in A$, $B \setminus \psi(a) \cup a$ is a basis.*

A quick proof in the case of finite rank matroids may be found as Proposition 6 [4, p. 337]. In Brualdi's proof of Lemma 1, it is shown directly that $|SA| \geq |A|$ for independent sets A, using some technical results of Asche [1].

Lemma 2 *For finite independent $C \subseteq E$, $|S_B C| \geq |C|$.*

In particular, we note that every basis is a regular set, so that $\mathcal{B}$ is the collection of maximal regular independent sets, and therefore, the regular independent sets also determine the matroid completely.

Lemma 3 *For $A \subseteq E$, every circuit in $A \cup B$ is in $A \cup S_B A$.*

Proof By a counterexample C with minimum $|C \cap (A \setminus B)|$. ∎

Definition For $A \subseteq E$, the *B-support incidence matrix* $N_B(A)$ of A is a not-necessarily-finite $(0, 1)$-matrix. The rows are indexed by A, the columns by $S_B A$, and we have

$$N_B(A)_{a,b} = \begin{cases} 1 & \text{if } b \in S_B(a), \\ 0 & \text{if } b \notin S_B(a). \end{cases}$$

When there is only one basis B involved, we write NA or $N(A)$ instead of $N_B(A)$, and call the matrix simply the *incidence matrix* of A. Each column of NA has at least one nonzero entry, as does each row for an element $x \in E$ which is not a zero of M.

Definition A *line* of a matrix is a row or a column. The *term rank* of a $(0, 1)$-matrix is the maximum size of a set of nonzero entries, with no two on a line (when the matrix is finite, this number is the same as the smallest number of lines required to cover all the ones in the matrix).

Lemma 4 *If $A \subseteq E$ is independent, then the term rank of NA is $|A|$.*

Proof If A is a basis, then use Lemma 1.

If A is any independent set, then A is contained in a basis C, and the usual argument suffices. ∎

Lemma 5 *A finite independent set $A \subseteq E$ is contained in a finite regular independent set $A' \subseteq E$.*

Proof It follows by adding elements of $SA \setminus A$, one at a time. ∎

Definition Let $R = \mathbb{Z}[X_{(a,b)}]$ be the polynomial ring in commuting indeterminates $\{X_{(a,b)} | a \in E \setminus B, \ b \in S(a)\}$. This ring is called the *formal polynomial ring* of the matroid M (note that it does depend on the basis B).

This ring is the main tool of investigation used in this paper.

Definition For $A \subseteq E$, the matrix $M(A)$, called the *formal incidence matrix* of A (see [6, 7]), is a rectangular matrix over R with rows indexed by A and columns by SA, with

$$M(A)_{a,b} = \begin{cases} 0 & \text{if } b \notin S(a), \\ 1 & \text{if } b = a \in B, \\ X_{(a,b)} & \text{if } b \neq a, b \in S(a). \end{cases}$$

For regular $A \subseteq E$, the matrix $M_0(A)$, called the *reduced* formal incidence matrix of A, has rows indexed by $A \setminus B$ and columns by $SA \setminus B$, with the same entries as in $M(A)$. We will often write MA for $M(A)$, and $M_0 A$ for $M_0(A)$. We will usually consider these matrices only for A finite. For $A \subseteq E$ finite, the *formal incidence polynomial* of A is $P(A) = \det M(A) = \pm \det M_0(A)$ [the row of $M(A)$ for an element of B has a one in one column and the rest zero].

Remark When A is finite, $M(A)$ is almost the formal incidence matrix of the (0, 1)-matrix $N(A)$. Since the matrix has a finite number of lines, it has the same rank over R as its rank over the field of fractions of R (see [6, 7]).

Lemma 6 (see [9]) *Rank $M(A)$ = term rank $N(A)$.*

Let I be the ideal of R generated by $P(A)$ for all finite regular dependent $A \subseteq E$, and let T be the semigroup in R generated by $\pm P(A)$ for all finite regular independent $A \subseteq E$.

Lemma 7 *The polynomial $1 \in T$; any $a \in E \setminus B$ which is not a zero of M and any $b \in S(a)$ have $X_{(a,b)} \in T$.*

Proof Let $A \subseteq B$. Then $SA = A$ and $M(A)$ is an $|A| \times |A|$ identity matrix, so $P(A) = 1$.

For the second part, let $a \in E \backslash B$ be a nonzero element of M. Then $C = \{a\} \cup S(a)$ is a circuit. For any $b \in S(a)$, $A = C \backslash \{b\}$ is independent and $SA = C \backslash \{a\}$. Thus, A is regular, and we see that $P(A)$ is $\pm X_{(a, b)}$. ∎

Proposition 1 *The ideal I contains $\det X$ for every $s \times s$ submatrix X of MA for $s > r = \text{rank } A$ for every finite dependent set $A \subseteq E$.*

Proof Let A be dependent with $r = \text{rank } A$. Let X be a matrix as in the statement of the proposition. Since A is dependent, there is a circuit $C \subseteq A$. Expanding $\det X$ by minors across all rows corresponding to elements of $A \backslash C$, we get a sum, with coefficients from R, of terms $\det Y$, for submatrices Y of MA all of whose rows are in C. If the columns of Y do not come from MC, then at least one is zero, so $\det Y$ is zero. If the columns of Y all come from MC, then Y is a $|C| \times |C|$ submatrix of MC.

It therefore suffices to show that I contains $\det X$ for every $|C| \times |C|$ submatrix X of MC for every circuit C. If C is regular, then there is only one such submatrix $X = MC$, and $\det X = PC$ is already in I. We may therefore assume that C is not regular, so that there are columns of MC outside X. If D is the set of elements of SC corresponding to columns outside X, then $C \cup D$ is dependent, and $S(C \cup D) = SC$, so that $C \cup D$ is regular. But we know that $D \subseteq B$, so $P(C \cup D) = \det X$ must be in I. ∎

Lemma 8 *If $D \subseteq A$ are both regular independent sets, then PD divides PA.*

Proof We consider the matrix MA, with rows arranged according to the elements of D first:

$$\begin{array}{c} D \\ A \backslash D \end{array} \begin{bmatrix} MD & 0 \\ \hline * & * \end{bmatrix},$$

where the $*$ denotes unspecified entries. The result is clear. ∎

Definition Let S be an entire ring (commutative ring with one, and without zero divisors), K its field of fractions.

An *S-representation* is a map $\psi : E \to V$ which maps zeros of M to zero, is injective on the set $E \#$ of nonzeros of M, and preserves independence: for any $A \subseteq E$, ψA is S-independent iff A is independent.

A matroid M on a set E is *S-linear*, or *linear over S*, if there is an S-representation $\psi : E \to V$, where V is a free S-module.

Remark If we consider V as being imbedded in a K-vector space KV, where K is the field of fractions of S, then a finite subset of V is K-independent iff it is S-independent. Therefore, a finite matroid is K-linear iff it is S-linear.

Let p be a rational prime number or zero, so that $\mathbb{Z}/p\mathbb{Z}$ is the prime ring (subring generated by one) in a field of characteristic p. If S is an entire ring of characteristic p, then $R_S = S \otimes_\mathbb{Z} R$ contains $R_p = R/pR$. Denote by I_S the ideal of R_S generated by I (i.e., the image of I under the obvious map $R \to R_S$), and by T_p the semigroup in R_p generated by T (i.e., the image of T under the map $R \to R_p$).

Theorem 1 (Vamos [8]; see [3] for a partial proof) *Let K be a field of characteristic p (a prime or zero), and M a matroid. If M is K-linear, then $T_p \cap I_K = \varnothing$. Conversely, if $T_p \cap I_K = \varnothing$, then M is F-linear for some extension field F of K.*

Proof If $\psi: E \to V$ is a K-representation, it is easy to show that $T_p \cap I_K = \varnothing$.

Conversely, suppose $T_p \cap I_K = \varnothing$. Then by a standard result of commutative algebra, there is a prime ideal P of R_K with $P \supseteq I_K$ and $T_p \cap P = \varnothing$. Let F be the field of fractions of the entire ring R_K/P, and let $\psi: R_K \to F$ be the quotient map. Note that F is an extension field of K, since $K \subseteq R_K$ must map isomorphically under ψ.

Let V be the F-vector space with basis consisting of symbols e_b for $b \in B$, and define $\psi: E \to V$ by $\psi(b) = e_b$ for $b \in B$, and

$$\psi(a) = \sum (b \in S_B(a))\psi(X_{(a,b)}) \cdot e_b$$

for $a \in E\backslash B$. This function will be shown to be an F-representation of M. We note that $\psi(a) = 0$ for a zero of M, since $S(a) = \varnothing$, and that $\psi(I_K) = 0$, $0 \notin \psi(T_p)$, by our choice of the ideal P. Since we know that for finite $A \subseteq E$, rank $\psi MA =$ rank ψA, it is sufficient to prove the following two lemmas. ∎

Lemma *If $A \subseteq E$ is finite and independent, then rank $\psi MA = |A|$.*

Proof If A is regular, then we already know this. If A is not regular, then A is contained in a finite regular independent set A'. We will use induction on $|A'\backslash A| \geq 1$.

Since $|A| < |A'|$, A is contained in an independent set $C \subseteq A'$, with $|C| = |A| + 1$. Also, ψMA is a submatrix of ψMC, and (by induction) ψMC has rank $|C|$.

Consider a $|C| \times |C|$ nonsingular submatrix X of ψMC, and its expansion by minors across the row corresponding to the element of $C\backslash A$. Since $\det X \neq 0$, not all of the cofactors can be zero. If we now choose a submatrix Y which gives a nonzero cofactor, then the columns of Y must correspond to elements of SA (since any column for an element outside SA is all zero), so Y is a submatrix of ψMA. Since Y has $|A|$ rows and columns, rank $\psi MA \geq |A|$, so we are done. ∎

Lemma *If $A \subseteq E$ is finite, dependent, then $\text{rank}_F \psi MA = \text{rank } A$.*

Proof Using the previous lemma, we have $\text{rank}_F \psi MA \geq \text{rank } A$, by choosing an independent subset C of A of size $\text{rank } A$. It therefore is sufficient to prove the reverse inequality. If $s > \text{rank } A$, and X is an $s \times s$ submatrix of MA, then $\det X \in I_K$, so $\psi \det X = 0$, and thus ψX is singular. Therefore, every $s \times s$ submatrix of ψMA has a zero determinant, so $\text{rank } \psi MA < s$. Taking $s = \text{rank } A + 1$, we get the result we need. ∎

Let M, R, T, I be as before. Let $J = \{n \in \mathbb{Z} \mid nT \cap I \neq \emptyset\}$, called the *representation spectrum* of M. Note that if $j \in J$ and $m \in \mathbb{Z}$, then $jm \in J$, since $t \in jT \cap I$ implies $mt \in jmT \cap I$.

Lemma $J = (0)$ iff $T \cap \mathbb{Q}I = \emptyset$, $J \neq \mathbb{Z}$ iff $T \cap I = \emptyset$.

Proof If $J \neq (0)$, then there is an element of $nT \cap I$, say, nt with $t \in T$. Since $t = (1/n)nt \in \mathbb{Q}I$, $T \cap \mathbb{Q}I \neq \emptyset$. Conversely, if $t \in T \cap \mathbb{Q}I$, then $t \in (m/n)I \subseteq (1/n)I$ for some integer $n \geq 1$, and $nt \in nT \cap I$. The last statement is obvious, since $J = \mathbb{Z}$ iff $1 \in J$. ∎

Lemma *If there is a characteristic p representation, then $J \subseteq p\mathbb{Z}$.*

Proof If p is a prime and $J \not\subseteq p\mathbb{Z}$, then there is an $nT \cap I \neq \emptyset$ with $p \nmid n$. Then $\bar{n}$ is invertible in $\bar{\mathbb{Z}} = \mathbb{Z}/p\mathbb{Z} \subseteq R_p$, so that $T_p \cap I_{GF(p)} \neq \emptyset$, and there cannot be a representation in any field of characteristic p. ∎

Corollary (Vamos [8], Rado [5]) *If there is a characteristic p representation for infinitely many primes p, then there is a characteristic zero representation (and hence for finite E a representation for all but finitely many primes).*

Proof Here $J \subseteq p\mathbb{Z}$ for infinitely many primes p, so $J = (0)$ and hence $T \cap \mathbb{Q}I = \emptyset$. There is therefore an extension field F of $\mathbb{Q}$ for which M is F-linear. By a theorem of Rado (see [5]), if E is finite, then there are characteristic p representations for all but finitely many primes p. ∎

Definition We know there is no characteristic p representation when $J \not\subseteq p\mathbb{Z}$. An *exceptional prime* is a prime p with $J \subseteq p\mathbb{Z}$ and no characteristic p representation.

We know there is no characteristic p representation iff $T_p \cap I_{GF(p)} \neq \emptyset$, i.e., iff there are elements $t \in T$, $i \in I$ with $t - i \in pR$. We therefore define $D = \{t - i \mid t \in T, i \in I\}$, called the *representation semigroup* of M. Since $D \cap pR \neq \emptyset$ iff $T_p \cap I_{GF(p)} \neq \emptyset$, a prime p is exceptional iff $J \subseteq p\mathbb{Z}$ and $D \cap pR \neq \emptyset$.

Lemma *D is a semigroup containing both T and $1 - I = \{1 - i \mid i \in I\}$.*

REPRESENTATIONS OF MATROIDS 273

Proof The last two assertions are clear, since $0 \in I$, $1 \in T$. If $t, t' \in T$ and $i, i' \in I$, then

$$(t - i)(t' - i') = tt' - it' - ti' + ii'$$

and $tt' \in T$, $it' + ti' - ii' \in I$, so the product $(t - i)(t' - i')$ is in D. ∎

Proposition 2 *Let M be a matroid on a finite set E. There are square-free integers m, l with $m \mid l$ for which: a prime p is exceptional iff $p \mid m$, the matroid M has a characteristic p representation iff $p \nmid l$ or $p \mid m$.*

Proof When $J \neq (0)$, there are only a finite number of primes p with $J \subseteq p\mathbb{Z}$, so there is a maximum square-free integer l for which $J \subseteq l\mathbb{Z}$. Among the primes $p \mid l$, there are only finitely many for which $D \cap pR \neq \emptyset$. Since $d \in D \cap pR$, $d' \in D \cap p'R$ implies $dd' \in D \cap pp'R$, there is a maximum square-free integer $m \mid l$ for which $D \cap mR \neq \emptyset$. Then there is no characteristic p representation for $p \nmid l$ (ordinary primes) or for $p \mid m$ (exceptional primes). For $p \mid l/m$, $D \cap pR = \emptyset$ and $J \subseteq p\mathbb{Z}$, so there is a characteristic p representation.

When $J = (0)$, there is a characteristic zero representation, so there fails to be a characteristic p representation for only a finite number of primes. The number m above still exists,

$$m = \prod (p \text{ prime} \mid J \subseteq p\mathbb{Z}, D \cap pR \neq \emptyset)$$
$$= \max\{\text{square-free } m \mid J \subseteq m\mathbb{Z} \text{ and } D \cap mR \neq \emptyset\}$$

and a prime p is exceptional iff $p \mid m$. We take $l = 0$ in this case, and state that by convention zero divides zero. ∎

Definition The pair (m, l) is called the *characteristic indicator* of M. These numbers gather together the information on representations of finite matroids, and present it conveniently.

We next illustrate these polynomial methods with two other applications. The first concerns representations of a matroid over a particular finite field, and the second the problem of representing a matroid over a field in such a way that an automorphism of the matroid can be represented as a linear transformation.

In both cases, the idea is that in addition to the equations required for matroid representability, the desired representation problem needs other equations to hold. These equations will be forced to hold by considering different ideals analogous to I in the polynomial ring.

Let q be a power of the prime p, and consider the representation of a matroid M over $K = GF(q)$. We use the same notation as in the main theorem. For respresentability of M over K, we should add equations that force each

$\psi(X_{(a,b)})$ into K. Since each $X_{(a,b)} \in T$, we know that the values will be non-zero, so we define $I_{(q)}$ to be the ideal generated by I_p and the polynomials $X_{(a,b)}^{q-1} - 1$ for all a, b [i.e., $a \in E \backslash B$ not a zero of M, and $b \in S_B(a)$].

Then $T_p \cap I_p = \emptyset$ implies the existence of a characteristic p representation, and $T_p \cap I_{(q)} = \emptyset$ implies the existence of a characteristic p representation for which each $\psi(X_{(a,b)})$ is a $(q-1)$-root of l. This is clearly enough to produce a K-representation. The converse is clear (i.e., that a K-representation of M implies that $T_p \cap I_{(q)} = \emptyset$).

The last application considers the representability of automorphisms of matroids. An *automorphism* of a matroid M is a permutation $g \in \text{Sym}(E)$ of the base set such that for all $A \subseteq E$, A is independent if Ag is independent (i.e., in the natural extension of g to the power set 2^E, g maps $\mathscr{E}$ into itself).

We seek conditions that imply that M is K-linear and that the automorphism g is the action on the image $\psi(E) \subseteq V$ of a matrix over K. We will actually determine when g can act on $\psi(E)$ as a linear transformation.

The automorphism g of the matroid M is said to be *compatible* with the K-representation $\psi: E \to V$ of M iff there is a linear transformation L on V for which $\psi(eg) = \psi(e)L$ for all $e \in E$. We write $\psi(g)$ for L and say that g is represented over K by $\psi(g)$.

It is clear that the condition that $g \in G$ be compatible with ψ is the same as

$$\psi(ag) = \psi(a)\psi(g)$$

for $a \in E$, and that this expression determines some polynomials in R which must be zero for any compatible representation. If we define $I(g)$ to be the ideal of R generated by I and these polynomials (there is one for each $a \in E$, $b \in B$), then a compatible representation will have $I(g) \cap T = \emptyset$. Moreover, if G is any group of automorphisms of M, and $I(G)$ is the ideal generated by $I(g)$ for $g \in G$, it can easily be shown that it suffices to take $I(g)$ only for a set of generators of G.

We assert that if $I(G) \cap T = \emptyset$, then M can be represented over an extension field of K with G compatibly represented. We note for this that we choose a prime ideal P of R which contains $I(G)$, and is disjoint from T. Then the quotient map $\psi: R \to R/P = F$ takes each element of $I(G)$ to zero, and yet maps T into the nonzero elements of F. Thus M is represented over the extension F of K, and it is easy to verify that G maps compatibly.

A characteristic indicator can be defined for any group G of automorphisms of a finite matroid M using $I(G)$ instead of I. It has similar properties.

REFERENCES

1. D. S. Asche, Minimal dependent sets, *J. Austral. Math. Soc.* **6** (1966) 259–262.
2. R. Brualdi, Comments on bases in dependence structures, *B. Austral. Math. Soc.* **1** (1969) 161–167.

3. C. B. Bruter, "Elements de la Theorie des Matroides," Lecture Notes in Mathematics, No. 387. Springer-Verlag, Berlin and New York, 1974.
4. T. Brylawski, Some properties of basic families of subsets, *Discrete Math.* **6** (1973) 333–341.
5. R. Rado, Note on independence functions, *Proc. London Math. Soc.* **7** (1957) 300–320.
6. H. B. Ryser, Indeterminates and incidence matrices, *Linear and Multilinear Algebra* **1** (1973) 149–157.
7. H. B. Ryser, The formal incidence matrix, *Linear and Multilinear Algebra* **3** (1975) 99–104.
8. P. Vamos, Representing matroids over fields, unpublished manuscript, 1971.
9. D. J. A. Welsh, "Matroid Theory," Academic Press, New York, 1976.

AMS 05B35

LANGUAGE/SYSTEMS GROUP
PATTERN ANALYSIS AND RECOGNITION CORPORATION
ROME, NEW YORK

STATE UNIVERSITY OF NEW YORK, COLLEGE AT UTICA/ROME
UTICA, NEW YORK

On Some Generalizations of Partial Geometry

RENU LASKAR

and

J. A. THAS

1. Introduction

In this paper we only consider finite structures. An *incident structure* π is a triple (P, L, I) where P and L are sets and $I \subseteq P \times L$. The elements of P and L are called *points* and *lines*, respectively, and I is called the *incidence relation* of π. We use the usual geometrical terminology. For example, a point p and a line l are *incident* if $(p, l) \in I$; two points p and p' are incident with a line l if $(p, l) \in I$ and $(p', l) \in I$; and two lines l and l' are incident with a point p if $(p, l) \in I$ and $(p, l') \in I$.

Bose [1] introduced an important class of incidence structures known as partial geometries. In recent years there have been many studies on partial geometries. In one direction, classifications of partial geometries in terms of their parameters have been studied, and constructions of such systems from known finite geometries given. Developments in another direction involve extensions of partial geometries to include broader classes of mathematical systems, as well as higher dimensional analogs of partial geometries.

Here, we attempt to describe some results in this area of investigation. However, we do not claim completeness in this survey. Some new examples of the three-dimensional analog of partial geometry are given.

2. Classes of Partial Geometries (Nondisjoint)

An (r, k, t)-*partial geometry* is an incidence structure satisfying the following axioms:

A1. Any two points are incident with at most one line.
A2. Each point is incident with r lines.
A3. Each line is incident with k points.
A4. If the point p is not incident with the line l, there exist exactly t points incident with l, which are collinear with p.

We will call such a system a 2-*dimensional* or *plane* partial geometry. The reader is referred to [1] for properties and characterization theorems. In an earlier paper Bruck [6] originated the concept of (r, k)-net and later he [7] studied further properties and embeddability conditions of an (r, k)-net in an affine plane of order k (which is a $(k + 1, k)$-net).

An (r, k)-*net* is an incidence structure satisfying the following conditions:

B1. Lines are partitioned into r disjoint nonempty parallel classes such that

  (a) each point is incident with exactly one line in each class;
  (b) two lines belonging to distinct classes intersect in exactly one point.

B2. Each line is incident with k points.

We will call such a system a *Bruck-net* or 2-*net*.

Let Π be a (r, k, t)-partial geometry. It is known that the following is true.

  (a) If $t = k$, then Π is a BIB design with $\lambda = 1$ (see [2, 17]).
  (b) If $t = k - 1$, then Π is a Bruck-net (see [2, 7, 17]).
  (c) If $t = 1$, then Π is a generalized quadrangle (introduced by Tits in [30]). For a detailed study of generalized quadrangles see [15].
  (d) For $1 < t < \min\{r - 1, k - 1\}$, all known examples are constructed with the aid of maximal arcs. In $PG(2, q)$ a *maximal n-arc* is defined to be a nonvoid set K of points meeting every line in n points or in no points. It is well known that if K is a maximal n-arc, K has $nq - q + n$ points, where $n | q$ if $n \ne q + 1$ [12]. We now describe some partial geometries arising from maximal arcs as given by Thas in [7, 9].

(i) Let K be a maximal n-arc with $1 < n < q$, of a projective plane S (not necessarily Desarguesian) of order q. Define an incidence structure Π as follows:

Π-points = all points of S not in K,

Π-lines = lines of S incident with n points of K.

Incidence is that of S.

The incidence structure Π forms a partial geometry with

$$r = q - \frac{q}{n} + 1, \quad k = q - n + 1, \quad \text{and} \quad t = q - \frac{q}{n} - n + 1.$$

(ii) Suppose there exists a maximal n-arc K with $1 < n < q$, in PG(2, q). Let PG(2, q) be embedded as a plane H in PG(3, q) = P. Define an incidence structure Π as follows:

Π-points = points of P not in H,

Π-lines = lines of P not contained in H and meeting K (necessary in a unique point).

Incidence is that of P.

The incidence structure Π forms a partial geometry with parameters

$$r = qn - q + n, \quad k = q, \quad \text{and} \quad t = n - 1.$$

As the existence of a maximal n-arc K in PG(2, q), with $n \leq q$, implies the existence of a maximal q/n − arc K' in PG(2, q), it follows that there exists a partial geometry with parameters

$$r = q(q - n + 1)/n, \quad k = q, \quad \text{and} \quad t = (q - n)/n.$$

For constructions of maximal arcs, see [13, 27].

3. Semipartial Geometries

A *semipartial geometry* is an incidence structure $S = (P, L, I)$ subject to the following axioms:

(i) Each point is incident with r lines; two distinct points are incident with at most one line.

(ii) Each line is incident with k points; two distinct lines are incident with at most one point.

(iii) If two points are not collinear, there are α ($\alpha > 0$) points collinear to both.

(iv) If a point p is not incident with a line l, then there is either no point incident with l which is collinear with p or there are exactly t ($t \geq 1$) points incident with l which are collinear with p.

Remarks

(1) Clearly $t \leq \min\{r, k\}$.
(2) If $t = k$, any two points are collinear and S is a 2-design.
(3) $t^2 \leq \alpha \leq rt$.
(4) A semipartial geometry with parameters r, k, t, α is a partial geometry if and only if $rt = \alpha$.
(5) A semipartial geometry with $t = 1$ is a partial quadrangle (introduced by Cameron [9]).

Debroey and Thas [10] and De Clerck and Thas [11] have studied semipartial geometries quite extensively and constructed many interesting examples from finite projective spaces. As a matter of fact, Thas has determined all partial geometries embeddable in $AG(n, k)$ for any n and Debroey and Thas have determined all the semipartial geometries embeddable in $PG(n, k - 1)$ for any n and in $AG(n, k)$ for $n = 2$ or 3. The reader is referred to [2, 8, 10, 11, 13, 25, 26, 28, 29].

4. Partial Geometric Design

Bose *et al.* [3] generalized the concept of a partial geometry to that of a partial geometric design. An incidence structure $\Pi = (P, L, I)$ is said to be an (r, k, t, c)-*partial geometric design* if the following axioms are satisfied:

(i) Each point is incident with r lines.
(ii) Each line is incident with k points.
(iii) $m_3(p, l) = t$ if p is not incident with l and $m_3(p, l) = c$ if p is incident with l, where $m_i(u, v)$ denotes the number of simple paths from u to v of length i in the bipartite graph $B(\Pi)$ of Π, where the vertices of $B(\Pi)$ are $P \cup L$ and (p, l) is an edge if and only if p is incident with l.

Clearly the dual of an (r, k, t, c)-partial geometric design Π is a (k, r, t, c)-partial geometric design Π^*, where points of Π^* are lines of Π and lines of Π^* are points of Π. A semilinear (a pair of points is incident with at most one line) partial geometric design is equivalent to a partial geometry. A partial geometric design is equivalent to a partial geometry if and only if $c = 0$.

Let Π be an (r, k, t, c)-partial geometric design. Let $G(\Pi)$ be the multigraph, whose vertices are points of Π, two vertices u_i, u_j being joined by m_{ij} edges if and only if the corresponding points in Π are incident with m_{ij} lines. The multigraph $G(\Pi)$ has the following properties:

(1) $G(\Pi)$ has exactly $(k/t)[(r-1)(k-1)+t-c] = v$ vertices,
(2) the degree of every vertex is $r(k-1)$,
(3) $\sum_{j=1, j\neq i}^{v} m_{ij}(m_{ij}-1) = rc$,
(4) for two vertices u_i and u_j with $m_{ij} = m \geq 1$, $\sum_{s=1, s\neq i,j}^{v} m_{is}m_{js} = m(k-2) + rt + m(c-t-r+1)$,
(5) $m_{ij} \leq r$,
(6) for $u_i \neq u_j$ with $m_{ij} = 0$, $\sum_{s=1; s\neq i,j}^{v} m_{is}m_{js} = rt$.

A multigraph G satisfying (1)–(6) is called a *pseudo (r, k, t, c)-partial geometric design graph*. In [3] the authors proved that if $c < r + t - 1$ and k is greater than a certain function of r, t, c, then there exists a unique (r, k, t, c)-partial geometric design Π such that $G \cong G(\Pi)$. Furthermore, the authors applied their results to prove an embedding theorem for a quasi-residual design. Since an (r, k, t)-partial geometry is an $(r, k, t, 0)$-partial geometric design, the characterization theorems for partial geometric design include those for partial geometry given by Bose [1]. These theorems have many applications. The reader is referred to [1, 3].

5. 3-nets

Laskar [18] introduced 3-nets as 3-dimensional analogs of Bruck-nets. A *3-net* is an incidence structure consisting of points, lines, and planes, together with an incidence relation subject to the following axioms:

(1) If a point p is incident with a line l, and l is incident with a plane Π, then p is incident with Π.
(2) Two intersecting lines are coplanar.
(3) Points and lines incident with a plane form a Bruck-net with parameters (k, n).
(4) Planes are partitioned into b parallel classes such that

   (a) two planes from two classes intersect in a unique line;
   (b) each point is in exactly one plane in each class;

(5) each line is in at least one plane and there exists a line contained in exactly r^* planes.

It is pointed out in [20] that not all of the parameters are independent; as a matter of fact, $r^* = k$ and $b = k^2 - k + 1$. It is shown in [18] that if we define two points of the geometry as first associates if incident with a line, second associates if incident with a plane but not incident with a line, and third associates otherwise, then the points of a 3-net form a 3-class association scheme (from which a partially balanced incomplete block design can be constructed).

Clearly an affine 3-space of order n is a 3-net. Another example of a 3-net is a cubic lattice graph [19]. A *cubic lattice graph* is one whose vertices are identified with ordered triplets (i, j, k); $i, j, k = 1, 2, \ldots, n$, with two vertices adjacent if and only if the corresponding triplets agree in two co-ordinates. If the triplets are taken as points, lines as triplets with one missing symbol [for example $(i, j, \cdot)$], and planes as triplets with two missing symbols [for example $(i, \cdot, \cdot)$], then a cubic lattice graph is a 3-net.

Laskar and Freeman [22] have constructed 3-nets from finite projective spaces as follows:

Let Σ be a $PG(3, p^t)$ and Π^* be a fixed $PG(2, p^t)$ contained in Σ. Let Π° be a projective subplane of Π^*, that is, $\Pi^\circ = PG(2, p^r)$ where $r|t$. Define an incidence structure N, where

N-points = points of Σ/Π^*,

N-lines = lines of Σ meeting Π^* in a point of Π°,

N-planes = planes of Σ intersecting Π^* in a line of Π°.

Incidence is the natural incidence. It can be checked easily that N forms a 3-net with parameters $k = p^r + 1$, $n = p^t$ where $r|t$.

Sprague [24] has given a characterization of 3-nets in the following:

Theorem [24] *Let N be a 3-net with parameters (k, n).*

(i) *If $k = 2$, then N is a cubic lattice graph.*

(ii) *If $k \geq 3$, then N is a 3-net constructed on finite projective spaces as just given.*

Sprague calls this a *Freeman–Laskar net*.

6. d-nets

The concept of a 3-net has been generalized to that of a d-net for any $d > 3$ by Dunbar and Laskar [14].

Let $d, k^*, n, \{b_j\}_{j=2}^d$ be integers. A system N consisting of nonempty finite collections of undefined objects called *j-dimensional sets* (or *j-sets*) where $0 \leq j \leq d - 1$, together with an incidence relation is a *d-net* if the following axioms hold:

(1) If a $(j - 1)$-set S_{j-1} is incident with a j-set which is incident with a $(j + 1)$-set S_{j+1}, then S_{j-1} is incident with S_{j+1}, for $1 \leq j \leq d - 2$.

(2) If two j-sets are both incident with a $(j - 1)$-set, they are both incident with a $(j + 1)$-set for $1 \leq j \leq d - 2$.

(3) Each j-set is incident with at least one $(j + 1)$-set, for $0 \leq j \leq d - 1$.

(4) The 0-sets and $(d-1)$-sets of N are called *points* and *hyperplanes*, respectively. Two hyperplanes are said to be *parallel* if they are not incident with a common point. The hyperplanes of N may be partitioned into b_d disjoint nonempty parallel classes such that the following is true:

(a) Two hyperplanes belonging to different classes are both incident with exactly one $(d-2)$-set S_{d-2}. Further, any set S_l (with $l \le d-3$) incident with both hyperplanes is incident with S_{d-2}.

(b) Each point of N is incident with exactly one hyperplane of each class.

(5) (a) The 0-sets and 1-sets incident with any 2-set form a (k^*, n)-net.

(b) For j such that $3 \le j \le d-1$, let S_j be any j-set. A finite j-net is formed by the k-sets incident with S_j, where $0 \le k \le j$, and b_j is the number of parallel classes into which the hyperplanes are partitioned (assuming that b_j is well defined).

Two points may be called *j*th *associates* if incident with a j-set but not incident with a $(j-1)$-set, $j = 1, 2, \ldots, d-1$. It is shown in [14] that the points of N form a d-class association scheme and that PBIB designs can be constructed. Clearly, a finite affine d-space is a d-net. Another example of a d-net is what may be called a *d-lattice graph*, whose vertices are identified with d-plets on n symbols, two vertices being adjacent if and only if the corresponding d-plets differ in exactly one co-ordinate. Such graphs were first studied in [23].

Constructions of d-nets based on $PG(d, p^t)$, generalizing analogous constructions of 3-nets, have been given by Freeman and Laskar [16] as follows:

Let Σ be a $PG(d, p^t)$, and let Π be a fixed hyperplane of Σ. Let $\Sigma^* = \Sigma - \Pi$. Let Π° be a subhyperplane of Π. In other words $\Pi^\circ = PG(d-1, p^r)$ with $r | t$. Define an incidence structure D where

D-points = points of Σ^*,

D-S_j-sets = j-dimensional subspaces of Σ not contained in Π and intersecting Π° in a $(j-1)$-dimensional subspace, $j = 1, 2, \ldots, d-1$.

Incidence is the natural incidence relation. It may be checked easily that D forms a d-net.

7. Partial Geometry of Dimension Three (or Partial 3-Space)

The concept of *partial geometry of dimension three*, a 3-dimensional analog of partial geometry, is due to Laskar and Dunbar [21] and is a system of undefined objects, points, lines, and planes, together with an incidence relation subject to the following conditions:

(i) If a point p is incident with a line l, and l is incident with a plane Π, then p is incident with Π.

(ii) (a) A pair of distinct planes is incident with at most one line.
(b) A pair of planes, if not incident with a line, is incident with at most one point.

(iii) The set of points and lines incident with a plane forms an (r, k, t)-partial geometry.

(iv) The set of lines and planes incident with a point p forms an (r, k^*, t^*)-partial geometry, where the points and lines of the geometry are the planes and lines through p, respectively, and the incidence is the natural incidence relation.

(v) Given a plane Π, and a line l not incident with Π, Π and l not intersecting in a point, there exist exactly u planes through l intersecting Π in a line and exactly $w - u$ planes through l intersecting Π in a point but not in a line.

(vi) Given a point p and a line l, p and l not coplanar, there exist exactly u^* points in l which are collinear with p, and $w^* - u^*$ points in l coplanar but not collinear with p.

(vii) Given a point p and a plane Π not containing p, there exist exactly x planes through p intersecting Π in a line.

If two points are called first associates if they are collinear, second associates if coplanar but not collinear, and third associates otherwise, it is shown in [21] that points of the system form a 3-class association scheme from which a PBIB design can be constructed. Also, given a partial 3-space Π, a dual partial 3-space Π^* is obtained by interchanging points and planes. It may be pointed out that the axioms (iii) and (iv) are duals of each other, the dual of (v) is (vi), and (vii) is self-dual.

8. Examples of Partial 3-Space

It is shown in [21] that a 3-net, a tetrahedral graph [4, 5], and a finite projective 3-space are examples of partial 3-spaces. We construct here some new examples. For definitions not given here, see [12].

(i) Let Q be a nonsingular hyperquadratic of index three in $PG(d, q)$, $d = 5, 6$, or 7. Then the points of Q together with the lines, planes of Q form a partial 3-space $Q(d, q)$ with the following parameters:

(a) $d = 5$

$r = k = t = q + 1$, $k^* = 2$, $t^* = 1$, $u = 0$, $w = 1$, $w^* = u^* = 1$,

$x = 1$.

(b) $d = 6$

$r = k = t = q + 1$, $k^* = q + 1$, $t^* = 1$, $u = 0$, $w = 1$,

$w^* = u^* = 1$, $x = 1$.

(c) $d = 7$

$r = k = t = q + 1$, $k^* = q^2 + 1$, $t^* = 1$, $u = 0$, $w = 1$, $x = 1$,

$w^* = u^* = 1$.

(ii) Let H be a nonsingular Hermitian primal (the set of absolute points of a unitary polarity) of the projective space $PG(d, q^2)$, $d = 5$ or 6. Then the points of H together with the lines and planes of H form a partial 3-space $H(d, q^2)$ with the following parameters:

(a) $d = 5$

$r = k = t = q^2 + 1$, $k^* = q + 1$, $t^* = 1$, $u = 0$, $w = 1$,

$u^* = w^* = 1$, $x = 1$.

(b) $d = 6$

$r = k = t = q^2 + 1$, $k^* = q^3 + 1$, $t^* = 1$, $u = 0$, $w = 1$,

$u^* = w^* = 1$, $x = 1$.

(iii) The points of $PG(5, q)$, together with the totally isotropic lines and planes with respect to a simplectic polarity form a partial 3-space with parameters.

$r = k = t = q + 1$, $k^* = q + 1$, $t^* = 1$, $u = 0$, $w = 1$,

$w^* = u^* = 1$, $x = 1$.

(iv) Consider a unital U of $PG(2, q^2)$. Suppose that $PG(2, q^2)$ is embedded in $PG(3, q^2)$. Define an incidence structure as follows:

Π-points = points of $PG(3, q^2)$ not contained in $PG(2, q^2)$,

Π-lines = lines of $PG(3, q^2)$ not contained in $PG(2, q^2)$ and containing no point of U,

Π-planes = planes of $PG(3, q^2)$ containing exactly one point of U.

Incidence is the natural incidence relation.

The incidence structure Π is a partial 3-space with parameters:
$$r = q^2, \quad k = q^2, \quad t = q^2 - 1, \quad k^* = q + 1, \quad t^* = q + 1, \quad u = q,$$
$$w = q, \quad u^* = q^2 - q - 1, \quad w^* = q^2, \quad x = q^3.$$

(v) Let K be a maximal n-arc, $1 \leq n < q$ of $PG(2, q)$. Let $PG(2, q)$ be embedded in $PG(3, q)$. Define Π as follows:

Π-points = points of $PG(3, q)$ not in $PG(2, q)$,

Π-lines = lines of $PG(3, q)$ not contained in $PG(2, q)$ and which contain no point of K,

Π-planes = planes of $PG(3, q)$ which do not contain a point of K.

Incidence is the natural one.

The incidence structure Π is a partial 3-space with parameters:
$$r = q + 1, \quad k = q, \quad t = q, \quad k^* = \frac{q}{n}, \quad t^* = \frac{q}{n}, \quad u = \frac{q}{n} - 1, \quad w = \frac{q}{n} - 1,$$
$$u^* = q + 1 - n, \quad w^* = q + 1 - n, \quad x = (q + 1)\left(\frac{q}{n} - 1\right).$$

(vi) Consider a subspace $PG(n - 2, q)$ of the projective space $PG(n + 1, q)$, $n \geq 3$. Define Π as follows:

Π-points = points of $PG(n + 1, q)$ not contained in $PG(n - 2, q)$,

Π-lines = lines of $PG(n + 1, q)$ having no common point with the $PG(n - 2, q)$,

Π-planes = planes of $PG(n + 1, q)$ having no common point with the $PG(n - 2, q)$.

Incidence is the natural one.

The structure Π is a partial 3-space with parameters:
$$r = k = t = q + 1, \quad k^* = q^{n-1}, \quad t^* = q, \quad u = 0, \quad w = q^2, \quad u^* = q,$$
$$w^* = q^*, \quad x = q^2.$$

The dual of this partial 3-space is isomorphic to the Freeman–Lasker net.

REFERENCES

1. R. C. Bose, Strongly regular graphs, partial geometries, and partially balanced design, *Pacific J. Math.* **13** (1963) 389–419.
2. R. C. Bose, Graphs and designs, finite geometric structures and their applications, *C.I.M.E. Bressanone* (1972) 3–104.
3. R. C. Bose, S. S. Shrikhande, and N. M. Singhi, Edge regular multigraphs and partial geometric designs, *Proc. Internat. Colloq. Combinatorial Theory, Accad. Naz. Lincei, Rome* (1973) 49–81.

4. R. C. Bose and R. C. Laskar, Characterization of tetrahedral graphs, *J. Combinatorial Theory* **3** (1967) 366–385.
5. R. C. Bose and R. Laskar, Eigenvalues of tetrahedral graphs, *Aequationes Math.* **4** (1970) 37–43.
6. R. H. Bruck, Finite nets I. Numerical invariants, *Canad. J. Math.* **3** (1951) 96–107.
7. R. H. Bruck, Finite nets II. Uniqueness and embedding, *Pacific J. Math.* **13** (1963) 421–457.
8. F. Buekenhout and C. Lefevre, Generalized quadrangles in projective spaces, *Arch. Math. (Basel)* **25** (1974) 540–552.
9. P. J. Cameron, Partial quadrangles, *Quart. Math. Oxford Ser* **25** (1974) 1–13.
10. I. Debroey and J. A. Thas, On semi partial geometries, *J. Combinatorial Theory*, to appear.
11. F. De Clerck and J. A. Thas, Partial geometries in finite projective spaces, *Arch. Math. (Basel)* **30** (1978) 537–540.
12. P. Dembowski, "Finite Geometries," Springer-Verlag, Berlin and New York, 1968.
13. R. H. F. Denniston, Some maximal arcs in finite projective planes, *J. Combinatorial Theory* **6** (1969) 317–319.
14. J. Dunbar and R. Laskar, Finite nets of dimension d., *Discrete Math.* **22** (1978) 1–24.
15. W. Feit and G. Higman, The nonexistence of certain generalized polygons, *J. Algebra* **1** (1964) 114–131.
16. J. W. Freeman and R. Laskar, Results on d-nets, *Proc. S. E. Conf.* (1977) 461–465.
17. D. G. Higman, Partial geometries, generalized quadrangles, and strongly regular graphs, *Atti Convegno Geometria Combinatoria Applicazioni, Perugia* (1971) 265–293.
18. R. Laskar, Finite nets of dimension three I, *J. Algebra* **32** (1974) 8–25.
19. R. Laskar, A characterization of cubic lattice graphs, *J. Combinatorial Theory* **3** (1967) 386–401.
20. R. Laskar and J. Dunbar, Note on "Finite Nets of Dimension Three I," Tech. Rep. No. 173. Clemson University, Clemson, South Carolina, 1974.
21. R. Laskar and J. Dunbar, Partial geometry of dimension three, *J. Combinatorial Theory Ser. A* **24** (1978) 187–201.
22. R. Laskar and J. W. Freeman, Further results on 3-nets, *Notices Amer. Math. Soc.* Jan. (1977) A-35.
23. H. A. Pellerin, Characterization problems in graph theory, Ph.D. Thesis, Clemson University, Clemson, South Carolina, 1972.
24. A. P. Sprague, A characterization of 3-nets, unpublished manuscript.
25. J. A. Thas and F. De Clerck, Partial geometries satisfying the axiom of Pasch, *Simon Stevin*, to appear.
26. J. A. Thas and P. De Winne, Generalized quadrangles in finite projective spaces, *J. Geometry*, to appear.
27. J. A. Thas, Construction of maximal arcs and partial geometries, *Geometriae Dedicala* **3** (1974) 61–64.
28. J. Thas, Combinatories of partial geometries and generalized quadrangles, *in* "Higher Combinatorics" (M. Aigner, ed.), pp. 183–199. Reidel, Boston, 1977.
29. J. A. Thas, Construction of partial geometries, *Simon Stevin* Jan. (1973) 95–98.
30. J. Tits, Sur la trialité et certains groupes qui s'en déduisent, *Publ. Math. I.H.E.S., Paris* **2** (1959) 14–60.

AMS 05B05, 05B25

Renu Lasker
CLEMSON UNIVERSITY
CLEMSON, SOUTH CAROLINA

J. A. Thas
SEMINAR OF HIGHER GEOMETRY
UNIVERSITY OF GHENT
GHENT, BELGIUM

GRAPH THEORY AND RELATED TOPICS

A Graph-Theoretical Approach to Embedding $(r, 1)$-Designs

D. McCARTHY, N. M. SINGHI,
and
S. A. VANSTONE

1. Introduction

An $(r, 1)$-*design* D consists of a set of *varieties* V and a collection of subsets (called *blocks*) of V such that each variety occurs in precisely r blocks, and each pair of distinct varieties occurs in precisely one block. D is said to be *trivial* if it contains a block containing all the varieties.

It is well known [2] that any nontrivial $(r, 1)$-design contains at most $r^2 - r + 1$ varieties and contains exactly this number of varieties if and only if it is a finite projective plane of order $r - 1$. McCarthy and Vanstone [1] have shown that any nontrivial $(r, 1)$-design having more than $n^2 - \alpha$ varieties is always embeddable in a finite projective plane of order $r - 1$ provided $r > 2\alpha^2 + 3\alpha + 2$. In this paper, we give a general theorem for embedding $(r, 1)$-designs within the framework of graph theory. As a particular case of this result, we deduce the design embedding just stated. In order to do this, we require several definitions.

For any graph G we define

(i) $d_1(x)$ to be the degree of vertex x.
(ii) $d_2(x)$ to be the number of paths of length 2 (2-paths) emanating from x.
(iii) $d_3(x, y)$ to be the number of 2-paths between 2 nonadjacent vertices x and y.
(iv) $d_4(x, y)$ to be the number of 2-paths between 2 adjacent vertices x and y.

Let D be an $(r, 1)$-design on v varieties. We define the $(r, 1)$-*graph* of D to be the graph, $G(D)$, obtained from D by associating a vertex with each block of D and joining two vertices with an edge if and only if the associated blocks have a variety in common. We define the *index* of $G(D)$ to be v. It is clear that an $(r, 1)$-graph inherits certain properties from the underlying $(r, 1)$-design.

Each variety in D corresponds to an r-clique in $G(D)$, and two r-cliques of this type have precisely one common vertex. So any graph G is an $(r, 1)$-graph of index v if and only if G can be decomposed into v r-cliques, any pair of which has precisely one common vertex. We consider some numerical conditions of $G(D)$. Suppose that the size (cardinality) of block B_i is k_i. If we let B_i also denote the corresponding vertex in $G(D)$, the following are easily verified.

$$d_1(B_i) = k_i(r - 1), \tag{1.1}$$

$$d_2(B_i) = k_i(r - 1)(v - k_i + r - 2), \tag{1.2}$$

$$d_3(B_i, B_j) = k_i k_j, \tag{1.3}$$

$$d_4(B_i, B_j) = (k_i - 1)(k_j - 1) + r - 2, \tag{1.4}$$

$$\sum_i k_i = vr. \tag{1.5}$$

We then define a *pseudo $(r, 1)$-graph* to be a graph G, with vertex set $V = \{B_1, B_2, \ldots, B_b\}$, which satisfies (1.1)–(1.5) for some integer v and some set of integers $\{k_1, k_2, \ldots, k_b\}$.

Certainly, (1.1)–(1.5) are necessary conditions for a graph to be an $(r, 1)$-graph, but they are not sufficient, as K_{43} exists while the corresponding $(7, 1)$-design, the finite projective plane of order six, does not exist. The two main results of this paper are stated as Theorems 1.1 and 1.2. Their proofs will be given in Sections 2 and 3, respectively.

Theorem 1.1 *A pseudo $(r, 1)$-graph of index v is an $(r, 1)$-graph of index v if*

(i) $r > v - 1$ *or*
(ii) $r \leq v - 1$ *and* $r > 2(v - r)^2 + 4(v - r) + 3$.

We let D_1 and D_2 represent $(r, 1)$-designs with variety sets V_1 and V_2, respectively, and block collections B_1 and B_2, respectively. We say that D_1 is *embedded* in D_2 if $V_1 \subseteq V_2$ and $B_1 \subseteq B_2$, where $B_1 \subseteq B_2$ implies containment up to multiplicity. If we let $G \backslash H$ denote the complement in a graph G of a subgraph H then the second main result may be stated as

Theorem 1.2 *An $(r, 1)$-design D_1 on v_1 varieties can be embedded in an $(r, 1)$-design E on v varieties if $G(E) \backslash G(D_1)$ is a pseudo $(r, 1)$-graph of index $v - v_1$ and*

  (i) $r > v - v_1 - 1$ or
  (ii) $r \leq v - v_1 - 1$ and $r > 2(v - v_1 - r)^2 + 4(v - v_1 - r) + 3$.

2. Proof of Theorem 1.1

We let G denote a pseudo $(r, 1)$-graph of index v with vertex set $V(G) = \{B_1, B_2, \ldots, B_b\}$ for which $d_1(B_i) = k_i(r - 1)$.

Lemma 2.1
$$\sum_{i=1}^{b} k_i(k_i - 1) = v(v - 1).$$

Proof Counting 2-paths through each vertex we obtain
$$\sum_{i=1}^{b} \tfrac{1}{2} k_i(r - 1)(k_i(r - 1) - 1),$$
and counting 2-paths which start at a vertex we have $\sum_{i=1}^{b} \tfrac{1}{2} d_2(B_i)$. Thus
$$\sum_{i=1}^{b} k_i(rk_i - k_i - 1) = \sum_{i=1}^{b} k_i(v - k_i + r - 2)$$
and since $\sum_{i=1}^{b} k_i = vr$ we obtain $\sum_{i=1}^{b} k_i(k_i - 1) = v(v - 1)$. ∎

We now suppose G contains a vertex X of degree $d(r - 1)$ through which there are d edge-disjoint r-cliques. We denote the vertex set of each r-clique by V_i^* ($i = 1, 2, \ldots, d$) and let $V_i = V_i^* \backslash \{X\}$ ($i = 1, 2, \ldots, d$).

Lemma 2.2
$$\sum_{\substack{j \\ B_j \in V_1 \cup \cdots \cup V_d}} (k_j - 1) = d(v - d).$$

Proof Since $d_2(X) = (v - d + r - 2)(r - 1)$, and
$$d_2(X) = \sum_{\substack{j \\ B_j \in V_1 \cup \cdots \cup V_d}} [d_1(B_j) - 1],$$
$$\sum_{\substack{j \\ B_j \in V_1 \cup \cdots \cup V_d}} [(r - 1)k_j - 1] = d(v - d + r - 2)(r - 1).$$

Thus
$$\sum_{\substack{j \\ B_j \in V_1 \cup \cdots \cup V_d}} k_j = d(v - d + r - 1)$$

and the result follows. ∎

Lemma 2.3 *If $d = 1$ or $d = 2$,*
$$\sum_{\substack{j \\ B_j \in V_i}} (k_j - 1) = v - d \qquad (i = 1, 2).$$

Proof For $d = 1$ the result follows directly from Lemma 2.2. For $d = 2$ we count the number of edges joining a vertex of V_1 to a vertex of V_2. For a vertex B_i in V_1 $d_4(X, B_i) = k_i - 1 + r - 2$. Since B_i is joined to the other $r - 2$ vertices in V_1, B_i is joined to precisely $k_i - 1$ vertices of V_2. Thus there are
$$\sum_{\substack{j \\ B_j \in V_2}} (k_j - 1)$$
edges from V_2 to V_1. So
$$\sum_{\substack{j \\ B_j \in V_1}} (k_j - 1) = \sum_{\substack{j \\ B_j \in V_2}} (k_j - 1).$$

Since
$$\sum_{\substack{j \\ B_j \in V_1 \cup V_2}} (k_j - 1) = 2(v - 2)$$

from Lemma 2.2, we have
$$\sum_{\substack{j \\ B_j \in V_1}} (k_j - 1) = \sum_{\substack{j \\ B_j \in V_2}} (k_j - 1) = v - 2. \qquad \blacksquare$$

We now consider one of the r-cliques with vertex set V_i^* and let B_j be any vertex not in V_i^*. We let l_j denote the number of vertices in V_i^* which are adjacent to B_j.

Lemma 2.4 *If*
$$\sum_{\substack{t \\ B_t \in V_i}} (k_t - 1) = v - d, \qquad l_j = k_j.$$

Proof We shall prove that
$$\sum_{\substack{j \\ B_j \notin V_i^*}} (l_j - k_j)^2 = 0.$$

We count the number of edges of type (B_t, B_j), where $B_t \in V_i^*$ and $B_j \notin V_i^*$. We obtain

$$\sum_{\substack{j \\ B_j \notin V_i^*}} l_j = \sum_{\substack{t \\ B_t \in V_i^*}} (r-1)(k_t - 1) = (r-1)(v-1)$$

since

$$\sum_{\substack{t \\ B_t \in V_i}} (k_t - 1) = v - d \quad \text{and} \quad d_1(X) = d(r-1).$$

Now

$$\sum_{\substack{j \\ B_j \notin V_i^*}} k_j = \sum_{j=1}^{b} k_j - \sum_{\substack{t \\ B_t \in V_i^*}} k_t = (r-1)(v-1)$$

so

$$\sum_{\substack{j \\ B_j \notin V_i^*}} l_j = \sum_{\substack{j \\ B_j \notin V_i^*}} k_j. \qquad (2.1)$$

We now count the number of paths of the type $B_t B_j B_l$ where $B_t, B_l \in V_i^*$ ($B_t \neq B_l$) and $B_j \notin V_i^*$. This is

$$\frac{1}{2} \sum_{\substack{j \\ B_j \notin V_i^*}} l_j(l_j - 1) = \frac{1}{2} \sum_{\substack{t \\ B_t \in V_i^*}} \sum_{\substack{l \\ B_l \in V_i^*, B_l \neq B_t}} [d_4(B_t, B_l) - (r-2)]$$

$$= \frac{1}{2} v(v-1) - \frac{1}{2} \sum_{\substack{t \\ B_t \in V_i^*}} k_t(k_t - 1)$$

$$= \frac{1}{2} \sum_{\substack{j \\ B_j \notin V_i^*}} k_j(k_j - 1) \quad \text{from Lemma 2.1.}$$

So

$$\sum_{\substack{j \\ B_j \notin V_i^*}} l_j(l_j - 1) = \sum_{\substack{j \\ B_j \notin V_i^*}} k_j(k_j - 1)$$

and from (2.1) we have

$$\sum_{\substack{j \\ B_j \notin V_i^*}} l_j^2 = \sum_{\substack{j \\ B_j \notin V_i^*}} k_j^2. \qquad (2.2)$$

We now count paths of the type $B_t B_j B_l$ in which $B_t \in V_i^*$, $B_j \notin V_i^*$, and B_l is any vertex other than B_j. Counting these paths through B_j we obtain

$$\sum_{\substack{j \\ B_j \notin V_i^*}} l_j(r-1)k_j$$

and counting the same paths originating at B_t we obtain

$$\sum_{\substack{t \\ B_t \in V_i^*}} \sum_{\substack{j \\ B_j \notin V_i^*;\, (B_t, B_j) \in E(G)}} (r-1)k_j.$$

Since

$$d_2(B_t) = \sum_{(B_t, B_j) \in E(G)} ((r-1)k_j - 1) = (r-1)k_t(v - k_t + r - 2),$$

we obtain

$$\sum_{\substack{j \\ (B_t, B_j) \in E(G)}} k_j = k_t(v - k_t + r - 1).$$

Now

$$\sum_{\substack{j \\ B_j \in V_i^*}} k_j = v + r - 1$$

so

$$\sum_{\substack{j \\ B_j \notin V_i^*;\, (B_t, B_j) \in E(G)}} k_j = (k_t - 1)(v - k_t + r - 1).$$

From the equations for the paths we now have

$$\sum_{\substack{j \\ B_j \notin V_i^*}} l_j k_j = (v + r - 1) \sum_{\substack{t \\ B_t \in V_i^*}} (k_t - 1) - \sum_{\substack{t \\ B_t \in V_i^*}} k_t^2 + \sum_{\substack{t \\ B_t \in V_i^*}} k_t.$$

Since

$$\sum_{\substack{t \\ B_t \in V_i^*}} (k_t - 1) = v - 1,$$

$$\sum_{\substack{j \\ B_j \notin V_i^*}} l_j k_j = \sum_{\substack{j \\ B_j \notin V_i^*}} k_j^2. \tag{2.3}$$

From (2.2) and (2.3),

$$\sum_{\substack{j \\ B_j \notin V_i^*}} (l_j - k_j)^2 = 0$$

and the lemma follows. ∎

Lemma 2.5 *A vertex $B_j \in V_i$ is adjacent to exactly $(d-1)(k_j - 1)$ vertices in $\bigcup_{k=1}^{d} V_k \setminus V_i$.*

Proof Since X is only adjacent to vertices in $\bigcup_{k=1}^{d} V_k$, and $d_4(B_j, X) = (k_j - 1)(d - 1) + r - 2$, the result follows. ∎

Lemma 2.6 *Let G be a pseudo $(r, 1)$-graph of index v. Let X be a vertex of degree $d(r - 1)$ through which there are d edge-disjoint r-cliques with vertex sets $V_1^*, V_2^*, \ldots, V_d^*$. We let G^* be the subgraph of G obtained by removing the edges of the r-cliques. If*

$$\sum_{\substack{j \\ B_j \in V_i}} (k_j - 1) = v - d \qquad \text{for} \quad i = 1, 2, \ldots, d,$$

G^* is a pseudo $(r, 1)$-graph of index $v - d$.

Proof We shall let $d_1^*(x)$, $d_2^*(x)$, $d_3^*(x, y)$, and $d_4^*(x, y)$ represent the appropriate functions in G^*. Since $d_1^*(X) = 0$, $d_1^*(B_i) = k_i^*(r - 1) = (k_i - 1)(r - 1)$ for $B_i \in \bigcup_{j=1}^{d} V_j$, and $d_1^*(B_i) = k_i^*(r - 1) = k_i(r - 1)$ for $B_i \notin \bigcup_{j=1}^{d} V_j$ condition (1.1) is satisfied in G^*. If $B_i \in V_k$ for some k, then

$$d_2^*(B_i) = d_2(B_i) - \sum_{\substack{j \\ B_j \neq B_i;\, B_j \in V_k^*}} [(r - 1)k_j - 1] - (r - 1)(d - 1)(k_i - 1)$$

follows from Lemma 2.5. Since

$$\sum_{\substack{j \\ B_j \in V_k}} (k_j - 1) = v - d$$

by hypothesis, we have

$$\begin{aligned}
d_2^*(B_i) &= (r - 1)k_i(v - k_i + r - 2) - (r - 1)(v - k_i + r - 2) \\
&\quad - (r - 1)(d - 1)(k_i - 1) \\
&= (r - 1)(k_i - 1)(v - d - (k_i - 1) + r - 2)
\end{aligned}$$

which satisfies (1.2) in G^*. If $B_i \notin V_k$ for some k, then by Lemma 2.4 B_i is adjacent to precisely k_i vertices in V_j^*, $j = 1, 2, \ldots, d$. Thus $d_2^*(B_i) = d_2(B_i) - d(r - 1)(k_i) = (r - 1)k_i(v - d - k_i + r - 2)$ which satisfies (1.2) in G^*. With Lemma 2.4 it is also possible to show that $d_3^*(B_i, B_j)$ and $d_4^*(B_i, B_j)$ satisfy (1.3) and (1.4), respectively, in G^*. Finally, it easily verified that $\sum_{i=1}^{b} k_i^* = (v - d)r$, so (1.5) is satisfied by G^*, and the proof is complete. ∎

We now determine conditions under which a pseudo $(r, 1)$-graph satisfies the hypothesis of Lemma 2.6 in the cases $d = 1$ and $d = 2$.

Lemma 2.7 *A pseudo $(r, 1)$-graph of index v contains a vertex of positive degree $d(r - 1)$ or less if $v < dr + 1$.*

Proof Assuming that $d_1(B_i) \geq (d + 1)(r - 1)$ for all i, $d_2(B_i) \geq k_i(r - 1)(dr + r - d - 2)$, but from (1.2), $d_2(B_i) = k_i(r - 1)(v - k_i + r - 2)$. Thus $v - k_i + r - 2 \geq dr + r - d - 2$, and so $v \geq dr + 1$ since $k_i \geq d + 1$. ∎

Lemma 2.8 *A vertex of degree $r - 1$ in a pseudo $(r, 1)$-graph is contained in an r-clique.*

Proof We let X denote a vertex of degree $r - 1$, and we consider a vertex B_i of degree $k_i(r - 1)$ which is adjacent to X. Since $d_4(B_i, X) = r - 2$, B_i is adjacent to each of the $r - 2$ other vertices which are adjacent to X. The result follows. ∎

Lemma 2.9 *Let G be a pseudo $(r, 1)$-graph of index v which contains no vertex of positive degree less than $d(r - 1)$. The maximum degree in G is $[v - (d - 1)(r - 1)](r - 1)$.*

Proof Let B_i be a vertex of maximum degree $k_i(r - 1)$. By hypothesis $d_1(B_j) \geq d(r - 1)$; therefore $d_2(B_i) = k_i(r - 1)(v - k_i + r - 2) \geq k_i(r - 1)[d(r - 1) - 1]$, and we have $k_i \leq v - (d - 1)(r - 1)$. ∎

We note that this implies $v - (d - 1)(r - 1) \geq d$ and for $d = 2, v \geq r + 1$.

Lemma 2.10 *Let G be a pseudo $(r, 1)$-graph of index v with no vertex of positive degree less than $d(r - 1)$. If $v \leq dr$ then any vertex of degree $d(r - 1)$ is adjacent to at least $d(dr - v)$ vertices of degree $d(r - 1)$.*

Proof Let B_i be a vertex of degree $d(r - 1)$ in G, and suppose B_i is adjacent to u vertices of degree $d(r - 1)$ and $d(r - 1) - u$ vertices of degree at least $(d + 1)(r - 1)$. Then the number of 2-paths from B_i is at least

$$u[d(r - 1) - 1] + [d(r - 1) - u][(d + 1)(r - 1) - 1].$$

So

$$d_2(B_i) = (r - 1)(d)(v - d + r - 2) \geq (u - 1)d(r - 1) + d(d + 1)(r - 1)^2 - u(d + 1)(r - 1)$$

which yields $u \geq d(dr - v)$.

For the following results, we let G be a pseudo $(r, 1)$-graph of index v with $v < 2r$. From Lemma 2.7, G contains a vertex of degree $r - 1$ and/or a vertex of degree $2(r - 1)$. We assume that G contains a vertex B of degree $2(r - 1)$ and no vertex of degree $r - 1$. Thus $v \geq r + 1$. The set of $2(r - 1)$ vertices adjacent to B we denote by U, and by Lemma 2.10, U contains a vertex B' of degree $2(r - 1)$. Since $d_4(B, B') = r - 1$, we partition the remaining $2r - 3$ vertices of U into sets T and T'; T contains the $r - 2$ vertices which are adjacent to B and are not adjacent to B', while T' contains the $r - 1$ vertices adjacent to both B and B'. ∎

Lemma 2.11 *A vertex $B^* \in T$ of degree $(\gamma + 1)(r - 1)$ with $\gamma \leq r - 3$ is adjacent to at least $r - \gamma - 3$ vertices of T.*

Proof Since $d_3(B', B^*) = 2\gamma + 2$ and $B'BB^*$ is a 2-path, B^* is adjacent to at most $2\gamma + 1$ vertices of T'. As $d_4(B, B^*) = \gamma + r - 2$, the result follows. ∎

Lemma 2.12 *The induced subgraph of T is an $(r-2)$-clique if $r > (v-r)^2 + 4(v-r) + 2$.*

Proof Assume there are two nonadjacent vertices $B_i, B_j \in T$ whose degrees are $(\gamma + 1)(r-1)$ and $(\delta + 1)(r-1)$, respectively. From Lemma 2.11, B_i and B_j are both adjacent to at least $(r-\gamma-3) + (r-\delta-3) - (r-4) = r - \gamma - \delta - 2$ vertices of T. Since $d_4(B_i, B_j) = (\gamma + 1)(\delta + 1)$ and $B_i BB_j$ is a 2-path, B_i and B_j are both adjacent to at most $\gamma\delta + \gamma + \delta$ vertices of T. Thus

$$r - \gamma - \delta - 2 \leq \gamma\delta + \gamma + \delta$$

and so

$$r \leq \gamma\delta + 2\delta + 2\gamma + 2$$

Now, if $r > \gamma\delta + 2\delta + 2\gamma + 2$ we have a contradiction, and B_i is adjacent to B_j. Setting γ and δ at their maximum values, obtained from Lemma 2.9 with $d = 2$, we obtain the stated result. ∎

Lemma 2.13 *T contains a vertex of degree $2(r-1)$ if $2v < 3r$.*

Proof Assuming that the degree of each vertex in T is at least $3(r-1)$ the number of 2-paths from B is at least $(r-2)(3r-4) + r(2r-3)$. So $d_2(B) = 2(r-1)(v+r-4) \geq (r-2)(3r-4) + r(2r-3)$, giving $3r \leq 2v$. The result follows. ∎

Lemma 2.14 *If $r > 2(v-r)^2 + 4(v-r) + 3$ there exists a vertex in T' which is adjacent to each vertex in T.*

Proof From Lemmas 2.12 and 2.13, T induces an $(r-2)$-clique and contains a vertex of degree $2(r-1)$, say B''.

We let S denote the $r-2$ vertices of U which are not adjacent to B''. Applying the proof technique of Lemma 2.12 to the set S we see that S induces an $(r-2)$-clique if $r > (v-r)^2 + 4(v-r) + 2$. Since S and T account for $2r - 4$ of the $2r - 2$ vertices adjacent to B, we let B_m denote one of the two remaining vertices with degree $k_m(r-1)$. Since $d_4(B, B_m) = k_m + r - 3$, B_m is adjacent to $k_m + r - 3$ vertices of $S \cup T \cup B_n$; thus B_m is adjacent to at least $(k_m + r - 4)/2$ vertices of S or T, say, T. We assume that some vertex $B_p \in T$, of degree $k_p(r-1)$, is not adjacent to B_m. Now B_m and B_p are both adjacent to at least $(k_m + r - 4)/2$ vertices of T, and since $d_3(B_m, B_p) = k_m k_p$, they are both adjacent to at most $k_m k_p - 1$ vertices of T (they are both adjacent to B). So $(k_m + r - 4)/2 \leq k_m k_p - 1$. If $(k_m + r - 4)/2 > k_m k_p - 1$ then B_m is adjacent to B_p. Using Lemma 2.9 with $d = 2$ we obtain the maximum value for k_m and k_p, which yields the desired result when used in the previous inequality. ∎

We let $T^* = T \cup B_m$. Using the same proof technique as in the previous lemma, it is clear that B_n is adjacent to each vertex in S or T^* if $r > 2(v - r)^2 + 4(v - r) + 3$. In the following results we will prove that B_n must be adjacent to each vertex in S.

For notational convenience we let $B = B_1$, $S = \{B_2, B_3, \ldots, B_{r-1}\}$, $T^* = \{B_r, B_{r+1}, \ldots, B_{2r-2}\}$, and $B_n = B_{2r-1}$. We assume that $T^* \cup \{B_{2r-1}\}$ induces an r-clique, while S induces an $(r-2)$-clique. We let $y_1 = \sum_{i=2}^{r-1} k_i$ and $y_2 = \sum_{i=r}^{2r-1} k_i$.

Lemma 2.15 $y_1 = v - 3$ and $y_2 = v + 2r - 3$.

Proof Since $d_2(B_1) = 2(r-1)(v+r-4) = \sum_{i=2}^{2r-1} [(r-1)k_i - 1]$ we obtain

$$y_1 + y_2 = \sum_{i=2}^{2r-1} k_i = 2(v + r - 3). \quad (2.4)$$

We now count the edges (B_i, B_j) for $2 \leq i \leq r-1$, $r \leq j \leq 2r-1$. These are given by $\sum_{i=2}^{r-1} [d_4(B_1, B_i) - (r-3)]$, and $\sum_{j=r}^{2r-1} [d_4(B_1, B_j) - (r-1)]$. With (1.4) we obtain

$$\sum_{i=2}^{r-1} k_i = \sum_{j=r}^{2r-1} (k_j - 2) = \sum_{j=r}^{2r-1} k_j - 2r.$$

Therefore, $y_2 - y_1 = 2r$; so with (2.4) we have

$$y_1 = \sum_{i=2}^{r-1} k_i = v - 3 \quad \text{and} \quad y_2 = \sum_{j=r}^{2r-1} k_j = v + 2r - 3$$

as required. ∎

We define x_l to be the number of vertices B_i ($1 \leq i \leq r-1$) which are adjacent to B_l ($r \leq l \leq b$).

Lemma 2.16 $x_l = k_l - 1$ ($r \leq l \leq b$) if $k_l \neq 0$; otherwise $x_l = 0$.

Proof We consider the following two sums:

$$S_1 = \sum_{l=r}^{b} x_l(x_l - k_l + 1) \quad \text{and} \quad S_2 = \sum_{l=r}^{b} (x_l - k_l)(x_l - k_l + 1).$$

We shall prove the result by showing that $S_1 = S_2 = 0$.

The number of edges of type (B_i, B_j) ($1 \leq i \leq r-1$; $r \leq j \leq b$) is given by

$$\sum_{j=r}^{b} x_j = \sum_{i=1}^{r-1} [(r-1)k_i - (r-2)] = (r-1)(v - r + 1). \quad (2.5)$$

EMBEDDING $(r, 1)$-DESIGNS

We now count the 2-paths of type $B_i B_t B_j$ where $1 \leq i, j \leq r - 1$, and $r \leq t \leq b$. These are given by

$$\sum_{\substack{i,j=1 \\ i \neq j}}^{r-1} [d_4(B_i, B_j) - (r - 3)] \quad \text{and} \quad \sum_{j=r}^{b} x_j(x_j - 1).$$

So

$$\sum_{j=r}^{b} x_j(x_j - 1) = \sum_{\substack{i,j=1 \\ i \neq j}}^{r-1} [(k_i - 1)(k_j - 1) + (r - 2) - (r - 3)]$$

$$= \sum_{i=1}^{r-1} (k_i - 1)\left[\sum_{j=1}^{r-1} (k_j - 1) - (k_i - 1)\right] + (r - 1)(r - 2)$$

$$= \sum_{i=1}^{r-1} (k_i - 1)(v - 3 + 2 - k_i - r + 2) + (r - 1)(r - 2)$$

$$= (v - r + 1)(v - r) - \sum_{i=1}^{r-1} k_i^2 + v - 1 + (r - 1)(r - 2).$$

Using (2.5) we obtain

$$\sum_{j=r}^{b} x_j^2 = (v + 1)(v - r) + r(r - 1) - \sum_{i=1}^{r-1} k_i^2. \tag{2.6}$$

The number of 2-paths of type $B_i B_j B_k$ ($1 \leq i \leq r - 1; r \leq j \leq b; 1 \leq k \leq b$) is given by

$$\sum_{j=r}^{b} x_j[(r - 1)k_j - 1] \quad \text{and} \quad \sum_{i=1}^{r-1} d_2(B_i) - \sum_{\substack{i,j=1 \\ i \neq j}}^{r-1} [(r - 1)k_j - 1].$$

So

$$\sum_{j=r}^{b} x_j(r - 1)k_j = \sum_{j=r}^{b} x_j + \sum_{i=1}^{r-1} (r - 1)k_i(v - k_i + r - 2)$$

$$- \sum_{i=1}^{r-1} \left(\sum_{j=1}^{r-1} [(r - 1)k_j - 1] - (r - 1)k_i + 1\right)$$

$$= (r - 1)(v - r + 1) + (r - 1)\sum_{i=1}^{r-1} k_i(v - k_i + r - 2)$$

$$- \sum_{i=1}^{r-1} (r - 1)(v - 1 - k_i) + (r - 1)(r - 2).$$

Thus

$$\sum_{j=r}^{b} x_j k_j = v - r + 1 + \sum_{i=1}^{r-1} [k_i(v - k_i + r - 2) - (v - 1 - k_i)] + r - 2$$

$$= (v + 1)(v - 1) - \sum_{i=1}^{r-1} k_i^2. \tag{2.7}$$

From Lemma 2.1

$$\sum_{i=1}^{b} k_i(k_i - 1) = v(v - 1)$$

so

$$\sum_{i=r}^{b} k_i(k_i - 1) = (v + 1)(v - 1) - \sum_{i=1}^{r-1} k_i^2. \tag{2.8}$$

Evaluating S_1 with (2.5), (2.6), and (2.7) we have $S_1 = 0$, while evaluating S_2 using (2.5), (2.6), (2.7), and (2.8) gives $S_2 = \sum_{l=r}^{b} (x_l - k_l)(x_l - k_l + 1) = 0$. $S_2 = 0$ shows that $x_l = k_l$ or $x_l = k_l - 1$, and this with $S_1 = 0$ shows that $x_l = k_l - 1$ or $x_l = k_l = 0$. ∎

Lemma 2.17 *Let G be a pseudo $(r, 1)$-graph of index v which contains a vertex B of degree $2(r - 1)$ and no vertex of degree $r - 1$. If $r > 2(v - r)^2 + 4(v - r) + 3$ the set U of $2(r - 1)$ vertices adjacent to B can be partitioned into two disjoint sets, each of cardinality $r - 1$, and each of which induces an $(r - 1)$-clique.*

Proof From Lemma 2.14 and the subsequent comments, if $r > 2(v - r)^2 + 4(v - r) + 3$ either U can be partitioned into an r-set and an $(r - 2)$-set which induce an r-clique and an $(r - 2)$-clique, respectively, or U can be partitioned into two $(r - 1)$-sets each of which induces an $(r - 1)$-clique. G contains b vertices and with no loss in generality we assume that $k_i = 0$ for $i > b - \alpha$ and $k_i \neq 0$ for $i \leq b - \alpha$ (if $\alpha = 0$ then $k_i \neq 0$ for any i). Assuming the former case to be true for U, we have from Lemma 2.16

$$\sum_{l=r}^{b-\alpha} (k_l - x_l) = b - \alpha - (r - 1).$$

Thus

$$b - \alpha = r - 1 + \sum_{l=r}^{b-\alpha} k_l - \sum_{l=r}^{b-\alpha} x_l$$

$$= r - 1 + vr - \sum_{l=1}^{r-1} k_l - (r - 1)(v - r + 1) = r^2 - r + 1. \tag{2.9}$$

So

$$\bar{k} = \frac{1}{b-\alpha} \sum_{i=1}^{b-\alpha} k_i = \frac{vr}{r^2 - r + 1}.$$

Since $r > 2(v - r)^2 + 4(v - r) + 3$, $v \le 2r - 2$, and so

$$\bar{k} = \frac{vr}{r^2 - r + 1} \le \frac{2r^2 - 2r}{r^2 - r + 1} < \frac{2r^2 - 2r + 2}{r^2 - r + 1} = 2.$$

But $\bar{k} < 2$ implies that some vertex has degree $r - 1$, and this contradicts the hypothesis. Thus U can be partitioned into two disjoint $(r-1)$-sets, each of which induces an $(r-1)$-clique. ∎

Theorem 1.1 *Let G be a pseudo $(r, 1)$-graph of index v. G is an $(r, 1)$-graph of index v if*

(i) $r > v - 1$ or
(ii) $r \le v - 1$ and $r > 2(v - r)^2 + 4(v - r) + 3$.

Proof For $r \le v - 1$, if $r > 2(v - r)^2 + 4(v - r) + 3$, $v < 2r + 1$, and so by Lemma 2.7 G contains a vertex of degree $r - 1$ or $2(r - 1)$.

If there is a vertex of degree $r - 1$, it is contained in an r-clique by Lemma 2.8. Then by Lemmas 2.3 and 2.6 removing the edges of the r-clique leaves a pseudo $(r, 1)$-graph of index $v - 1$. If there is no vertex of degree $r - 1$, then the vertex of degree $2(r - 1)$ is contained in two edge-disjoint r-cliques by Lemma 2.17. Removing the edges from these two r-cliques we obtain a pseudo $(r, 1)$-graph of index $v - 2$ using Lemmas 2.3 and 2.6. Applying this argument iteratively to the subsequent pseudo $(r, 1)$-graphs we see that G can be decomposed into a set of v edge-disjoint r-cliques. In the case $v < r + 1$, the previous decomposition of G is also valid since G and the subsequent pseudo $(r, 1)$-subgraphs contain a vertex of degree $r - 1$ by Lemma 2.7. Since there are $\binom{k_i}{2}$ distinct pairs of edge-disjoint r-cliques through a vertex of degree $k_i(r - 1)$, and $\sum_{i=1}^{b} \binom{k_i}{2} = \binom{v}{2}$, any two r-cliques have precisely one vertex in common. Thus we see that G is an $(r, 1)$-graph of index v. ∎

3. Proof of Theorem 1.2 and an Application

Theorem 1.2 *An $(r, 1)$-design D_1 on v_1 varieties is embeddable in an $(r, 1)$-design E on v varieties if $G(E) \backslash G(D_1)$ is a pseudo $(r, 1)$-graph of index $v - v_1$ and*

(i) $r > v - v_1 - 1$ or
(ii) $r \le v - v_1 - 1$ and $r > 2(v - v_1 - r)^2 + 4(v - v_1 - r) + 3$.

Proof Since $G(E)\backslash G(D_1)$ is a pseudo $(r, 1)$-graph, and $r > v - v_1 - 1$ or $r \leq v - v_1 - 1$ and $r > 2(v - v_1 - r)^2 + 4(v - v_1 - r) + 3$, $G(E)\backslash G(D_1)$ is an $(r, 1)$-graph of index $v - v_1$ by Theorem 1.1. We let D_2 be the associated $(r, 1)$-design on $v - v_1$ varieties. Now $G(D_1) \subseteq G(E)$, so there is a natural correspondence between the blocks of D_1 and D_2. We shall let V_1 and V_2 denote the variety sets of D_1 and D_2, respectively. We also let $B_1, B_2, \ldots, B_b$ and $\bar{B}_1, \bar{B}_2, \ldots, \bar{B}_b$ denote the blocks of D_1 and D_2, respectively, where B_i and $\bar{B}_i$ are related by the correspondence just mentioned. Then we define $D_1 \cup D_2$ to be the incidence structure defined on the variety set $V_1 \cup V_2$, and with blocks $B_1 \cup \bar{B}_1, B_2 \cup \bar{B}_2, \ldots, B_b \cup \bar{B}_b$. Certainly pairs of varieties from D_1 or D_2 occur precisely once in the blocks of $D_1 \cup D_2$. Since adjacent vertices in D_1 are nonadjacent in D_2, no pair of varieties v_i, v_j ($v_i \in V_1$, $v_j \in V_2$) occurs more than once in the blocks of $D_1 \cup D_2$. There are $\binom{v}{2}$ pairs in the blocks of E; since $|V_1 \cup V_2| = v$ and no pair of varieties occurs twice in $D_1 \cup D_2$, $D_1 \cup D_2$ is an $(r, 1)$-design on v varieties. Thus $D_1 \cup D_2 = E$ and the proof is complete. ∎

Using this theorem we will determine some results on embedding $(r, 1)$-designs into finite projective planes. These results can also be found in McCarthy and Vanstone [1].

Lemma 3.1 *Let D be an $(r, 1)$-design of index $r^2 - r + 1 - v$. Then $H = K_{r^2 - r + 1} \backslash G(D)$ is a pseudo $(r, 1)$-graph of index v.*

Proof Certainly $G(D) \subseteq K_{r^2 - r + 1}$. We let $B_1, B_2, \ldots, B_{r^2 - r + 1}$ denote the vertices of $G(D)$, where we add vertices of degree zero if necessary. The four functions $d_1(x), d_2(x), d_3(x, y), d_4(x, y)$ are distinguished in $G(D)$ and H by the superscripts G and H, respectively. Without loss of generality we let $d_1^{(G)}(B_i) = (r - k_i)(r - 1)$ for $i = 1, 2, \ldots, r^2 - r + 1$. It is immediate that

$$d_1^{(H)}(B_i) = k_i(r - 1). \tag{3.1}$$

Now $d_2^{(G)}(B_i) = (r - 1)(r - k_i)(r^2 - r - 1 - v + k_i)$ from (1.2). Also

$$d_2^{(G)}(B_i) = \sum_{\substack{j \\ (B_i, B_j) \in E(G)}} [d_1^{(G)}(B_j) - 1]$$

$$= \sum_{\substack{j \\ (B_i, B_j) \in E(G)}} (r - 1)(r - k_j) - (r - 1)(r - k_i).$$

Therefore

$$\sum_{\substack{j \\ (B_i, B_j) \in E(G)}} (r - k_j) = (r - k_i)(r^2 - r - v + k_i)$$

EMBEDDING $(r, 1)$-DESIGNS

and so
$$\sum_{\substack{j \\ (B_i, B_j) \in E(G)}} k_j = (r - k_i)(v - k_i). \tag{3.2}$$

Since $G(D)$ is an $(r, 1)$-graph of order $r^2 - r + 1 - v$,
$$\sum_{i=1}^{r^2-r+1} (r - k_i) = (r^2 - r + 1 - v)r$$

and so
$$\sum_{i=1}^{r^2-r+1} k_i = vr. \tag{3.3}$$

So
$$d_2^{(H)}(B_i) = \sum_{\substack{j \\ (B_i, B_j) \in E(H)}} [d_1^{(H)}(B_j) - 1] = \sum_{\substack{j \\ (B_i, B_j) \notin E(G)}} (r-1)k_j - (r-1)k_i$$
$$= (r - 1)[vr - k_i - (r - k_i)(v - k_i) - k_i]$$

from (3.2) and (3.3).

This simplifies to give
$$d_2^{(H)}(B_i) = (r - 1)k_i(v - k_i + r - 2). \tag{3.4}$$

Since nonadjacent vertices in H are adjacent in $G(D)$, we have by the principle of inclusion–exclusion
$$d_3^{(H)}(B_i, B_j) = r^2 - r - 1 - [d_1^{(G)}(B_i) - 1] - [d_1^{(G)}(B_j) - 1] + d_4^{(G)}(B_i, B_j)$$
$$= k_i k_j. \tag{3.5}$$

Similarly, adjacent vertices in H are nonadjacent in $G(D)$, so
$$d_4^{(H)}(B_i, B_j) = r^2 - r - 1 - d_1^{(G)}(B_i) - d_1^{(G)}(B_j) + d_3^{(G)}(B_i, B_j)$$
$$= (k_i - 1)(k_j - 1) + r - 2. \tag{3.6}$$

Now (3.1), (3.4), (3.5), (3.6), and (3.3) satisfy (1.1)–(1.5), respectively, and the proof is complete. ∎

Since K_{r^2-r+1} is the $(r, 1)$-graph of the finite projective plane of order $r - 1$ we obtain the following

Theorem 3.1 *Let D be an $(n + 1, 1)$-design on v_1 varieties. Then*

(i) *for $v_1 > n^2 - 1$, D can be embedded in a finite projective plane of order n.*

(ii) *for $v_1 = n^2 - \alpha$ ($\alpha \geq 1$), D can be embedded in a finite projective plane or order n if $n > 2\alpha^2 + 4\alpha + 2$.*

Proof We let $n = r - 1$. From Lemma 3.1 and (i) of Theorem 1.2, with $v = r^2 - r + 1 = n^2 + n + 1$ we obtain the previous (i). Then letting $v_1 = n^2 - \alpha$ ($\alpha \geq 1$) and using (ii) of Theorem 1.2 and Lemma 3.1 we obtain the previous (ii). ∎

REFERENCES

1. D. McCarthy and S. A. Vanstone, Embedding $(r, 1)$-designs in finite projective planes, *Discrete Math.* **19** (1977) 67–76.
2. R. G. Stanton and R. C. Mullin, Inductive methods for balanced incomplete block designs, *Ann. Math. Statist.* **37** (1966) 1348–1354.

AMS 05B30

D. McCarthy
DEPARTMENT OF COMPUTER SCIENCE
UNIVERSITY OF MANITOBA
WINNIPEG, MANITOBA

N. M. Singhi
SCHOOL OF MATHEMATICS
TATA INSTITUTE OF FUNDAMENTAL RESEARCH
COBALA, BOMBAY, INDIA

S. A. Vanstone
ST. JEROME'S COLLEGE
UNIVERSITY OF WATERLOO
WATERLOO, ONTARIO

Chromatic Enumeration for Triangulations

J. D. McFALL

Introduction

We shall use the same terminology as [1, Section 1]. In [1] Tutte derives an equation for the chromatic enumeration of rooted planar near-triangulations of the 2-sphere, and in [2] and [3] the equation is solved for λ equal to particular Beraha numbers. The equation remains unsolved for arbitrary λ.

By keeping track of the valencies of both endpoints of the root-edge of each triangulation we use a basic recursive relation between chromatic polynomials of particular maps to derive an equation for the chromatic enumeration of rooted planar triangulations of the 2-sphere. There is a marked resemblance between Eq. (5) of [4], for the dichromatic enumeration of rooted planar maps, and Eq. (20) of this paper.

An Equation for the Chromatic Enumeration of Rooted Planar Triangulations

Introduce the generating function

$$F = F(u_1, u_2, z, \lambda) = \sum_T u_1^{n_1(T)} u_2^{n_2(T)} z^{t(T)} P(T, \lambda) \qquad (1)$$

where the sum is taken over all combinatorially distinct rooted planar triangulations T (henceforth called triangulations), $n_1(T)$ is the valency of the root-vertex of T, $n_2(T)$ is the valency of the non-root-vertex of the root-edge of T, $t(T)$ is the total number of triangles of T, and $P(T, \lambda)$ is the chromatic polynomial of the graph corresponding to T.

Regarding F as a power series in z with coefficients in the ring of polynomials, over Z, in u_1, u_2 and λ we write

$$F = \sum_{k \geq 0} f_k z^k \tag{2}$$

where $f_k = f_k(u_1, u_2, \lambda) \in Z[u_1, u_2, \lambda]$ for all $k \geq 0$. Associated with a triangulation T is a triangulation $\tilde{T}$ obtained by reversing the direction of the root-edge of T. Since $P(T, \lambda) = P(\tilde{T}, \lambda)$ and $t(T) = t(\tilde{T})$ each f_k is symmetric in u_1 and u_2.

It is convenient to define

$$S = S(u_1, z, \lambda) = F(u_1, 1, z, \lambda) \tag{3}$$

and

$$R = R(u_2, z, \lambda) = F(1, u_2, z, \lambda),$$

and to write

$$S = \sum_{k \geq 0} s_k z^k \quad \text{and} \quad R = \sum_{k \geq 0} r_k z^k.$$

Only even powers of z occur in F and

$$f_0 = s_0 = r_0 = 0, \qquad f_2 = \lambda(\lambda - 1)(\lambda - 2) u_1{}^2 u_2{}^2$$

and

$$s_2 = \lambda(\lambda - 1)(\lambda - 2) u_1{}^2, \qquad r_2 = \lambda(\lambda - 1)(\lambda - 2) u_2{}^2. \tag{4}$$

As shown in Fig. 1a, for a triangulation T we denote by E the root-edge, by F the (unbounded) root-face, by V the root-vertex, by W the non-root-vertex of E, by U the remaining unnamed vertex adjacent to F, and by A the edge VU. The third vertex of the bounded triangle adjacent to A shall be denoted by Y. Of course it may be that $Y = W$. Let $\mathcal{K}$ be the collection of all triangulations T for which $Y \neq W$.

Given $T \in \mathcal{K}$ there are associated planar maps $\theta_A(T)$, $\phi_A(T)$, and $\psi_A(T)$. The map $\theta_A(T)$ is obtained from T by deleting A and joining Y to W by an edge in the root-face of T, as shown in Fig. 1b. Then $\theta_A(T)$ is a triangulation with root-face, root-edge, and root-vertex the same as those for T. As shown in Fig. 1c, $\phi_A(T)$ is obtained from T by deleting A and UW, and identifying U with V and UY with VY. Taking the root-vertex to be $V(=U)$, the root-edge to be VW, and the root-face to be the unbounded triangle, $\phi_A(T)$ is a

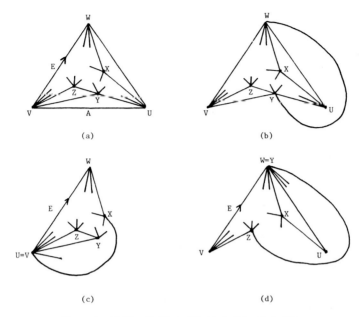

Figure 1 (a) $T \in \mathcal{K}$. (b) $\theta_A(T)$. (c) $\phi_A(T)$. (d) $\psi_A(T)$.

triangulation possibly having a loop at V. Lastly, $\psi_A(T)$ is obtained from T by deleting A and VY, and identifying Y with W and YU with WU, as shown in Fig. 1d. The root-vertex and root-edge of $\psi_A(T)$ are the same as in T and the root-face is the unbounded triangle of $\psi_A(T)$. Therefore, $\psi_A(T)$ is a triangulation possibly having a loop at W.

These maps are related by the equation (see [5])

$$P(T, \lambda) - P(\theta_A(T), \lambda) = P(\psi_A(T), \lambda) - P(\phi_A(T), \lambda). \tag{5}$$

Defining

$$F_1 = \sum_{T \in \mathcal{K}} u_1^{n_1(T)} u_2^{n_2(T)} z^{t(T)} P(T, \lambda), \tag{6}$$

$$F_2 = \sum_{T \in \mathcal{K}} u_1^{n_1(T)} u_2^{n_2(T)} z^{t(T)} P(\theta_A(T), \lambda), \tag{7}$$

$$F_3 = \sum_{T \in \mathcal{K}} u_1^{n_1(T)} u_2^{n_2(T)} z^{t(T)} P(\phi_A(T), \lambda), \tag{8}$$

and

$$F_4 = \sum_{T \in \mathcal{K}} u_1^{n_1(T)} u_2^{n_2(T)} z^{t(T)} P(\psi_A(T), \lambda) \tag{9}$$

by (5)
$$F_1 - F_2 + F_3 - F_4 = 0. \tag{10}$$

Each of the functions in (10) can be expressed in terms of F, S, and R.

Considering F_1, the triangulations T not in $\mathscr{K}$ have $Y = W$. As shown in Fig. 2 T may have digons adjacent to the outer edges VW and UW. Assuming neither digon degenerates to a link-map T defines triangulations T_1 and T_2.

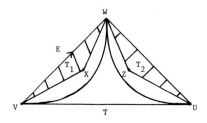

Figure 2

Specifically T_1 has root-vertex V, root-edge E, root-face an unbounded triangle with edges E, WX, and XV, and the interior of T_1 is the interior of the digon with the outer edge other than E deleted. The triangulation T_2 has root-vertex U, root-edge UW, root-face an unbounded triangle with edges UW, WZ, and ZU. The interior of T_2 is the interior of the digon with outer edge other than UW deleted.

Allowing for the cases when one or both digons are link-maps,

$$F_1 = F - \sum_{T \notin \mathscr{K}} u_1^{n_1(T)} u_2^{n_2(T)} z^{t(T)} P(T, \lambda)$$

$$= F - \lambda(\lambda - 1)(\lambda - 2) u_1^2 u_2^2 z^2 - \sum_{T_1} (u_1^{n_1(T_1)+2} u_2^{n_2(T_1)+2} z^{t(T_1)+2}$$

$$\times (\lambda - 2) P(T_1, \lambda)) - \sum_{T_2} (u_1{}^2 u_2^{n_2(T_2)+2} z^{t(T_2)+2} (\lambda - 2) P(T_2, \lambda))$$

$$- \sum_{(T_1, T_2)} \left(u_1^{n_1(T_1)+2} u_2^{n_2(T_1)+n_2(T_2)+2} z^{t(T_1)+t(T_2)+2} \right.$$

$$\left. \times \frac{(\lambda - 2)}{\lambda(\lambda - 1)} P(T_1, \lambda) P(T_2, \lambda) \right)$$

$$= (1 - (\lambda - 2) u_1^2 u_2^2 z^2) F - \frac{(\lambda - 2)}{\lambda(\lambda - 1)} u_1{}^2 u_2{}^2 z^2 FR$$

$$- (\lambda - 2) u_1{}^2 u_2{}^2 z^2 R - \lambda(\lambda - 1)(\lambda - 2) u_1{}^2 u_2{}^2 z^2. \tag{11}$$

From Fig. 1b it follows that

$$F_2 = u_1 u_2^{-1} \sum_{T \in \mathcal{K}} u_1^{n_1(\theta_A(T))} u_2^{n_2(\theta_A(T))} z^{t(\theta_A(T))} P(\theta_A(T), \lambda). \qquad (12)$$

Given an arbitrary triangulation N with $U \neq V$ (labeled as in Fig. 3a) there is a unique triangulation $T \in \mathcal{K}$, as shown in Fig. 3b, such that $\theta_A(T) = N$. Let $\mathcal{L}$ be the collection of all triangulations N for which $U = V$ (as labeled in Fig. 3a). That is, $\mathcal{L}$ is the collection of all triangulations N which are not of the form $\theta_A(T)$ for some $T \in \mathcal{K}$. For each $N \in \mathcal{L}$ there are possibly two digons adjacent with VW and VY, shown in Fig. 3c, and hence two triangulations N_1 and N_2 analogous to Fig. 2. When the digons are nondegenerate N_1 has root-vertex V, root-edge E, and root-face an unbounded triangle with edges VW, WX, and XV. The triangulation N_2 has root-vertex V, root-edge VY, and root-face an unbounded triangle with edges VZ, ZY, and YV. Allowing for either or both digons to be link-maps, (12) can be written as

$$F_2 = u_1 u_2^{-1} \left\{ F - \sum_{N \in \mathcal{L}} u_1^{n_1(N)} u_2^{n_2(N)} z^{t(N)} P(N, \lambda) \right\}$$

$$= u_1 u_2^{-1} \left\{ F - \lambda(\lambda - 1)(\lambda - 2) u_1^2 u_2^2 z^2 - \sum_{N_1} (u_1^{n_1(N_1)+2} u_2^{n_2(N_1)+2} \right.$$

$$\times z^{t(N_1)+2} (\lambda - 2) P(N_1, \lambda)) - \sum_{N_2} (u_1^{n_1(N_2)+2} u_2^2 z^{t(N_2)+2} (\lambda - 2) P(N_2, \lambda))$$

$$- \sum_{(N_1, N_2)} \left(u_1^{n_1(N_1)+n_1(N_2)+2} u_2^{n_2(N_1)+2} z^{t(N_1)+t(N_2)+2} \frac{(\lambda - 2)}{\lambda(\lambda - 1)} \right.$$

$$\times P(N_1, \lambda) P(N_2, \lambda) \right\} $$

$$= u_1 u_2^{-1} \left\{ (1 - (\lambda - 2) u_1^2 u_2^2 z^2) F - \frac{(\lambda - 2)}{\lambda(\lambda - 1)} u_1^3 u_2 z^2 FS \right.$$

$$- (\lambda - 2) u_1^3 u_2 z^2 S - \lambda(\lambda - 1)(\lambda - 2) u_1^3 u_2 z^2 \right\}. \qquad (13)$$

Comparing (13) and (11),

$$F_1(u_1, u_2, z, \lambda) = \frac{u_1}{u_2} F_2(u_2, u_1, z, \lambda). \qquad (14)$$

The term $u_1^{n_1(T)} u_2^{n_2(T)} z^{t(T)} P(T, \lambda)$ in F_1 appears on the right side of (14) as $u_1^{n_2(T')+1} u_2^{n_1(T')-1} z^{t(T')} P(\theta_A(T'), \lambda)$ where $T' = \widetilde{\theta_A(T)}$. Conversely, the term $u_1^{n_2(T)+1} u_2^{n_1(T)-1} z^{t(T)} P(\theta_A(T), \lambda)$ in $(u_1/u_2) F_2(u_2, u_1, z, \lambda)$ appears in F_1 as $u_1^{n_1(T'')} u_2^{n_2(T'')} z^{t(T'')} P(T'', \lambda)$ where $T'' = \widetilde{\theta_A(T)}$.

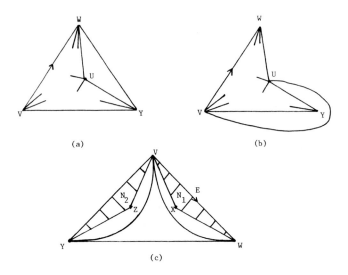

Figure 3 (a) $N = \theta_A(T)$. (b) T. (c) $N \in \mathcal{L}$.

In the sum F_3 we agree to omit all those terms corresponding to $T \in \mathcal{K}$ for which $\phi_A(T)$ has a loop. From Fig. 1c,

$$F_3 = u_2 z^2 \sum_{T \in \mathcal{K}} (u_1^{n_1(T)} u_2^{n_2(\phi_A(T))} z^{t(\phi_A(T))} P(\phi_A(T), \lambda)). \tag{15}$$

Given an arbitrary triangulation N, as in Fig. 4a, for each edge VY with $Y \neq W$ there is a unique $T \in \mathcal{K}$, shown in Fig. 4b, such that $\phi_A(T) = N$. As

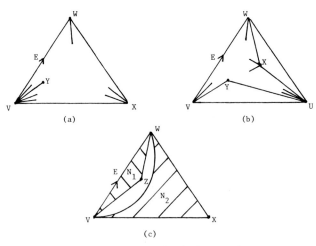

Figure 4 (a) $N = \phi_A(T)$. (b) T. (c) N.

Y varies over edges VY of N such that $Y \neq W$, $n_1(T)$ varies from three to $n_1(N) + 1$, missing those numbers corresponding to edges VY with $Y = W$ and equaling each number at most once.

When $Y = W$, in N, there will be a digon adjacent to E (we do not try to split the root-edge), and N defines triangulations N_1 and N_2 as in Fig. 4c. The root-vertex of N_1 is V, the root-edge is E, and the root-face is an unbounded triangle with edges VW, WZ, and ZV. The interior of N_1 is the interior of the digon with the outer edge other than E deleted. The root-vertex of N_2 is V, the root-edge is the outer edge of the digon other than E, and the root-face is an unbounded triangle with edges VX, XW, and the just defined root-edge. Any triangulation can occur as N_1 or N_2 and given N_1 and N_2 we can construct N. Hence (15) becomes

$$F_3 = u_2 z^2 \left\{ \sum_N \left(\sum_{j=3}^{n_1(N)+1} (u_1{}^j) u_2^{n_2(N)} z^{t(N)} P(N, \lambda) \right) \right.$$

$$\left. - \sum_{(N_1, N_2)} \left(u_1^{n_1(N_1) + 2} u_2^{n_2(N_1) + n_2(N_2)} z^{t(N_1) + t(N_2)} \frac{P(N_1, \lambda) P(N_2, \lambda)}{\lambda(\lambda - 1)} \right) \right\}$$

$$= u_2 z^2 \left\{ \frac{u_1{}^3}{u_1 - 1} \sum_N ((u_1^{n_1(N) - 1} - 1) u_2^{n_2(N)} z^{t(N)} P(N, \lambda)) - \frac{u_1{}^2}{\lambda(\lambda - 1)} FR \right\}$$

$$= \frac{u_1{}^2 u_2 z^2}{u_1 - 1} F - \frac{u_1{}^2 u_2 z^2}{\lambda(\lambda - 1)} FR - \frac{u_1{}^3 u_2 z^2}{u_1 - 1} R. \tag{16}$$

Lastly, we rewrite F_4, looking at Fig. 1d, as

$$F_4 = u_1{}^2 z^2 \sum_{T \in \mathcal{K}} u_1^{n_1(\psi_A(T))} u_2^{n_2(T)} z^{t(\psi_A(T))} P(\psi_A(T), \lambda), \tag{17}$$

where we agree to omit terms corresponding to those $T \in \mathcal{K}$ for which $\psi_A(T)$ has a loop.

Given a triangulation N, for each edge WU (as labeled in Fig. 5a), such that $U \neq V$ there is a unique triangulation $T \in \mathcal{K}$, shown in Fig. 5b, satisfying

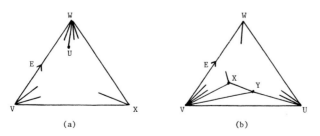

Figure 5 (a) $S = \psi_A(T)$. (b) T.

$\psi_A(T) = N$. Splitting all edges WU in N, where $U \ne V$, yields triangulations T such that $n_2(T)$ takes each number from two to $n_2(N)$ inclusive, missing exactly those numbers corresponding to edges WU, in N, with $U = V$. Corresponding to edges UV in N (excepting the root-edge) there are triangulations N_1 and N_2 as shown in Fig. 4c and described in the derivation of the expression for F_3. Therefore,

$$F_4 = u_1^2 z^2 \left\{ \sum_N \left(\sum_{j=2}^{n_2(N)} (u_2{}^j) u_1^{n_1(N)} z^{t(N)} P(N, \lambda) \right) \right.$$
$$\left. - \sum_{(N_1, N_2)} \left(u_1^{n_1(N_1) + n_1(N_2)} u_2^{n_2(N_1) + 1} z^{t(N_1) + t(N_2)} \frac{P(N_1, \lambda) P(N_2, \lambda)}{\lambda(\lambda - 1)} \right) \right\}$$
$$= u_1^2 z^2 \left\{ u_2^2 \sum_N \frac{u_2^{n_2(N)-1} - 1}{u_2 - 1} u_1^{n_1(N)} z^{t(N)} P(N, \lambda) - \frac{u_2}{\lambda(\lambda - 1)} FS \right\}$$
$$= \frac{u_1^2 u_2 z^2}{u_2 - 1} F - \frac{u_1^2 u_2 z^2}{\lambda(\lambda - 1)} FS - \frac{u_1^2 u_2^2 z^2}{u_2 - 1} S. \tag{18}$$

From (18) and (16),

$$F_3(u_1, u_2, z, \lambda) = \frac{u_1}{u_2} F_4(u_2, u_1, z, \lambda) \tag{19}$$

The term $u_1^{n_1(T)} u_2^{n_2(T)} z^{t(T)} P(\phi_A(T), \lambda)$ in F_3 appears on the right side of (19) as $u_1^{n_2(T') + 1} u_2^{n_1(T') - 1} z^{t(T')} P(\psi_A(T'), \lambda)$ where $T' = \overline{\theta_A(T)}$. Conversely the term $u_1^{n_2(T) + 1} u_2^{n_1(T) - 1} z^{t(T)} P(\psi_A(T), \lambda)$ in $(u_1/u_2) F_4(u_2, u_1, z, \lambda)$ appears as $u_1^{n_1(T'')} u_2^{n_2(T'')} z^{t(T'')} P(\phi_A(T''), \lambda)$ in F_3, where $T'' = \overline{\theta_A(T)}$.

Substituting (11), (13), (16), and (18) into (10) and rearranging we get an equation for the chromatic enumeration of triangulations,

$$\frac{u_1 - u_2}{u_1^2 u_2^2 z^2} F = \lambda(\lambda - 1)(\lambda - 2)(u_1 - u_2) + (\lambda - 2)(u_1 - u_2)F$$
$$+ \frac{(\lambda - 2)u_1 + 1}{\lambda(\lambda - 1)} FS - \frac{(\lambda - 2)u_2 + 1}{\lambda(\lambda - 1)} FR$$
$$+ \frac{u_2 S - F}{u_2 - 1} - \frac{u_1 R - F}{u_1 - 1} + (\lambda - 2)(u_1 S - u_2 R). \tag{20}$$

Having specified $f_0 = s_0 = r_0 = 0$, (20) uniquely determines F. Also $(u_2 S - F)/(u_2 - 1)$ and $(u_1 R - F)/(u_1 - 1)$ are power series in z with coefficients in $Z[u_1, u_2, \lambda]$ and $u_1 - u_2$ divides each coefficient of the power series in z on the right side of (20).

Differentiating (20) with respect to λ and setting $\lambda = 0$ and $\lambda = 1$ yields nothing. However, for $\lambda = 2$ we get the equation

$$F' = 2u_1{}^2 u_2{}^2 z^2$$
$$+ \frac{u_1{}^2 u_2{}^2 z^2}{u_1 - u_2} \left\{ \frac{(u_2 - u_1)F' + (u_1 - 1)u_2 S' - (u_2 - 1)u_1 R'}{(u_1 - 1)(u_2 - 1)} \right\}, \quad (21)$$

where the primes indicate taking the derivative of the coefficients of powers of z and setting $\lambda = 2$. Using the Waterloo computer, tables of the coefficients have been calculated up to the coefficient of z^{26}, but (21) has not been solved in general.

The Connection with Near-Triangulations

There is a close connection between our function S (or R) and the function l introduced by Tutte in [1].
Introduce

$$L = L(u_1, z, \lambda) = \sum_N u_1^{n_1(N)} z^{q(N)} P(N, \lambda), \quad (22)$$

where N ranges over all combinatorially distinct rooted planar near-triangulations having root-face a digon, $n_1(N)$ is the valency of the root-vertex of N, $q(N)$ is the number of *interior* triangles of N, and $P(N, \lambda)$ is the chromatic polynomial of the graph corresponding to N.

Given such an N, not equal to the link-map, there is an associated triangulation T, as in Fig. 6b. We obtain T by deleting the root-edge of N. The root-vertex of T is V, the root-edge is the non-root-edge of the root-face of

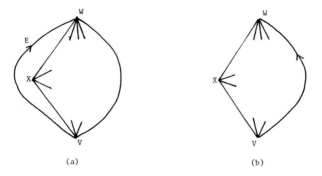

Figure 6 (a) N. (b) T.

N, and the root-face is an unbounded triangle with edges VW, WX, and XV. Therefore,

$$L = u_1 \lambda(\lambda - 1) + u_1 \sum_T u_1^{n_1(T)} z^{t(T)} P(T, \lambda)$$

$$= u_1 \lambda(\lambda - 1) + u_1 S. \qquad (23)$$

The function l introduced in [1, Section 2] is related to L by the equation $L(y, z, \lambda) = l$. This connection with near-triangulations allows for the checking of our tables against some of the formulas in [1, Section 5].

ACKNOWLEDGMENT

The author wishes to thank Professor Tutte for many stimulating conversations concerning chromatic enumeration.

REFERENCES

1. W. T. Tutte, Chromatic sums for rooted planar triangulations: The cases $\lambda = 1$ and $\lambda = 2$, *Canad. J. Math.* **25** (1973) 426–447.
2. W. T. Tutte, Chromatic sums for rooted planar triangulations II: The case $\lambda = \tau + 1$, *Canad. J. Math.* **25** (1973) 657–671.
3. W. T. Tutte, Chromatic sums for rooted planar triangulations, III: The case $\lambda = 3$, *Canad. J. Math.* **25** (1973) 780–790.
4. W. T. Tutte, Dichromatic sums for rooted planar maps, *Proc. Symp. Pure Math.* **19** (1971) 235–245.
5. W. T. Tutte, "Lectures on Chromatic Polynomials," Mimeo Ser. No. 600.25. Department of Statistics, University of North Carolina, Chapel Hill, 1970.

AMS 05C30

DEPARTMENT OF COMBINATORICS AND OPTIMIZATION
UNIVERSITY OF WATERLOO
WATERLOO, ONTARIO

A Covering Problem
in Binary Spaces of Finite Dimension

R. C. MULLIN

and

R. G. STANTON

1. Introduction

Let $V_k(2)$ denote the vector space of k-tuples over GF(2). An (m, k) *cover of* $V_k(2)$ (or simply an (m, k) *cover*) is a subset of $V_k(2) - \{0\}$ which contains k linearly independent vectors and which has non-null intersection with each $k - m$-dimensional subspace of $V_k(2)$. As a nondegeneracy condition, we require that m satisfy the inequality $1 \leq m \leq k - 1$. It is assumed henceforth that m and k are so related.

For convenience we shall consider the vectors in such a cover as columns of a binary matrix. Let $M(k, n)$ denote the set of all $k \times n$ binary matrices which have rank k and which contain no zero column. Let $C_m(k, n)$ be the subset of matrices in $M(k, n)$ whose columns form an (m, k) cover. An (m, k) cover C is *minimal* if no proper subset of C is also an (m, k) cover. Let $\underline{C}_m(k, n)$ be the subset of matrices in $M(k, n)$ which are minimal (m, k) covers. The following sets of integers are of interest. Let

$$Q_m = \{n - k : \underline{C}_m(k, n) \text{ is not empty}\}.$$

It will be shown that $Q_1 = \{1\}$. We conjecture that Q_m is finite for $m = 1, 2, \ldots$, and that $Q_2 = \{4, 6, 9\}$. This will be discussed in greater detail later. The main purpose of this paper is to determine the covering number $\alpha(m, k)$ which is defined to be the least value of n for which $C_m(k, n)$ is nonempty. Similarly we define $\beta(m, k)$ to be the least value of n_0 such that for $n > n_0$ any member of $M(k, n)$, all of whose columns are distinct, is an (m, k) cover; that is, for $n > \beta(m, k)$ any member of $M(k, n)$, all of whose columns are distinct, is necessarily an (m, k) cover. The number $\beta(m, k)$ is more easily obtained than its counterpart $\alpha(m, k)$. It can be obtained directly from the following lemma. Subsequently, we refer to a set of vectors which contain a subset of k linearly independent vectors as a set of *rank* or *dimension* k.

Lemma 1.1 *Let W denote an l-dimensional subspace of $V_k(2)$ in which l does not exceed $k - 1$. Then $\overline{W}$, the complement of W in $V_k(2)$, is a set of rank k.*

Proof Let us assume that the largest linearly independent sets in $\overline{W}$ contain at most $t \le k - 1$ independent vectors, and let X be a set of t such vectors. Now, under linear combination, X must generate all of $\overline{W}$, otherwise $\overline{W}$ would contain a vector independent from X contrary to the definition of t. Moreover, X generates zero, which lies in W. Hence

$$2^{k-1} \ge 2^t > 2^k - 2^l, \qquad 2^l > 2^k - 2^{k-1} = 2^{k-1},$$

which contradicts the fact that l does not exceed $k - 1$. ∎

Theorem 1.2 $\beta(m, k) = 2^k - 2^{k-m}$.

Proof Let W denote a $k - m$-dimensional subspace of $V_k(2)$. Let $\overline{W}$ denote the complement of W in $V_k(2)$ as in the preceding. Then clearly the matrix whose columns are the vectors of $\overline{W}$ belongs to $M(k, n_0) - C_m(k, n_0)$, where $n_0 = 2^k - 2^{k-m}$. Equally clearly each member of $M(k, n)$ for $n > n_0$ is an (m, k) cover. ∎

2. Bounds for $\alpha(m, k)$

In this section an upper bound for $\alpha(m, k)$ is given. We show later that this bound can always be met. It is convenient to refer to a subspace of $V_n(2)$ as *punctured* if the zero vector is removed from it.

Theorem 2.1 $\alpha(m, k) \le k + 2^{m+1} - m - 2$.

Proof Let W be a punctured $m + 1$-dimensional subspace of $V_n(2)$. Since $W \cup \{0\}$ must meet every subspace V of dimension $k - m$ in a subspace of positive dimension, W meets every such V. Now W contains a set X of rank $m + 1$ and this set can be extended to a basis of $V_k(2)$ by adjoining a set Y of $k - m - 1$ linearly independent vectors to it. The set $W \cup Y$ is an (m, k)-covering of cardinality $k + 2^{m+1} - m - 2$. ∎

3. A Characterization of Covers

For subsequent work, the idea of the Boolean sum is useful. Let $u = (u_1, u_2, \ldots, u_n)$ and $v = (v_1, v_2, \ldots, v_n)$ be two binary vectors. Then the Boolean sum $u \oplus v$ of u and v is the vector $(u_1 \oplus v_1, u_2 \oplus v_2, \ldots, u_n \oplus v_n)$ where $0 \oplus 0 = 0$, $0 \oplus 1 = 1 \oplus 0 = 1 \oplus 1 = 1$. Clearly this operation is associative and commutative.

Until this point there has been no advantage in viewing the vectors as column vectors of a matrix. The purpose in doing so becomes evident in the next theorem.

Theorem 3.1 *The columns of a $k \times n$ binary matrix M of rank k form an (m, k)-cover of $V_k(2)$ if and only if there does not exist a set X of m vectors in $R(M)$, the row space of M, whose Boolean sum is j, the all ones vector of dimension n.*

Proof We first show that if the columns of M are an (m, k)-cover, then no such set X exists. Indeed, assume that such a set exists, and has rank $t \leq m$. Therefore, there exists a set Y of t linearly independent vectors in $V_k(2)$ which are not simultaneously orthogonal to any column of M; that is, no column of M lies in the orthogonal complement Y' of the space generated by Y. But Y' is a space of dimension $k - t$, and since $t \leq m$, Y' contains a subspace of dimension $k - m$ which is not covered (met) by the columns of M, contradicting the fact that M was assumed to be a cover. Conversely, given that no such set X exists, every set of m independent vectors in $V_k(2)$ is simultaneously orthogonal to some column of M, and hence the columns of M meet the set of orthogonal complements of all m-dimensional subspaces of $V_k(2)$; that is, the columns of M meet all $k - m$ dimensional subspaces of $V_k(2)$. ∎

The previous theorem characterizes those members of $M(k, n)$ which belong to $C_m(k, n)$ in terms of row spaces. We refer to a set of m vectors whose Boolean sum is j as an *m-Boolism*.

It is desirable to give an even more direct characterization where possible. In doing so, the following definition and lemma are useful. Two $k \times n$ binary matrices are said to be *combinatorially equivalent* if one can be obtained from the other by elementary row operations and/or column permutations.

Lemma 3.2 *Let M and N be combinatorially equivalent $k \times n$ binary matrices. Then M belongs to $C_m(k, n)$ if and only if N belongs to $C_m(k, n)$.*

Proof Since combinatorial equivalence clearly preserves rank, the existence of zero columns, and the existence of m-Boolisms in the corresponding row spaces, the result is immediate. ∎

We note that since an elementary row operation on a $k \times n$ binary matrix may be viewed as an invertible linear operator L of $V_k(2)$ applied simultaneously to the columns of M, and that since for any permutation π of the columns of M and any elementary row operation L we have $\pi(L(M)) = L(\pi(M))$, clearly for any pair of combinatorially equivalent $k \times n$ matrices $M = (C_1, C_2, \ldots, C_n)$ and $N = (D_1, D_2, \ldots, D_n)$ there exists a map $U = QT$ where Q is a permutation of degree n and L is an invertible linear transformation of $V_k(2)$ such that $D_i = QT(C_i), i = 1, 2, \ldots, n$. We refer to U as a *CE* map.

Lemma 3.3 *Combinatorial equivalence preserves linear combinations. More precisely, if $k \times n$ matrices M and N are combinatorially equivalent under a CE map U, then for any subset X of columns of M, the relation $\sum_{C \in X} C = 0$ holds if and only if $\sum_{C \in U(X)} C = 0$ holds.*

Proof Let $U = QT$ where Q is a permutation and T is an invertible linear operator as previously described. Then $\sum_{C \in X} C = 0$ if and only if $T(\sum_{C \in X} C) = 0$; that is, if and only if $\sum_{C \in X} T(C) = 0$. Since the permutation can be viewed as merely relabeling the terms in the last sum, the lemma follows. ∎

The *weight* of a binary k-tuple is the number of nonzero elements it contains. A column of even weight is said to be *even*. These concepts are useful in the following.

Theorem 3.4 (Tutte) *A member M of $M(k, n)$ belongs to $C_1(k, n)$ if and only if, for any given set of X of linearly independent columns of M, there exists a column of M which is the sum of an even number of columns of X.*

Proof Let us first assume that M belongs to $C_1(k, n)$. Since M then has rank k, there is a set X of linearly independent columns in M. Let N be a matrix combinatorially equivalent to M and which has the form

$$N = (I_k, S)$$

where I_k is an identity matrix of order k which corresponds to the set X under some CE map. Clearly S must be nonempty since the row space of I_k contains j, and this would contradict Lemma 3.2. Now if all members of S have an odd weight, then the sum of all rows is j. Hence some column of S has an even weight. This column is then a sum of an even number of columns of I_k, and by Lemma 3.3, some column of M is the sum of an even number of members of X.

Conversely, let us assume that M contains a set X of linearly independent columns and that some column is the sum of an even number of vectors in X. Then there exists a corresponding combinatorially equivalent matrix $N = (I_k, S)$ where S contains a column (say column t) of even weight. Since any vector in the row space of N which is not a sum of all rows of N

must have at least one zero component among the first k and since the sum of all rows of N has a zero in column t, the row space of N does not contain j, and hence M belongs to $C_1(k, n)$. ∎

Corollary 1 *If in any member of $M(k, n)$ for some set of k linearly independent vectors X there is a column of M which consists of the sum of an even number of columns of X, then for all sets Y of k independent columns there is a column of M which is a sum of an even number of columns of M.*

Corollary 2 *For $k \geq 2$, $\alpha(1, k) = k + 1$.*

Theorem 3.5 (i) $\alpha(2, k) = k + 4$ for $k \geq 3$.
(ii) *A member M of $M(k, k + 4)$ is a member of $C_2(k, k + 4)$ if and only if some set of seven columns of M form a punctured subspace of dimension three in $V_k(2)$.*

Proof As in Theorem 2.1 it is clear that if M contains a punctured 3-dimensional subspace, then M belongs to $C_2(k, k + 4)$. We now show that if $M \in C_2(k, n)$, then $n \geq k + 4$, with equality only if M contains a punctured 3-dimensional subspace. Let us assume, without loss of generality, that $M = (I_k, S)$, where S is not empty since by Theorem 3.4, $\alpha(2, k) \geq k + 1$. Let us assume that S has t columns, that is, $t = n - k$. Note that the row space $R(M)$ of M contains a vector whose first k components are ones. Further rank $S < t$, since otherwise one could find a vector in $R(M)$ whose last t components are also ones, and $R(M)$ would contain a 2-Boolism. Therefore $t \geq 2$, and if $t = 2$, then both columns of S are identical, which contradicts the fact that $t \geq 2$ in a smallest minimal cover. Hence $t \geq 3$. Assume $t = 3$. Since $j \notin R(M)$, at least one column must be even, and as before S contains no repeated column. This implies that rank $S = 2$ and either one or three columns of S has even weight. Assume first that only one column, without loss of generality the last, has even weight. Then $j' = (1, 1, 1, \ldots, 1, 0)$ belongs to $R(M)$. Further, since S has no zero column, some row r_i of M has a one in the last column. Then $\{r_i, j'\}$ is a 2-Boolism. Therefore if $t = 3$, all columns of S must have even weight. In this case the vector $j'' = (1, 1, 1, \ldots, 1, 0, 0, 0)$ belongs to $R(M)$. Thus no subset of rows of S can sum to $(1, 1, 1)$. Therefore no row of S can have weight three, nor can $(1, 0, 0)$, $(0, 1, 0)$, and $(0, 0, 1)$ all occur as rows of S. Hence some row of S has weight two. Without loss of generality we may assume that this row is $(1, 1, 0)$ and by combinatorial equivalence it can be taken to be the first row of S. Now a row $(0, 0, 1)$ becomes impossible and since S has no zero column, we may assume without loss of generality that S has a second row $(1, 0, 1)$. Let r_1 and r_2 denote the first and second rows of M, respectively. Then $j'' + r_1 = (0, 1, 1, 1, \ldots, 1, 1, 0)$ and $j'' + r_2 = (1, 0, 1, 1, 1, \ldots, 1, 0, 1)$ form a 2-Boolism in $R(M)$, and hence $t > 3$.

Now consider the case $t = 4$. Again the row space of M denoted by $R(M)$ contains a vector whose first k components are ones and thus no subset of rows of S can sum to $(1, 1, 1, 1)$. In particular, no row of S has weight four. Let us assume that the largest weight of a row in S is two. Again, since not all rows can have weight one we may assume without loss of generality that the first row of S is $(1, 1, 0, 0)$. This excludes $(0, 0, 1, 1)$ as a row of S. Nor can the remaining rows of S all belong to $\{(0, 0, 1, 0), (0, 0, 0, 1)\}$. Thus we may assume that the second row of S is $(1, 0, 0, 1)$. Again $(0, 0, 1, 1)$ is impossible, as are now $(0, 1, 1, 0)$ or $(0, 0, 1, 0)$ leaving only $(1, 0, 1, 0)$ as a third row with a one in position three. But then the first three rows of S sum to $(1, 1, 1, 1)$. Therefore S must contain a row of weight three, and without loss of generality we may assume that S has $(1, 1, 1, 0)$ as a first row. This excludes $(0, 0, 0, 1)$ as a row. Since the last column of S is not a zero column, without loss of generality we may assume that the second row of S is either (a) $(1, 0, 0, 1)$, or (b) $(1, 1, 0, 1)$. Now let $v = (1, 1, 1, \ldots, \varepsilon_1, \varepsilon_2, \varepsilon_3, \varepsilon_4)$ be the vector obtained by summing all rows of M. Let r_1 and r_2 denote the first and second rows of M, respectively. Since $\{r_1, v\}$ is not a 2-Boolism, $\varepsilon_4 = 0$. Also since S has rank ≤ 3 and since S has no repeated columns, S has rank three. Now consider case (a). It follows at once that if C_1, C_2, C_3 and C_4 are the columns of S that either $C_1 + C_2 + C_4 = 0$ or $C_1 + C_3 + C_4 = 0$. Since the parity of the weight of a sum of binary vectors is the binary sum of the parities of the weights of those vectors, we have immediately that either $\varepsilon_1 + \varepsilon_2 = 0$ or $\varepsilon_1 + \varepsilon_3 = 0$. Because of the symmetry of C_2 and C_3, the two cases are equivalent. Hence we need only consider the case in which $\varepsilon_1 + \varepsilon_2 = 0$. If $\varepsilon_1 = \varepsilon_2 = 0$, then $\{v + r_1, v + r_2\}$ is a 2-Boolism. If $\varepsilon_1 = \varepsilon_2 = 1$, then $\{v, v + r_1 + r_2\}$ is a 2-Boolism. Thus the vector $(1, 0, 0, 1)$ cannot be a row of S. Nor by symmetry neither can $(0, 1, 0, 1)$ or $(0, 0, 1, 1)$.

Hence we may consider the second row of S to be $(1, 1, 0, 1)$. Since this row is present, neither $(1, 0, 1, 0)$ nor $(0, 1, 1, 0)$ can be a row of S. Therefore if S has a row of weight two, it must be $(0, 0, 1, 1)$. As before this implies that either $C_1 + C_3 + C_4 = 0$ or $C_2 + C_3 + C_4 = 0$, the two solutions being equivalent under symmetry. We assume therefore that $C_1 + C_3 + C_4 = 0$. Since $\varepsilon_3 + \varepsilon_4 = 0$ because neither $\{v, r_1\}$ nor $\{v, r_2\}$ is a 2-Boolism, we have $\varepsilon_1 = 0$. Hence $\{v + r_1, v + r_3\}$ is a 2-Boolism. Hence the only possible rows of S are the four vectors of weight three and two vectors of weight one, namely, $(1, 0, 0, 0)$ and $(0, 1, 0, 0)$, the latter being equivalent under symmetry. Since S has rank three and no repeated columns, neither of these are possible. Hence S admits only row vectors of weight three. Nor can all four of these be present, since S has rank three.

Recall that we have shown that without loss of generality, we can assume that the first rows of S are $(1, 1, 1, 0)$ and $(1, 1, 0, 1)$. Now suppose that one of these rows is repeated. Again, without loss of generality, we may assume that it is (1, 1, 1, 0) and that it occurs in the third row of S. Again let v be the sum of all rows of M. Again, since S has rank ≤ 3, if $v = (1, 1, \ldots, 1, \varepsilon_1, \varepsilon_2, \varepsilon_3, \varepsilon_4)$, then $\varepsilon_3 = \varepsilon_4$. If $\varepsilon_3 = 0$, then $\{v + r_1 + r_2, v + r_3\}$ is a 2-Boolism. If $\varepsilon_3 = 1$, then $\{v, v + r_2\}$ is a 2-Boolism. Similarly no other row of weight three can be repeated as a row of S. Therefore S has either two or three rows of weight three, all such rows being distinct. But the case of two such rows only is impossible, since in this case, assuming the configuration to have the first two rows being (1, 1, 1, 0) and (1, 1, 0, 1), respectively, then $v = (1, 1, 0, 0, 1, 1)$, and $\{v, r_1\}$ is a 2-Boolism. Hence we may assume that the first three rows of S are (1, 1, 1, 0), (1, 1, 0, 1), and (1, 0, 1, 1), respectively, in which case the first three columns of M together with S form a punctured 3-dimensional subspace of $V_k(2)$. ∎

At this point it is natural to ask if all (2, k)-covers contain a punctured subspace of dimension three. Prior to answering this, we develop a criterion for minimal covers. An (m, k)-cover of C is *minimal* if no proper subset of C is also an (m, k)-cover. Clearly all (2, k)-covers contain a punctured subspace of dimension three if and only if all minimal (2, k)-covers contain $k + 4$ vectors.

Suppose M belongs to $C_m(k, n)$. We say that a column D of M is *B-essential* if there exists a set $X(D)$ of m vectors in $R(M)$ which simultaneously vanish in the coordinate position corresponding to D and which do not vanish simultaneously in any other coordinate position. A column D of M is *R-essential* if M', the matrix formed by removing D from M, has rank $k - 1$. A column is *essential* if it is either B-essential or R-essential or both. Otherwise it is *inessential*.

Lemma 3.6 *Let M belong to $C_m(k, n)$. Then M is minimal if and only if every column of M is essential.*

The proof is straightforward. ∎

The matrix in Fig. 1 belongs to $C_2(4, 10)$. It is readily verified that every column is essential (in fact, B-essential), hence it is minimal. Hence it contains no punctured subspace of dimension three.

$$\begin{pmatrix} 1 & 0 & 0 & 0 & 1 & 1 & 0 & 0 & 0 & 1 \\ 0 & 1 & 0 & 0 & 1 & 0 & 1 & 0 & 1 & 0 \\ 0 & 0 & 1 & 0 & 0 & 1 & 1 & 1 & 0 & 0 \\ 0 & 0 & 0 & 1 & 0 & 0 & 0 & 1 & 1 & 1 \end{pmatrix}$$

Figure 1 A (2, 4) cover with no punctured 3-space.

It is useful to characterize $C_{k+1}(k, k+1)$. Before doing so, we note that while consideration of (k, m)-covers with $m \le k$ is omitted by the constraints on m in Section 1, it is easily seen that none exist should this constraint on m be removed.

Lemma 3.7 *A matrix M belongs to $C_{k-1}(k, n)$ if and only if it contains all possible nonzero binary columns of length k. Such a cover is minimal if and only if it has no repeated columns.*

Proof Such a cover must meet all punctured 1-dimensional subspaces, that is, all vectors. ∎

Corollary $\alpha(k - 1, k) = 2^k - 1$.

4. Further Results on $\alpha(m, k)$

The following lemma is sufficient to discuss the asymptotic behavior of $\alpha(m, k)$ for fixed m and large k.

Lemma 4.1 *For $k > m$, $\alpha(m, k + 1) \le \alpha(m, k) + 1$.*

Proof Suppose that M belongs to $C_m(k, \alpha(m, k))$. Then clearly $M^* = M \dotplus 1$ has rank $k + 1$, where $\dotplus$ represents the direct sum of matrices and the one is viewed as the identity matrix of order one. Suppose that $R(M^*)$ contains an m-Boolism B. Then the restriction of B to the row space of M is an m-Boolism in $R(M)$, contrary to the definition of M. Hence M^* belongs to $C_m(k + 1, \alpha(m, k) + 1)$. ∎

Suppose that m is such that for all $k \ge m + 1$, the equality $\alpha(m, k + 1) = \alpha(m, k) + 1$ holds. Then clearly, for that value of m, $\alpha(m, k) = k + 2^{m+1} - m - 2$ for all $k \ge m + 1$. We shall show that this is always the case.

An (m, k)-cover M is said to be *reducing* if it is minimal, has rank $k \ge m + 2$, and has $n \le \alpha(m, k - 1)$ columns. A binary matrix M is *m-complete* if it has m rows and its columns contain the set of all binary m-tuples.

Lemma 4.2 *Let M be a $k \times n$ reducing (m, k)-cover. Then every nonzero vector r of $R(M)$, the row space of M has weight 2^m.*

Proof Let r and M be as described in the hypothesis. Then r can be embedded in a basis X for M which has a column (without loss of generality the first), which has weight one, that one occurring in r. Let M' be a $(k - 1) \times n$ matrix containing all rows of X except r. By the minimality of M, M' has precisely one column of zeros. Let M'' be the matrix derived from M' by removing its zero column. Since M'' has fewer than $\alpha(m, k - 1)$ columns, rank $k - 1$, and no zero column, its row space contains an m-Boolism B which corresponds to a subset of m rows of $R(M)$ of the form

$$M^* = \begin{pmatrix} 0 & & \\ 0 & & \\ \vdots & B & \\ 0 & & \end{pmatrix}$$

since $R(M)$ has no m-Boolism.

Now let α be any subset of rows of M^*. Let $M^*(\alpha)$ denote the matrix obtained from M^* by adding r to each row of M which occurs in α. Since $R(M)$ contains no m-Boolism, this implies that there exists a column $D(\alpha)$ of M^* which has ones in the rows of α, zeros in the remaining rows of M^*, and a one in the coordinate position of r which corresponds to $D(\alpha)$. This establishes the fact tht M^* contains a submatrix N' which is m-complete, and whose corresponding coordinate entries in r are all ones. Hence r has weight at least 2^m. ∎

Theorem 4.3 $\alpha(m, k) = 2^{m+1} + k - m - 2$ for $k \geq m + 1$.

Proof We have shown that this result is true for $m = 1$ and 2. Let us assume that the theorem is false, and that μ is the least value of m for which the theorem fails. Further, let v be the least value of $v \geq m + 1$ for which $\alpha(\mu, k) < 2^{\mu+1} + k - \mu - 2$. Clearly $v \geq \mu + 2$. By definition of μ and v there exists a (μ, v)-reducing cover M. Let r be any row of $R(M)$ the row space of M which has the property that the restriction of $R(M)$ to the positions in which r vanishes has rank $v - 1$ (this is possible in the row space of any rank v matrix if $v > 1$). Clearly r is not the zero vector. Now M is combinatorially equivalent to a matrix M^* whose first row is r and which has the form

$$\begin{pmatrix} 1 1 \cdots 1 & \vdots & 0 0 \cdots 0 \\ N & \vdots & W \end{pmatrix}.$$

Since the row space of M^* has no μ-Boolism, the row space of W has no $(\mu - 1)$-Boolism. Moreover W has no zero column, and W has rank $v - 1$. Hence W is a $(\mu - 1, k - 1)$-cover, and therefore W has at least $2^\mu + v - \mu - 2$ columns. Therefore M^* has at least $2^\mu + 2^\mu + v - \mu = 2^{\mu+1} - v + \mu - 2$ columns, contradicting the fact that M is a reducing matrix. ∎

5. Tangential Blocks

In this section, the problem of Tutte is reformulated in terms of vector spaces and the language of this paper in order to make comparisons of this paper with the results of Tutte [2] and its later developments by Datta ([1]; see also Datta's paper in this volume, p. 121).

Let m be a positive integer and q an integer greater than m. A set B of vectors of $V_q(2) - \{0\}$ is an *m-block* if its dimension is at least $m + 1$ and it includes at least one vector from each subspace of $V_q(2)$ of dimension $q - m$. In particular, if the dimension of B (which we write as dim B) is q, then B is a (q, m)-cover. Conversely, every (q, m)-cover is an m-block in $V_q(2)$. An m-block B is *minimal* if no proper subset of B is a k block. Tutte has shown that every minimal m-block of dimension k in $V_q(2)$ can also be viewed as a (minimal) m-block in $V_k(2)$. Hence it is proper to speak of minimal m-blocks as objects independent of the containing space. Let C be a non-null subset of an m-block B. A *tangent* of C in B is any (q, m)-space in $V_q(2)$ which contains all the vectors of C but no vector in B which is independent of them. B is a *tangential m-block* if every non-null subset of B, of dimension not exceeding $q - k$, has a tangent in B. Tutte has shown that every tangential m-block is minimal, and suggests the conjecture that there are only three tangential 2-blocks, called the Fano, Desargues, and Petersen blocks, which are of dimensions 3, 4, and 6, respectively and which contain 7, 10, and 15 vectors, respectively. The relation between k-blocks and covers is illustrated in the next section.

6. The Sets Q_m

The sets Q_m defined in Section 1 can be redefined as follows. Let $N(m, k)$ denote the set of minimal (m, k)-covers. If $M \in N(m, k)$, let $|M|$ denote the number of columns of M. Then

$$Q_m = \{|M| - k : M \in N(m, k), k = m + 1, m + 2, \ldots\}.$$

By Theorem 3.4, $Q_1 = \{1\}$. By Theorem 4.3, the least number of Q_m is $2^{m+1} - m - 2$. As mentioned earlier, it is reasonable to conjecture that Q_m is finite for all m. In particular, we discuss Q_2. Because of the Fano, Desargues, and Petersen tangential 2-blocks, $\{4, 6, 9\} \subseteq Q_2$. Using techniques similar to those employed in the proof of Theorem 3.5, van Rees has shown that $5 \notin Q_2$. We show now that if Q_2 is finite, then it can be established whether or not the Fano, Desargues, and Petersen 2-blocks are the only tangential 2-blocks. To do so, we introduce irreducible covers, which correspond to minimal m-blocks in Tutte's theory.

Let M be an (m, k)-cover. We say that M is *reducible* (to M_1) if M is combinatorially equivalent to a matrix M' of the form

$$M' = \begin{pmatrix} M_1 & 0 \\ \hline 0 & M_2 \end{pmatrix},$$

where zero denotes a matrix of zeros, M_2 is a matrix of rank k_1, where $1 \le k_1 \le k - m - 1$, and M_1 is an $(m, k - k_1)$-cover. Otherwise M is *irreducible*. If M_1 is a minimal $(m, k - k_1)$-cover, then we say that M is *minimally reducible*.

Lemma 6.1 *If a minimal matrix M is reducible to a matrix M_1, then M_1 is also minimal.*

Proof If M_1 is not minimal, then it is possible to remove some column of M_1 to produce a proper $(m, k - k_1)$ subcover. Removing the corresponding row of M produces a proper (m, k) subcover of M contradicting its minimality. ∎

Lemma 6.2 *If a minimal (m, k)-cover M contains a column C which is R-essential but not B-essential, then M is reducible.*

Proof By row reduction of column C, M is equivalent to a matrix M' of the form

$$\begin{pmatrix} M_1 & \vdots & 0 \\ & \vdots & 0 \\ & \vdots & 0 \\ \hdashline w & \vdots & 1 \end{pmatrix}.$$

If $w = 0$, then the result follows. If $w \ne 0$, then w belongs to the row space of M, since $M - \{C\}$ has rank $k - 1$. Hence M' is combinatorially equivalent to

$$M'' = \begin{pmatrix} M_1 & \vdots & 0 \\ & \vdots & 0 \\ & \vdots & 0 \\ \hdashline 0 \; 0 \; 0 & \vdots & 1 \end{pmatrix}.$$
∎

Corollary *In an irreducible minimal (m, k)-cover M, every column is B-essential.*

Lemma 6.3 *Let M be an irreducible minimal (m, k)-cover with $m \ge 1$ in the canonical form $M = (I_n S)$. Then*

(i) *S has no zero row, and*
(ii) *S has no pairs of identical rows.*

Proof That S has no zero row is clear from irreducibility. To show that S has no pairs of identical rows, we proceed as follows. Without loss of generality (because of combinatorial equivalence), we may assume that rows one and two, denoted by r_1 and r_2, respectively, of S are identical. As noted above, both columns one and two, denoted by C_1 and C_2, respectively, are essential in M. Since C_1 is essential, there is a set of m vectors, $v_1, v_2, \ldots, v_m$ in the row space of M such that the matrix

$$A = \begin{pmatrix} v_1 \\ v_2 \\ \vdots \\ v_m \end{pmatrix}$$

has column one as its only column of zeros. Note that if any row in A is replaced by the sum of all rows, we obtain another set of m columns whose only zero column is the first. Hence we can assume that the second column has at least two ones. By interchanging the roles of r_1 and r_2 in the formation of A, we obtain A', a matrix consisting of members of the row space of M which differs from A only inasmuch as columns one and two are interchanged. Let r be a row of A which has a one in column two, and let r' be the corresponding row in A'. Replacing r by r', we have a set of m rows in the row space whose Boolean sum is j, contradicting the fact that M is an (m, k)-cover. ∎

Theorem 6.4 *Let γ be any positive integer. Then*

(i) $\gamma \in Q_m$ *if and only if there exists an irreducible minimal (m, k)-cover for some $k \leq 2^\gamma - 1$ which contains $k + \gamma$ vectors; and*

(ii) *any minimal (m, k)-cover with $k > 2^\gamma - 1$ containing $k + \gamma$ vectors is reducible.*

Proof If $\gamma \in Q_m$, there exists a minimal (m, k)-cover M with $k + \gamma$ columns. Either M is irreducible or it is not. If M is not irreducible, then

$$M \sim \begin{pmatrix} M_1 & \vdots & 0 \\ \hdashline 0 & \vdots & M_2 \end{pmatrix},$$

where M_1 is irreducible. Moreover, M_2 is square, otherwise some column of M_2 is dependent on the remaining columns of M_2, and the removal of the corresponding column from M yields a proper subcover, contradicting the minimality of M. Hence M_1 is an (m, k_1)-cover with $k_1 + \gamma$ vectors. Repeating this operation, we obtain an irreducible minimal (m, k^*)-cover with $k^* + \gamma$ vectors for some integer k^*. Therefore there exists an irreducible minimal (m, k)-cover M with $k + \gamma$ columns for some integer k.

Assume that M is in the canonical form $(I_k S)$. By Lemma 6.3, S, which is k by γ, has no repeated rows and no zero row. Hence k does not exceed $2^\gamma - 1$. Conversely, if such a cover exists, then $\gamma \in Q(m)$. ∎

Theorem 6.5 *If $Q(m)$ is finite, then the set of minimal m-blocks is also finite, and conversely.*

Proof First assume that $Q(m)$ is finite with maximum entry μ. The set of all minimal irreducible (m, k)-covers is finite, since they occur in a space of dimension $2^\mu - 1$ or lower. However as noted earlier, every minimal m-block B of dimension k can be viewed as an (m, k)-cover M. It is clearly a minimal (m, k)-cover. Moreover it is an irreducible (m, k)-cover, since if some reduction to a cover M_1 were possible, then as Tutte has shown [2, p. 26] M_1 would be an m-block in $V_k(2)$, contradicting the minimality of B. Hence the set of minimal m-blocks is finite. Conversely, let us assume that the set of minimal m-blocks is finite. Each irreducible minimal (m, k)-cover is an m-block in $V_k(2)$. Suppose that this m-block B is not a minimal block. Then X, a proper subset of B, is also an m-block in $V_k(2)$. Let M be the matrix corresponding to B, and let C be a column of M in $B \backslash X$. By row reduction on C, M is combinatorially equivalent to M' where

$$M' = \begin{pmatrix} & 0 \\ A & 0 \\ & 0 \\ w & 1 \end{pmatrix}.$$

Since B is a minimal (m, k)-cover, the rank of $B \backslash X$ is $k - 1$. Hence w belongs to the row space of A, and M is reducible, a contradiction. Thus B is a minimal m block, and the set of irreducible minimal m blocks is finite. Since each member of $Q(m)$ corresponds to some irreducible minimal (m, k)-cover, Q_m is also finite. ∎

This proof establishes the equivalence of irreducible minimal (m, k)-covers and minimal m-blocks of dimension k. Further if Q_2 is finite, the method of Theorem 6.5 produces a finite algorithm for determining all tangential 2-blocks.

The techniques employed by Datta ([1]; see also Datta's paper in this volume, p. 121) and Tutte [2] are useful in looking for irreducible minimal (m, k)-covers. We note that many of the techniques and results employed here are also valid in $GF(q)$.

REFERENCES

1. B. T. Datta, On tangential 2-blocks, *Discrete Math.* **15** (1976) 1–22.
2. W. T. Tutte, On the algebraic theory of graph colourings, *J. Combinatorial Theory* **1** (1966) 15–50.

AMS 05B30

R. C. Mullin
UNIVERSITY OF WATERLOO
WATERLOO, ONTARIO

R. G. Stanton
UNIVERSITY OF MANITOBA
WINNIPEG, MANITOBA

The Tutte Polynomial and Percolation

J. G. OXLEY

and

D. J. A. WELSH

1. Introduction

In 1947 Tutte made the first systematic study of the following natural set of functional equations defined on the class of all graphs:

$$f(G) = f(H) \quad \text{if} \quad G \text{ is isomorphic to } H,$$
$$f(G) = f(G_e') + f(G_e''), \tag{1}$$
$$f(G_1 + G_2) = f(G_1)f(G_2),$$

where G_e', G_e'' respectively denote the graphs obtained from G by deletion and contraction of the edge e, and $G_1 + G_2$ denotes the disjoint union of the graphs G_1 and G_2. He showed in particular that subject to "boundary conditions" specifying f on bridges and loops, f is uniquely determined and is what is now commonly called the *Tutte polynomial* of the cycle or polygon matroid of the graph.

The connection between the Tutte polynomial of a graph and the solution to the general percolation model on that graph has been noticed by Essam [4], Fortuin and Kasteleyn [5], and Temperley and Lieb [9]. In this paper we have aimed at extending this relationship in two directions. In the first place, motivated by the percolation model, we study a slightly more general set of equations than (1). This extends Brylawski's extension to matroids [2] of Tutte's work [10, 11]. In Section 3 we consider a more general percolation model on arbitrary clutters which reduces to the classical case when the clutters are the edge sets of paths in graphs. In Section 4 we show that the theory of the Tutte polynomial does not extend to arbitrary clutters and that matroids are the limiting structures for which a Tutte polynomial can be sensibly defined. The final section gives a brief account of other results obtainable for clutter percolation and relates the problem to classical results of extremal set theory.

We use $|X|$ to denote the cardinality of the set X, and write $X \cup e$ to denote $X \cup \{e\}$ and $X \backslash e$ for $X \backslash \{e\}$ for any singleton set $\{e\}$.

The matroid terminology is that of [12]. If M is a matroid on a finite set S with rank function ρ, its *rank generating function* $R(M; x, y)$ is defined by

$$R(M; x, y) = \sum_{A \subseteq S} x^{\rho S - \rho A} y^{|A| - \rho A}.$$

Its *Tutte polynomial* $T(M; x, y)$ is defined by

$$T(M; x, y) = R(M; x - 1, y - 1).$$

If $e \in S$, we use M_e' and M_e'' to denote the matroids $M|(S \backslash e)$ and $M \cdot (S \backslash e)$, respectively. The matroid of rank zero on a singleton set we call *the loop* and denote by L; its dual *the coloop* we denote by L^*.

2. A Matroid Equation

Consider the following problem. Does there exist a unique real function f defined on the class of all matroids and satisfying the following conditions?

If matroids M and N are isomorphic, then

$$f(M) = f(N). \tag{1}$$

For fixed nonzero real numbers a and b and any matroid M,

$$f(M) = af(M_e') + bf(M_e'') \tag{2}$$

provided e is neither a loop nor a coloop of M.

For any pair of matroids M_1, M_2 on disjoint sets

$$f(M_1 + M_2) = f(M_1) f(M_2). \tag{3}$$

Theorem 1 *There is a unique real-valued function f satisfying* (1), (2), *and* (3) *together with the boundary conditions*

$$f(L^*) = x, \qquad (4)$$

$$f(L) = y. \qquad (5)$$

This function is given for any matroid M on S by

$$f(M) = a^{|S|-\rho S} b^{\rho S} T(M; b^{-1}x, a^{-1}y), \qquad (6)$$

where T is the Tutte polynomial of M.

Proof It is easy to check that f as defined in (6) satisfies (1)–(5). The uniqueness follows from the uniqueness of the Tutte polynomial (see [12, Chap. 15]). ∎

Example: Random Submatroids Let M be a matroid on a finite set S and suppose that each element of S has, independently of all other elements, a probability $q = 1 - p$ of being deleted from S. The resulting restriction minor $\omega(M)$ of M is what we call a *random submatroid* of M, corresponding in the obvious way to a random graph when M is the cycle matroid of the complete graph. Suppose $P(p; M)$ is the probability that $\omega(M)$ has the same rank as M. Then provided e is neither a loop nor a coloop of M

$$P(p; M) = qP(p; M_e') + pP(p; M_e'')$$

and

$$P(p; M_1 + M_2) = P(p; M_1)P(p; M_2).$$

Also

$$P(p; M) = \begin{cases} p & \text{if } M \text{ is a coloop,} \\ 1 & \text{if } M \text{ is a loop.} \end{cases}$$

Hence by Theorem 1

$$P(p; M) = q^{|S|-\rho S} p^{\rho S} T(M; 1, q^{-1}). \qquad (7)$$

By a similar argument if $r(M; \theta) = \mathscr{E}(\theta^{\rho(\omega(M))})$ denotes the probability generating function of the rank of a random submatroid of M we have that when e is neither a loop nor a coloop

$$r(M; \theta) = qr(M_e'; \theta) + p\theta\, r(M_e''; \theta),$$
$$r(L^*; \theta) = q + p\theta, \qquad r(L; \theta) = 1,$$

and hence

$$r(M; \theta) = q^{|S|-\rho S}(p\theta)^{\rho S} T\!\left(M; \frac{q}{p\theta}+1, \frac{1}{q}\right).$$

3. The Percolation Model for Clutters

Classical percolation theory as introduced by Broadbent and Hammersley [1] is concerned with the flow of liquid through randomly dammed graphs. We shall define a percolation model which has a wider scope than this but which clearly contains the classical model as a special case.

Let S be a finite set and let $\mathscr{A} = (A_i : i \in I)$ be a family of subsets of S with the property that for $i \neq j$, $A_i \not\subseteq A_j$. Such a family is called a *clutter* or *Sperner family*. Suppose that each element of S is, independently of all other elements, painted white with probability p or black with probability $q = 1 - p$. This defines a probability space Ω of possible realisations and we call this space the *percolation model* on $\mathscr{A}$. The classical model is the special case where S is the edge set of a finite graph and $\mathscr{A}$ is some collection of paths.

For given $\mathscr{A}$ and p we define the percolation probability $P(\mathscr{A}; p)$ to be the probability that some member of $\mathscr{A}$ has all its members painted white. Thus

$$P(\mathscr{A}; p) = \sum p^{|X|} q^{|S \setminus X|},$$

where the sum is over all subsets X of S which contain some member of $\mathscr{A}$. Thus if $|S| = n$ and we let u_k denote the number of k-subsets of S which contain a member of $\mathscr{A}$, we naturally define the *upper polynomial* $U(\mathscr{A}; z)$ by

$$U(\mathscr{A}; z) = \sum_{k=0}^{n} u_k z^k$$

and we notice that $P(\mathscr{A}; p) = q^n U(\mathscr{A}; p/q)$.

Example Let G be a finite connected graph and take $\mathscr{A}$ to be the collection of edge sets of spanning trees of G. Then $P(\mathscr{A}; p)$ is just the probability that a random subgraph of G is connected. This is clearly the same as the probability that a random submatroid of the cycle matroid $M(G)$ has full rank and hence by (2.7)

$$P(\mathscr{A}; p) = q^{|E|-|V|+1} p^{|V|-1} T(M(G); 1, q^{-1}),$$

where E and V are, respectively, the edge and vertex sets of G.

The following definitions are fairly standard. If $\mathscr{A}$ is a clutter on S and T is a subset of S let

$$\mathscr{A} | T = \{A_i : A_i \in \mathscr{A}, A_i \subseteq T\},$$
$$\mathscr{A} \cdot T = \{\text{minimal sets of the form } A_i \cap T : A_i \in \mathscr{A}\},$$

and if $T = S \setminus e$ write

$$\mathscr{A} | T = \mathscr{A}'_e, \qquad \mathscr{A} \cdot T = \mathscr{A}''_e.$$

The *direct sum* $\mathcal{A}_1 + \mathcal{A}_2$ of two clutters on disjoint sets is the collection of sets $\{A_1 \cup A_2 : A_1 \in \mathcal{A}_1, A_2 \in \mathcal{A}_2\}$. Their *union* $\mathcal{A}_1 \cup \mathcal{A}_2 = \{A : A \in \mathcal{A}_1 \text{ or } A \in \mathcal{A}_2\}$. We emphasize that these operations only apply to clutters on disjoint sets, hence whenever we write $\mathcal{A} + \mathcal{B}$ or $\mathcal{A} \cup \mathcal{B}$ it will implicitly be assumed that $\mathcal{A}$ and $\mathcal{B}$ are on disjoint sets. The *blocker* $\mathcal{A}^*$ of $\mathcal{A}$ is the collection of minimal sets X such that $X \cap A_i \neq \emptyset$ for all $A_i \in \mathcal{A}$. Edmonds and Fulkerson [3] show that $(\mathcal{A}^*)^* = \mathcal{A}$ and it is easy to show that

$$(\mathcal{A}|T)^* = \mathcal{A}^* \cdot T, \qquad (\mathcal{A} \cdot T)^* = \mathcal{A}^* | T.$$

An element e of S is *essential* for $\mathcal{A}$ if e belongs to each $A_i \in \mathcal{A}$ and is *redundant* if e belongs to no A_i. The following results are easy to see.

If e is redundant,

$$P(\mathcal{A}; p) = P(\mathcal{A}_e'; p). \tag{1}$$

If e is essential,

$$P(\mathcal{A}; p) = pP(\mathcal{A}_e''; p). \tag{2}$$

If e is neither redundant nor essential,

$$P(\mathcal{A}; p) = q\, P(\mathcal{A}_e'; p) + p\, P(\mathcal{A}_e''; p), \tag{3}$$

$$P(\mathcal{A}_1 + \mathcal{A}_2; p) = P(\mathcal{A}_1; p) P(\mathcal{A}_2; p). \tag{4}$$

If $S = \{e\}$, then

$$P(\mathcal{A}; p) = \begin{cases} p & \text{if } e \text{ is essential and } \mathcal{A} \text{ is nonempty,} \\ 1 & \text{if } e \text{ is redundant and } \mathcal{A} \text{ is nonempty.} \end{cases} \tag{5}$$

4. On (Not) Extending the Tutte Polynomial

In view of (3.3)–(3.5) it is natural to ask whether the theory of the Tutte polynomial can be extended to arbitrary clutters. First note that on the singleton set $S = \{e\}$ there are three clutters, namely,

$$E = \{\{e\}\}, \qquad R = \{\emptyset\},$$

and the empty clutter. We call E *the essential clutter* and R *the redundant clutter*; they are unique up to isomophism. Then if a and b are nonzero real numbers we ask if there exists a function $f(\mathcal{A}; x, y)$ of two real variables x and y defined on the class of all finite nonempty clutters such that the following rules are satisfied.

If $\mathcal{A}$ and $\mathcal{B}$ are isomorphic clutters,

$$f(\mathcal{A}; x, y) = f(\mathcal{B}; x, y). \tag{1}$$

If e is neither essential nor redundant for $\mathscr{A}$, then

$$f(\mathscr{A}; x, y) = a\, f(\mathscr{A}_e'; x, y) + b\, f(\mathscr{A}_e''; x, y), \qquad (2)$$

$$f(\mathscr{A}_1 + \mathscr{A}_2; x, y) = f(\mathscr{A}_1; x, y) f(\mathscr{A}_2; x, y), \qquad (3)$$

$$f(E; x, y) = x, \qquad f(R; x, y) = y. \qquad (4)$$

The main result of this section shows that Theorem 2.1 is best possible in the following sense.

Theorem 1 *If a and b are fixed nonzero real numbers then a function $f(\mathscr{A}; x, y)$ satisfying (1)–(4) is uniquely defined if and only if $\mathscr{A}$ is the collection of bases of a matroid.*

Proof From (3) and (4) it is clear that

$$f(\mathscr{A}; x, y) = \begin{cases} x\, f(\mathscr{A}_e''; x, y) & \text{if } e \text{ is essential,} \\ y\, f(\mathscr{A}_e'; x, y) & \text{if } e \text{ is redundant.} \end{cases} \qquad (5)$$

We use induction on the size n of the ground set S. Let $\mathscr{S}_n$ be the collection of clutters on sets of size n. The theorem is true when $n = 1$; suppose true for all $k < n$. Let $\mathscr{A} \in \mathscr{S}_n$ and suppose f is uniquely defined for $\mathscr{A}$. If e is essential for $\mathscr{A}$, then by (5)

$$f(\mathscr{A}; x, y) = x\, f(\mathscr{A}_e''; x, y)$$

and since f is uniquely defined on $\mathscr{A}$, it must be on $\mathscr{A}_e''$, and hence by the induction hypothesis $\mathscr{A}_e''$ is the set of bases of a matroid on $S \setminus e$. Hence $\mathscr{A}$ is the free single element extension of $\mathscr{A}_e''$. A similar argument gives the result when $\mathscr{A}$ has a redundant element. Hence we may suppose each element of S is neither redundant nor essential for $\mathscr{A}$.

Let A_1, A_2 be distinct members of $\mathscr{A}$ and let $e \in A_1 \setminus A_2$. If $A_1 \cup A_2 \neq S$ let $h \in S \setminus (A_1 \cup A_2)$. Then

$$f(\mathscr{A}; x, y) = a\, f(\mathscr{A}_h'; x, y) + b\, f(\mathscr{A}_h''; x, y).$$

As $f(\mathscr{A}; x, y)$ is uniquely defined, so is $f(\mathscr{A}_h'; x, y)$ and since $\mathscr{A}_h' \in \mathscr{S}_{n-1}$, it is the set of bases of a matroid on $S \setminus h$. But $A_1, A_2 \in \mathscr{A}_h'$, hence there exists $g \in A_2 \setminus A_1$ such that $(A_2 \setminus g) \cup e \in \mathscr{A}_h'$ and hence is in $\mathscr{A}$.

If $A_1 \cup A_2 = S$ and $A_1 \cap A_2 \neq \emptyset$, we choose $h \in A_1 \cap A_2$ and then since $f(\mathscr{A}; x, y)$ is uniquely defined, $f(\mathscr{A}_h''; x, y)$ must be uniquely defined. But $\mathscr{A}_h'' \in \mathscr{S}_{n-1}$ and is by the induction hypothesis the set of bases of a matroid on $S \setminus h$. Since $A_1 \setminus h, A_2 \setminus h$ are members of $\mathscr{A}_h''$ there exists $g \in (A_2 \setminus h) \setminus (A_1 \setminus h)$ such that $(A_2 \setminus \{h, g\}) \cup e \in \mathscr{A}_h''$. Hence either $(A_2 \setminus g) \cup e \in \mathscr{A}$ or $(A_2 \setminus \{h, g\}) \cup e \in \mathscr{A}$. Suppose the latter, then $e \in ((A_2 \setminus \{h, g\}) \cup e) \cap A_1$ so that $A_1 \setminus e$ and $A_2 \setminus \{h, g\} \in \mathscr{A}_e''$. Hence by the induction hypothesis since $\mathscr{A}_e''$ is the set of bases of a matroid, $|A_1 \setminus e| = |A_2 \setminus \{h, g\}|$, that is, $|A_2| = |A_1| + 1$. But since $A_1 \setminus h$ and $A_2 \setminus h \in \mathscr{A}_h''$ this is a contradiction and hence $(A_2 \setminus g) \cup e \in \mathscr{A}$.

Finally suppose that $A_1 \cup A_2 = S$ and $A_1 \cap A_2 = \emptyset$. If $u \in A_1$, then $A_1 \setminus u \in \mathscr{A}_u''$ and there exists $A_2' \subseteq A_2$ such that $A_2' \in \mathscr{A}_u''$. Thus by induction $|A_2'| = |A_1 \setminus u|$. Hence either

  (i) $A_2' = A_2$ and $|A_1| = |A_2| + 1$; or
  (ii) $A_2' \neq A_2$ and $A_2' \cup u \in \mathscr{A}$ and $|A_2| > |A_2'| = |A_1| - 1$.

Choose $v \in A_2$. By the same argument there exists $A_1' \subseteq A_1$ such that $A_1' \in \mathscr{A}_v''$ and $|A_1'| = |A_2| - 1$ and either

  (iii) $A_1' = A_1$ and hence $|A_1| = |A_2| - 1$; or
  (iv) $A_1' \neq A_1$ and $A_1' \cup v \in \mathscr{A}$ and $|A_1| > |A_1'| = |A_2| - 1$.

First note that (i) and (iii) cannot both hold. Suppose that (i) and (iv) hold. If $A_1' = \emptyset$, then $A_2 = \{v\}$ and, by (i), $|A_1| = 2$. Since $A_1 \cup A_2 = S$ this forces $S = \{u, v, w\}$ and $\mathscr{A} = (\{v\}, \{u, w\})$ and it is easy to check that f is not uniquely defined for this clutter. Thus we may assume $A_1' \neq \emptyset$. Choose $c \in A_1'$. Then $\mathscr{A}_c''$ is the set of bases of a matroid on $S \setminus c$. Now since $(A_1' \cup v) \setminus c$ and $A_1 \setminus c \in \mathscr{A}_c''$ we have $|(A_1' \cup v) \setminus c| = |A_1 \setminus c|$, hence $|A_1'| = |A_1| - 1$. Then by (iv), $|A_1| = |A_2|$, contrary to (i).

If (ii) and (iii) hold, then interchanging the roles of A_1 and A_2, A_1' and A_2', and u and v in the preceding argument again gives a contradiction. Hence (ii) and (iv) hold and hence $|A_1| = |A_2|$. It follows that $A_2' = A_2 \setminus z$ for some z in A_2 and hence $(A_2 \setminus z) \cup u \in \mathscr{A}$. As $A_1 \cap A_2 = \emptyset$, $z \in A_2 \setminus A_1$. Hence if we let $u = e$ and $g = z$, then we obtain the required result. This completes the proof of the theorem. ∎

Now suppose we replace conditions (2)–(4) by their dual conditions, where here we mean dual in the blocking sense. Conditions (2) and (3) become:

If $\{e\} \notin \mathscr{A}$ and e is not redundant for $\mathscr{A}$, then

$$f(\mathscr{A}; x, y) = af(\mathscr{A}/e; x, y) + bf(\mathscr{A} \setminus e; x, y), \qquad (2^*)$$

$$f(\mathscr{A}_1 \cup \mathscr{A}_2; x, y) = f(\mathscr{A}_1; x, y) f(\mathscr{A}_2; x, y). \qquad (3^*)$$

Since the dual of the essential clutter E is itself but the dual of the redundant clutter R is the empty clutter, Z, the dual form of Theorem 1 reads as follows.

Theorem 1* *If a and b are fixed nonzero real numbers then a function $f(\mathscr{A}; x, y)$ satisfying (1), (2*), (3*) and*

$$f(E) = x, \qquad f(Z) = y \qquad (4^*)$$

is uniquely defined for a finite clutter $\mathscr{A} \neq \{\emptyset\}$ if and only if $\mathscr{A}$ is the collection of circuits of a matroid.

Proof The proof follows from the fact that the blocking clutter of the collection of bases of a matroid is the set of circuits of the dual matroid. ∎

5. The Clutter Polynomial

In spite of Theorem 4.1 we notice that when $x = p$, $y = 1$, $a + b = 1$, and $b = p$, then by (3.1)–(3.5) the percolation probability $P(\mathscr{A}; p)$ satisfies (4.1)–(4.4) for all $p \in [0, 1]$. This prompts the question, for which points (a, b, x, y) of $\mathbb{R}^4$ is there a uniquely defined function satisfying (4.1)–(4.4) for all finite nonempty clutters? Apart from the percolation probability we notice that if $x = y = a + b$ then $f(\mathscr{A}) = (a + b)^{|S|}$ is a solution where S is the set supporting $\mathscr{A}$. Our next theorem shows that apart from this and other more trivial cases the essentially unique solution is the percolation probability. These trivial solutions are the following:

the plane $\{(a, b, 0, 0)\}$, $\quad f(\mathscr{A}) = 0$, $\hfill (1)$

the plane $\{(a, 0, 0, y)\}$, $\quad f(\mathscr{A}) = \begin{cases} y^{|S|} & \text{if } \mathscr{A} = \{\varnothing\}, \\ 0 & \text{otherwise,} \end{cases}$ $\hfill (2)$

the plane $\{(0, b, x, 0)\}$, $\quad f(\mathscr{A}) = \begin{cases} x^{|S|} & \text{if } \mathscr{A} = \{S\}, \\ 0 & \text{otherwise,} \end{cases}$ $\hfill (3)$

the plane $\{(0, 0, x, y)\}$, $\quad f(\mathscr{A}) = \begin{cases} x^{|T|} y^{|S \setminus T|} & \text{if } \mathscr{A} = \{T\} \\ & \text{for some } T \subseteq S, \\ 0 & \text{otherwise} \end{cases}$ $\hfill (4)$

the plane $\{(0, b, x, b)\}$, $\quad f(\mathscr{A}) = x^{e(\mathscr{A})} b^{|S| - e(\mathscr{A})}$, $\hfill (5)$

where $e(\mathscr{A})$ is the number of essential elements of $\mathscr{A}$,

the plane $\{(a, b, a + b, a + b)\}$, $\quad f(\mathscr{A}) = (a + b)^{|S|}$. $\hfill (6)$

Each of these solutions has a certain degeneracy about it inasmuch as the solution says very little about the structure of the clutter $\mathscr{A}$. Our next theorem says that there is only one "nondegenerate" solution.

Theorem 1 *The sets of points (a, b, x, y) of $\mathbb{R}^4$ for which there is a uniquely defined function f satisfying (1)–(4) in Section 4 for all finite nonempty clutters are given by the six degenerate cases (1)–(6) together with the plane $\{(a, b, b, a + b)\}$ which gives rise to the solution*

$$f(\mathscr{A}; x, y) = a^{|S|} U(\mathscr{A}; ba^{-1})$$

where U is the upper polynomial of the clutter $\mathscr{A}$ on S.

Sketch of Proof Let $\mathscr{A}_1$ be the clutter $\{\{1, 2\}, \{2, 3\}, \{3, 4\}\}$ on $\{1, 2, 3, 4\}$. Evaluating f in two different orders by the rules (4.1)–(4.4) shows that in order that f should be uniquely defined for $\mathscr{A}_1$ we must have

$$a^2 x^2 + abxy + abx^2 + b^2 y^2 = ax^2 y + abxy + b^2 y^2. \hfill (7)$$

Similarly with $\mathscr{A}_2$, the clutter $\{\{1\}, \{2, 3\}\}$ on $\{1, 2, 3\}$, we get the identity
$$ax^2 + by^2 = axy + abx + b^2y. \tag{8}$$
A third example $\mathscr{A}_3 = \{\{1, 2\}, \{3\}, \{4\}\}$ gives the constraint
$$a^2x^2 + aby^2 + by^3 = a^2xy + aby^2 + a^2bx + ab^2y + b^2y^2. \tag{9}$$
From constraints (7)–(9) it is not difficult to check that the only solutions possible are those listed, and these in fact are solutions. ∎

6. Further Results on Clutter Percolation

We close this paper by outlining some of the main results of [7]. Our original motivation in studying $P(\mathscr{A}; p)$ was to see how much of the known form of the classical percolation probability depended on the fact that the clutter $\mathscr{A}$ in question was the set of paths of a regular crystal lattice. For example, if $P_n(p)$ denotes the probability that at least n points of a lattice are joined by an open path to the origin then all the curves of $P_n(p)$ for various lattices obtained by simulation procedures (see, e.g., Shante and Kirkpatrick [8] or Welsh [13]) suggest that $d^2P_n(p)/dp^2$ has a single root in $(0, 1)$. Although we have not been able to show that this is true for arbitrary clutters, indeed it does not even seem to have been proved for the clutters of paths in crystal lattices, we have obtained the following properties of $P(\mathscr{A}; p)$ when $\mathscr{A}$ is a clutter on a finite set.

Provided neither $\mathscr{A}$ nor $\mathscr{A}^*$ has a member of size 1,
$$P''(p) \text{ has at least one root in } (0, 1). \tag{1}$$

It is also easy to see that for any clutter $\mathscr{A}$,
$$P(\mathscr{A}; p) + P(\mathscr{A}^*; q) = 1. \tag{2}$$

We have also been able to obtain best possible upper and lower bounds for the percolation probability in terms of the cardinalities of the members of the clutter. More precisely we prove:

Theorem 1 *If $a_1, a_2, \ldots, a_k$ are positive integers and if $\mathscr{A} = (A_i : 1 \leq i \leq k)$ is any clutter such that $|A_i| = a_i$ $(1 \leq i \leq k)$, then*
$$\sum_{j=1}^{k} p^{a_j}q^{j-1} \leq P(\mathscr{A}; p) \leq 1 - \prod_{i=1}^{k}(1 - p^{a_i}).$$

Moreover both bounds can be achieved by the clutters of edge sets of paths joining two fixed vertices in a graph.

The right hand inequality is a straightforward deduction from the FKG inequality (see [6]). The extremal graph is a set of k disjoint paths joining two fixed vertices. The lower bound is more interesting. The extremal clutter is again the clutter of paths joining two fixed points in a graph of $\max(a_i) + k - 1$ edges and it has the further property that of all clutters with this *profile* $(a_1, a_2, \ldots, a_k)$ it is a clutter with the minimum possible number of blockers.

Notice also that by taking $p = q = \frac{1}{2}$ we get the following corollary which is reminiscent of theorems of Kruskal–Katona type on the existence of Sperner families.

Corollary *If $\mathscr{A}$ is a clutter $(A_1, A_2, \ldots, A_k)$ of subsets of an n set and $u(\mathscr{A}) = \{X: X \supseteq A_i, 1 \leq i \leq k\}$ then if $|A_i| = a_i$ $(1 \leq i \leq k)$,*

$$|u(\mathscr{A})| \geq \sum_{i=1}^{k} 2^{n - a_i - i + 1}.$$

ACKNOWLEDGMENT

The first author is most grateful to CSIRO (Australia) for its generous financial support.

REFERENCES

1. S. R. Broadbent and J. M. Hammersley, Percolation processes I. Crystals and mazes, *Proc. Cambridge Philos. Soc.* **53** (1957) 629–641.
2. T. H. Brylawski, A decomposition for combinatorial geometries, *Trans. Amer. Math. Soc.* **171** (1972) 235–282.
3. J. Edmonds and D. R. Fulkerson, Bottleneck extrema, *J. Combinatorial Theory* **8** (1970) 299–306.
4. J. W. Essam, Graph theory and statistical physics, *Discrete Math.* **1** (1971) 83–112.
5. C. M. Fortuin and P. W. Kasteleyn, On the random cluster model I. Introduction and relation to other models, *Physica* **57** (1972) 536–564.
6. C. M. Fortuin, P. W. Kasteleyn, and J. Ginibre, Correlation inequalities on some partially ordered sets, *Comm. Math. Phys.* **22** (1971) 89–103.
7. J. G. Oxley and D. J. A. Welsh, On some percolation results of J. M. Hammersley, *J. Appl. Probability*, to appear.
8. V. K. S. Shante and S. Kirkpatrick, An introduction to percolation theory, *Advances in Phys.* **20** (1971) 325–356.
9. H. N. V. Temperley and E. H. Lieb, Relations between the percolation and colouring problem and other graph theoretical problems associated with regular planar lattices: Some exact results for the percolation problem, *Proc. Roy. Soc. London Ser. A* **322** (1971) 251–280.
10. W. T. Tutte, A ring in graph theory, *Proc. Cambridge Philos. Soc.* **43** (1947) 26–40.
11. W. T. Tutte, A contribution to the theory of chromatic polynomials, *Canad. J. Math.* **6** (1954) 80–91.

12. D. J. A. Welsh, "Matroid Theory," London Mathematical Society Monographs, No. 8. Academic Press, New York, 1976.
13. D. J. A. Welsh, Percolation and related topics, *Sci. Progr.* (*London*) **64** (1977) 67–85.

AMS 60K35, 05B35

MERTON COLLEGE
OXFORD, ENGLAND

Sums of Circuits

P. D. SEYMOUR

1. Introduction

Ford and Fulkerson's max-flow min-cut theorem [2] is a statement about the feasible values of flows from a source to a sink in a directed graph with capacitated edges. It is easy to see that we may restrict ourselves to flows which are sums of flows along directed paths from source to sink without invalidating the theorem, and expressed in that form it becomes the following.

(1.1) *Let s, t be distinct vertices of a directed graph $G = (V, E)$, and let $\mathscr{P}$ be the collection of all directed paths from s to t. Let $c : E \to \mathbb{Q}^+$ and $q \in \mathbb{Q}^+$ be given; then the following are equivalent:*

 (i) *for some $\alpha : \mathscr{P} \to \mathbb{Q}^+$, $\sum_{P \in \mathscr{P}} \alpha(P) f_P \leq c$, and $\sum_{P \in \mathscr{P}} \alpha(P) \geq q$;*
 (ii) *for each $X \subseteq V$ with $s \in X$ and $t \notin X$, $c(\partial^+(X)) \geq q$.*

[Some conventions and terminology: graphs may have multiple edges, but not loops; $\mathbb{Q}^+$ is the set of non-negative rational numbers; for any subgraph H of a graph $G = (V, E)$ (or simply a subset $H \subseteq E$), $f_H : E \to \mathbb{Q}^+$ is the zero-one function which takes the value one just on the edges of H; for $X \subseteq V$, $\partial(X)$ is the set of edges of G with just one end in X, and when G is directed, $\partial^+(X)$ is the subset of $\partial(X)$ directed out; and finally, for any map $f : Y \to Z$, if $X \subseteq Y$ then $f(X)$ denotes $\sum_{x \in X} f(x)$.]

There is a closely-related statement about sums of paths between two vertices of an *undirected* graph.

(1.2) *Let s, t be distinct vertices of a graph $G = (V, E)$ and let $\mathscr{P}$ be the collection of all paths between s and t. Let $c: E \to \mathbb{Q}^+$ and $q \in \mathbb{Q}^+$ be given; then the following are equivalent:*

  (i) *for some $\alpha: \mathscr{P} \to \mathbb{Q}^+$, $\sum_{P \in \mathscr{P}} \alpha(P) f_P \leq c$, and $\sum_{P \in \mathscr{P}} \alpha(P) \geq q$;*
  (ii) *for each $X \subseteq V$ with $s \in X$ and $t \notin X$, $c(\partial(X)) \geq q$.*

One can derive this as a corollary of (1.1). But the reader should notice the crucial point that if $P_1, P_2 \in \mathscr{P}$ have a common edge e, then $\alpha(P_1)$ and $\alpha(P_2)$ are *added* [when checking $\sum_{P \in \mathscr{P}} \alpha(P) f_P(e) \leq c(e)$] even when P_1 and P_2 use e in opposite senses. In (1.2), therefore, we are no longer talking about flows; because to obtain a flow one would need to cancel paths using edges in opposite senses.

Let us now return to the directed case (1.1), and consider the graph G' obtained from G by adding a new edge e_0 directed from t to s. The directed paths in $\mathscr{P}$ correspond to the directed circuits of G' which use e_0; and we may regard (1.1) as a condition for when there is a "non-negative rational sum of directed circuits" f such that $f(e_0) \geq q$ and $f(e) \leq c(e)$ for each $e \in E$. Hoffman [4] extended this, permitting both an upper and a lower bound on each edge.

(1.3) (Hoffman's Circulation Theorem) *Let $G = (V, E)$ be a directed graph and let $\mathscr{C}$ be its collection of directed circuits. Let $u, l: E \to \mathbb{Q}^+$ satisfy $u \geq l \geq 0$; then the following are equivalent:*

  (i) *there exists $\alpha: \mathscr{C} \to \mathbb{Q}^+$ such that $u \geq \sum_{C \in \mathscr{C}} \alpha(C) f_C \geq l$;*
  (ii) *for each $X \subseteq V$, $u(\partial^+(X)) \geq l(\partial^+(V - X))$.*

The main theorem of this paper is the following analogous extension of (1.2).

(1.4) *Let $G = (V, E)$ be a graph, and let $\mathscr{C}$ be its collection of circuits. Let $u, l: E \to \mathbb{Q}^+$ satisfy $u \geq l \geq 0$; then the following are equivalent:*

  (i) *there exists $\alpha: \mathscr{C} \to \mathbb{Q}^+$ such that $u \geq \sum_{C \in \mathscr{C}} \alpha(C) f_C \geq l$;*
  (ii) *for each cut B and $e \in B$, $l(e) \leq u(B - \{e\})$.*

[A *cut* is a nonempty set of edges of the form $\partial(X)$ for some $X \subseteq V$].

One might suspect this of being a corollary of Hoffman's theorem, in the same way that (1.2) is a corollary of (1.1), but this is not so. We shall see later (Section 3) that (1.4) does not work in integers, while (1.3) does; and in Section 4 that (1.4) does not extend to regular matroids, while (1.3) does. A further, more fundamental, difference between the two theorems appears when we take $l \equiv u$; (1.3) becomes obvious, but (1.4) remains nontrivial. Indeed, the case $l \equiv u$ is essentially the whole problem.

2. Proof of the Theorem

Suppose that $u, l: E \to \mathbb{R}^+$ satisfy $u \geq l \geq 0$; we say that the pair (u, l) is *cut-admissible* (*for G*) if for each cut B and each $e \in B$ we have $l(e) \leq u(B - \{e\})$. The first step is to reduce the problem to the case $l \equiv u$.

(2.1) *If (u, l) is cut-admissible for $G = (V, E)$, then there exists $p: E \to \mathbb{Q}^+$ such that $u \geq p \geq l$ and such that (p, p) is also cut-admissible.*

Proof We proceed by induction on the number of edges e with $u(e) > l(e)$. If this number is zero we may take $p \equiv u$; and so we assume that $u(e_0) > l(e_0)$ for some $e_0 \in E$.

For any $t \in \mathbb{Q}^+$ with $u(e_0) \geq t \geq l(e_0)$, we define $u_t, l_t: E \to \mathbb{Q}^+$ as follows: $u_t(e_0) = l_t(e_0) = t$, and for $e \neq e_0$, $u_t(e) = u(e)$ and $l_t(e) = l(e)$. Certainly we have $u \geq u_t \geq l_t \geq l$, and $u_t(e_0) = l_t(e_0)$; so if we can choose t such that (u_t, l_t) is cut-admissible, then the conclusion will follow by induction.

We may assume that taking $t = u(e_0)$ does not work; thus if $t_1 = \min(u(B - \{e_0\}): B$ a cut containing $e_0)$, then $l(e_0) \leq t_1 < u(e_0)$. Choose a cut B_1 containing e_0 such that $t_1 = u(B_1 - \{e_0\})$. We claim that if we take $t = t_1$ then (u_t, l_t) is cut-admissible. To check this we need only show that for each cut B containing e_0 and each $e \in B - \{e_0\}$, we have $l_t(e) \leq u_t(B - \{e\})$, that is,

$$l(e) \leq u(B - \{e\}) + t_1 - u(e_0) \tag{$*$}$$

But if $e \in B_1$, then $l(e) \leq u(e) \leq u(B_1 - \{e_0\}) = t_1$ and $u(e_0) \leq u(B - \{e\})$, and so $(*)$ holds in this case. On the other hand, if $e \notin B_1$, then there is a cut $B_2 \subseteq B \cup B_1$, with $e \in B_2$ and $e_0 \notin B_2$; and so

$$l(e) \leq u(B_2 - \{e\}) \leq u(B - \{e_0, e\}) + u(B_1 - \{e_0\})$$
$$= u(B - \{e\}) + t_1 - u(e_0)$$

as required. ∎

Notes (i) We have proved in addition that if l, u are integer-valued, then p may be chosen integer-valued.

(ii) The only property of graphs used here is the fact that if B, B_1 are cuts and $e_0 \in B \cap B_1$ and $e \in B - B_1$, then there is a cut $B_2 \subseteq B \cup B_1$ with $e \in B_2$ and $e_0 \notin B_2$. This is a matroid property, and the argument works equally well in matroid terms.

(iii) One can prove Hoffman's theorem (1.3) by the same approach.

The next lemma is complicated to state and easy to prove, but it is worth stating separately because we shall use it twice.

(2.2) *Let $G = (V, E)$ be a graph, and $\mathscr{C}$ be its collection of circuits. Suppose that we have a map $p_0: E \to \mathbb{Z}^+$ and a cut B, such that $p_0(e_0) = p_0(B - \{e_0\})$. Let X_1, X_2 be the sets of vertices on opposite sides of B. For*

$i = 1, 2$, let G_i be obtained from G by deleting all edges with both ends in X_i and identifying all the vertices of X_i. Let $\mathscr{C}_i$ be the collection of circuits of G_i, and let p_i be the restriction of p_0 to $E(G_i)$. Suppose that for $i = 1, 2$ there exists $\alpha_i \colon \mathscr{C}_i \to \mathbb{Z}^+$ such that $\sum_{C \in \mathscr{C}} \alpha_i(C) f_C = p_i$. Then there exists $\alpha \colon \mathscr{C} \to \mathbb{Z}^+$ such that $\sum_{C \in \mathscr{C}} \alpha(C) f_C = p_0$.

(Here $\mathbb{Z}^+$ is the set of non-negative integers.)

Proof By hypothesis, there is a list of circuits from $\mathscr{C}_i$ (with possible repetition) using each edge of G_i $p_i(e)$ times. Let $B = \{e_0, e_1, \ldots, e_k\}$; then B is a cut of G_i, and $p_i(e_0) = p_i(B - \{e_0\})$; and so each circuit in this list uses either no edge in B or precisely two edges in B one of which is e_0. There are thus $p_0(e_j)$ circuits in the list using e_j and e_0 ($j = 1, \ldots, k$) and no other circuits in the list which meet B. We combine the lists for $i = 1, 2$, pairing the circuits which meet B in the obvious way, to make a list of circuits of G using edge e precisely $p_0(e)$ times. This completes the proof. ∎

Notes (i) The integrality of p and α is not required for the application in this section (although it is in the next); however, some such device is needed in any case to make the "pairing" operation work in the proof.

(ii) Observe that if (p_0, p_0) is cut-admissible for G and p_i, G_i are constructed as in (2.2) then (p_i, p_i) is cut-admissible for G_i ($i = 1, 2$). This is because identifying vertices does not introduce new cuts.

The next lemma is due to Rick Giles and myself, and is the most tricky single step in the proof of (1.4).

(2.3) Let $G = (V, E)$ be a non-null bridgeless graph, and let $\phi \colon V \to E$ map each vertex to some incident edge. Then there is a circuit C^* such that $\phi(v) \in E(C^*)$ for each $v \in V(C^*)$.

(A *bridge* is a cut of one edge.)

Proof We proceed by induction on $|V|$, and assume that the result is true for any graph G' with $|V(G')| < |V|$. G is non-null; choose a vertex x_1. Let x_2 be the other end of $\phi(x_1)$. For $i = 2, 3, \ldots$, define x_{i+1} inductively as follows:

If $\phi(x_i) \neq \phi(x_{i-1})$, let x_{i+1} be the end of $\phi(x_i)$ different from x_i.

If $\phi(x_i) = \phi(x_{i-1})$, let x_{i+1} be the other end of some edge incident with x_i but different from $\phi(x_i)$. [Such an edge exists because $\phi(x_i)$ is not a bridge of G.] G is finite, and so there is repetition in this sequence. Choose r, s with $r < s$ such that $x_1, \ldots, x_{s-1}$ are all distinct but $x_r = x_s$. Then the sequence $x_r, x_{r+1}, \ldots, x_s$ gives a circuit C of G such that for $r + 1 \leq i \leq s - 1$, $\phi(x_i) \in E(C)$. We may assume that $\phi(x_r) \notin E(C)$, for otherwise the theorem is satisfied with $C^* = C$. Thus x_{r+1} is not an end of $\phi(x_r)$; and so $r \geq 2$, and $\phi(x_{r-1}) = \phi(x_r)$. $x_{r-1} \notin V(C)$ because $x_1, \ldots, x_{s-1}$ are all distinct.

SUMS OF CIRCUITS

Let $G' = (V', E')$ be the result of contracting all the edges with both ends in $V(C)$, all the vertices of C becoming identified in a vertex v_0, say. Define $\phi': V' \to E'$ as follows:

$$\phi'(v_0) = \phi(x_r),$$
$$\phi'(x) = \phi(x) \, (x \in V' - \{v_0\}).$$

G' is non-null and bridgeless, and so by induction there is a circuit C' of G' such that $\phi'(v) \in E(C')$ for each $v \in V(C')$. If $v_0 \notin V(C')$ the theorem is satisfied by taking $C^* = C'$, and so we assume that $v_0 \in V(C')$. Then $\phi(x_r) = \phi'(v_0) \in E(C')$, and so $x_{r-1} \in V(C')$. Let e be the other edge of C' incident with v_0, and let z be its other end. Now $e \in E(G)$, and e is incident with z and some vertex x_t say in $V(C)$.

If $x_t = x_r$ then C' is a circuit of G and we may take $C^* = C'$ to satisfy the theorem. We therefore assume that $x_t \neq x_r$. Let P be the path from x_r to x_t which is a section of C and which uses the edge $\phi(x_t)$. (So the vertices of P are either $x_r, x_{r+1}, \ldots, x_{t-1}, x_t$ or $x_r, x_{s-1}, \ldots, x_{t+1}, x_t$.) In G, the edges of C' give another path from x_r to x_t; let C^* be the composition of this and P. Then C^* satisfies the theorem. ∎

This implies the following (irrelevant) curiosity, due to Kotzig [9].

(2.4) Corollary *If G is a non-null bridgeless graph with a 1-factor, then G has at least two 1-factors.*

Proof Let F be a 1-factor. For each $v \in V(G)$, define $\phi(v)$ to be the edge in F incident with v. From (2.3) there is a circuit C^* such that $\phi(v) \in E(C^*)$ for each $v \in V(C^*)$. Thus C^* consists of edges alternately in and not in F; and so $F \triangle E(C^*)$ is another 1-factor. ($\triangle$ denotes symmetric difference.) ∎

Now to complete the proof of (1.4). Because of (2.1) it is only necessary to show the following.

(2.5) *Let $G = (V, E)$ be a graph, with collection of circuits $\mathscr{C}$. Let $p: E \to \mathbb{Q}^+$ be such that (p, p) is cut-admissible. Then there exists $\alpha: \mathscr{C} \to \mathbb{Q}^+$ such that $\sum_{C \in \mathscr{C}} \alpha(C) f = p$.*

Proof We use induction on $|V| + |E|$, and assume that the result is true for any pair G', p' with $|V(G')| + |E(G')| < |V| + |E|$. Thus we may assume that G is connected.

For any map $p': E \to \mathbb{Q}^+$, let $n(p')$ denote the number of pairs (B, e) of a cut B and an edge $e \in B$ such that $p'(e) = p'(B - \{e\})$. $n(p')$ is bounded above, and we assume with a second induction that the result is true for G and any p' with $n(p') > n(p)$.

(1) The result is true for G and any p' such that $p'(e) = 0$ for some edge $e \in E$.

This follows from our first induction, by considering the restriction of p' to the edges of the graph obtained by deleting e.

(2) The result is true for G and p' if there is a cut B with at least two vertices on either side, and an edge $e \in B$ such that $p'(e) = p'(B - \{e\})$.

For $i = 1, 2$ define X_i, G_i, $\mathscr{C}_i$, as in (2.2), and let p_i' be the restriction of p' to $E(G_i)$. (p_i', p_i') is cut-admissible for G_i, and so there exists $\alpha_i' : \mathscr{C}_i \to \mathbb{Q}^+$ such that $\sum_{C \in \mathscr{C}} \alpha_i(C) f_C = p_i'$, by induction. Choose an integer $N > 0$ such that $N\alpha_1'$ and $N\alpha_2'$ are both integer-valued. Put $\alpha_i = N\alpha_i'$ and $p_i = Np_i'$ ($i = 1, 2$), and $p_0 = Np'$. Then the hypotheses of (2.2) are satisfied, and so there exists $\alpha : \mathscr{C} \to \mathbb{Z}^+$ such that $\sum_{C \in \mathscr{C}} \alpha(C) f_C = Np'$. Put $\alpha' = (1/N)\alpha$; and then α' shows that the result is true for G, p'.

(3) The result is true for G and p.

We may assume that $|V| \geq 2$. For each vertex v, define $\phi(v)$ to be an edge e incident with v with $p(e)$ maximum. Since (p, p) is cut-admissible, and since by (1) we may assume that $p(e) \neq 0$ ($e \in E$), it follows that G is bridgeless; and so by (2.3) there is a circuit C^* of G such that if $v \in V(C^*)$ then $\phi(v) \in E(C^*)$. For $t \in \mathbb{Q}^+$, define $p_t : E \to \mathbb{Q}$ by $p_t = p - tf_{C^*}$. Choose t as large as possible such that $p_t \geq 0$ and (p_t, p_t) is cut-admissible. (This is possible since $t = 0$ satisfies these conditions, and the set of such t is closed above.) If p_t is expressible in the form $\sum_{C \in \mathscr{C}} \alpha(C) f_C$ then so is p, because $p = p_t + tf_{C^*}$. We claim that it follows from one of our inductive hypotheses that p_t is so expressible.

Observe that if B is a cut and $e \in B$, and $p(e) = p(B - \{e\})$, then $p_t(e) = p_t(B - \{e\})$ [because (p_t, p_t) is cut-admissible]; thus $n(p_t) \geq n(p)$. From the maximality of t, one of the following is true:

(i) $\phi_t(e) = 0$ for some $e \in E$. Then the result holds for G, ϕ_t by (1).

(ii) For some cut B with at least two vertices on either side, and for some $e \in B$, $p_t(e) = p_t(B - \{e\})$. In this case the result holds for G, by (2).

(iii) For some $v \in V$ and some $e \in \partial(\{v\})$, $p_t(e) = p_t(\partial(\{v\}) - \{e\})$ and C^* passes through v but does not use e. Then

$$p(e) \leq p(\phi(v)) < p(\partial(\{v\}) - \{e\})$$

by definition of ϕ; it follows that $n(p_t) > n(p)$ and the result follows for G, p_t by our second inductive hypothesis. This completes the proof of (1.4). ∎

3. Integral Constraints and Integral Sums

It is easy to see that if u, l in (1.3) are integer-valued then α may be chosen to be integer-valued as well. This is not so with (1.4); for a counterexample, let G consist of three parallel edges, with $u \equiv l \equiv 1$. But this is rather too trivial a counterexample to convince one that integrality is difficult; perhaps we have just chosen the wrong conjecture. A more reasonable conjecture would be the relevant one which comes in the general format "given u, l with $u \geq l \geq 0$, either there exists a suitable α with $u \geq \sum_{C \in \mathscr{C}} \alpha(C) f_C \geq l$, or there is a bad cut." Here by a "bad cut" we mean a cut B such that the restriction of u, l to B could not be the restriction of any u', l' to B in any graph G' for which B is a cut and for which there is a suitable α' satisfying $u' \geq \sum_{C \in \mathscr{C}'} \alpha(C) f_C \geq l'$. [(1.3), (1.4), and the integral form of (1.3) may all be recast in this format; and indeed, so may most of the other theorems of this paper.] We are led to the conjecture "given integral u, l with $u \geq l \geq 0$, then there exists $\alpha: \mathscr{C} \to \mathbb{Z}^+$ such that $u \geq \sum_{C \in \mathscr{C}} \alpha(C) f_C \geq l$ if and only if (u, l) is cut-admissible and for each cut B there is an even number between $u(B)$ and $l(B)$." But unfortunately this also is false; for a counterexample, take the graph of three parallel edges, and replace one of the edges by two edges e_1, e_2 in series; define $l(e_1) = 0$, $u(e_2) = 2$, and otherwise $l \equiv 1 \equiv u$. Thus this problem remains unsolved.

However, we can go much further in the special case $l \equiv u$ ($\equiv p$, say). In that case, the "bad cut" format leads us to the conjecture "given $p: E \to \mathbb{Z}^+$, there exists $\alpha: \mathscr{C} \to \mathbb{Z}^+$ such that $p = \sum_{C \in \mathscr{C}} \alpha(C) f_C$ if and only if (p, p) is cut-admissible and for each cut B, $p(B)$ is even." We shall prove below that this is true for planar graphs (assuming the four color theorem). But once again it is not true in general; for a counterexample, take G to be the Petersen graph, and let F be some 1-factor; define $p(e) = 2$ ($e \in F$), and $p(e) = 1$, otherwise. However, the following weaker conjecture is still open.

(3.1) **Conjecture** *If $p: E \to \mathbb{Z}^+$ and (p, p) is cut-admissible and $p(e)$ is even for each $e \in E$, then there exists $\alpha: \mathscr{C} \to \mathbb{Z}^+$ such that $p = \sum_{C \in \mathscr{C}} \alpha(C) f_C$.*

In view of (2.1), this is equivalent to

(3.2) **Conjecture** *If $u, l: E \to \mathbb{Z}^+$ satisfy $u \geq l \geq 0$, and (u, l) is cut-admissible, then there exists $\alpha: \mathscr{C} \to \mathbb{Q}^+$, half-integer valued, such that $u \geq \sum_{C \in \mathscr{C}} \alpha(C) f_C \geq l$.*

These are true, a fortiori, for planar graphs, but I have made no progress with a general proof. Indeed, the following very special case of (3.1) (when $p \equiv 2$) is still unsolved:

(3.3) **Conjecture** *Let G be a bridgeless graph; then there is a list of circuits of G, with each edge in precisely two of them.*

It is rather easy to see that the smallest counterexample to (3.3) must be a cubic, non-3-edge-colourable, bridgeless graph which is not the Petersen graph. Also, (3.3) holds for any cubic graph which can be embedded on some surface in such a way that the perimeters of regions are circuits (i.e., have no repeated edges) and so the following folklore conjecture (probably due to Tutte) is relevant:

(3.4) Conjecture *Any bridgeless cubic graph can be embedded on a surface of its own genus in such a way that the perimeters of all regions are circuits.*

Now for the main theorem of this section.

(3.5) *Let $G = (V, E)$ be a planar graph, with collection of circuits $\mathscr{C}$. Given $p: E \to \mathbb{Z}^+$, the following are equivalent:*

(i) *there exists $\alpha: \mathscr{C} \to \mathbb{Z}^+$ such that $p = \sum_{C \in \mathscr{C}} \alpha(C) f_C$;*
(ii) *(p, p) is cut-admissible and for each cut B, $p(B)$ is even.*

Proof We proceed by induction on $n(G, p)$, which is defined to be $\sum_{e \in E} (p(e) + 1)^2$. Clearly we may assume that G is connected. Moreover, we may assume that $p(e) \neq 0$ ($e \in E$); because if $p(e) = 0$ we may delete e from G, and this operation decreases $n(G, p)$.

(1) Suppose that there is a cut B with more than one vertex on either side, and $e \in B$ such that $p(e) = p(B - \{e\})$. B is a union of minimal cut-sets; choose a minimal cut-set $B_1 \subseteq B$ with $e \in B_1$. Now

$$p(e) \leq p(B_1 - \{e\}) = p(B - \{e\}) - p(B - B_1) = p(e) - p(B - B_1)$$

and so $p(B - B_1) = 0$. Since p is nonzero everywhere, we have $B = B_1$, and so B is a minimal cut-set. Let X_1, X_2 be the sets of vertices on either side of B. Then the subgraphs of G induced by X_1, X_2 are connected, and thus the graphs G_1, G_2 defined in (2.2) may be obtained by contracting edges, and hence are planar. Put $p_0 = p$, and define $\mathscr{C}_i$, p_i as in (2.2) ($i = 1, 2$). By induction the theorem holds for the pairs G_i, p_i ($i = 1, 2$), and hence by (2.2) for G, p. We therefore assume that there is no such cut B.

(2) Suppose that $v \in V$ is such that $p(e) = 1$ for each edge e incident with v. We may assume that v is adjacent with two distinct vertices, for otherwise the result follows by deleting v and using induction. We can therefore choose edges e_1, e_2 incident with v, with their other ends w_1, w_2 (say) distinct, such that G' is planar, where G' is the graph obtained by deleting e_1, e_2 and adding a new edge e_0 with ends w_1, w_2. Define $p': E(G') \to \mathbb{Z}^+$ by $p'(e_0) = 1$, and $p'(e) = p(e)$ ($e \in E - \{e_1, e_2\}$). It is easy to see that $p'(B)$ is even for each cut B of G'; we wish to show also that (p', p') is cut-admissible (for G'). Suppose, for a contradiction, that B is a cut of G' and $e \in B$, and $p'(e) > p'(B - \{e\})$. Since $p'(B)$ is even, it follows that $p'(e) \geq p'(B - \{e\}) + 2 \geq 2$, and hence that $e \neq e_0$. Choose $X \subseteq V(G')$ with $v \in X$ such that B is the set of edges of G' with just one end in X. Now

$$p'(B) \geq p(\partial(X)) - 2 \quad \text{and} \quad p'(e) = p(e) \leq p(\partial(X) - \{e\});$$

comparing these with

$$p'(e) \geq p'(B - \{e\}) + 2$$

we deduce that equality holds in all three inequalities. Thus in particular, $p(e) = p(\partial(X) - \{e\})$, and so by the assumed nonapplicability of (1), we deduce that $X = V - \{v\}$. But then e is incident with v, since $e \in \partial(X)$; this is impossible, since $p(e) > 1$. We therefore assume that there is no such $v \in V$.

Choose some edge e_0 with $p(e_0) > 1$, such that $p(e_0)$ is minimum. Let e_0 have ends x, y. Add a new edge e_1, parallel to e_0, forming G'; and define $p': E(G') \to \mathbb{Z}^+$ by

$$p'(e_1) = 1, \quad p'(e_0) = p(e_0) - 1, \quad p'(e) = p(e) \quad (e \in E - \{e_0\}).$$

Then $n(G', p') < n(G, p)$ since $p(e_0) \geq 2$, and so by induction there is a list L' of circuits of G' using each edge e of G' $p'(e)$ times (it is easy to see that (p', p') is cut-admissible, etc.). We may assume that the circuit with two edges e_0, e_1, occurs in L', for otherwise we could replace e_1 by e_0 wherever it occurred, and obtain a suitable list of circuits for G, p.

It follows that there is a list L of circuits of G using e_0 $p(e_0) - 2$ times, and every other edge e $p(e)$ times. Put $L = L_1 \cup L_2$, where the circuits in L_1 do not use e_0 and the ones in L_2 do.

For distinct $u, v \in V$, write $u \to v$ if either some circuit of L_1 passes through u and v, or there is some circuit of L_2 which passes through y, x, v, u in that order (or the reverse). Let $U \subseteq V$ be the set of vertices $u \in V$ such that there is a sequence

$$x = v_0 \to v_1 \to \cdots \to v_k = u$$

for some $k \geq 0$. Suppose, for a contradiction, that $y \notin U$. It follows that no circuit of L_1 uses an edge in $\partial(U)$, and any circuit of L_2 uses only one edge of $\partial(U)$ different from e_0. Thus

$$p(e_0) - 2 = p(\partial(U) - \{e_0\})$$

which is impossible since (p, p) is cut-admissible.

Thus $y \in U$, and there is a sequence

$$x = v_0 \to v_1 \to \cdots \to v_k = y$$

for some $k \geq 0$. Choose such a sequence with k minimum. For $i = 1, \ldots, k$, choose a circuit $C_i \in L$ using v_{i-1} and v_i, such that if $C_i \in L_2$ then y, x, v_i, v_{i-1} occur in that order (or the reverse) on C_i. Let $q'(e)$ be the number of distinct C_i that use e. Define

$$q(e_0) = q'(e_0) + 2, \quad q(e) = q'(e) \quad \text{otherwise.}$$

q' is an "integer sum of circuits" and so (q', q') is cut-admissible and $q'(B)$ is even for every cut B. It follows that $q(B)$ is even for every cut B; and we claim that (q, q) is cut-admissible as well. It is only necessary to check that if $X \subseteq V - \{y\}$ and $x \in X$, then $q(e_0) \le q(\partial(X) - \{e_0\})$. But if this is false, then

$$q'(e_0) = q'(\partial(X) - \{e_0\})$$

and so each C_i which intersects $\partial(X)$ uses e_0 and just one other edge in $\partial(X)$. This contradicts the fact that for some i, $v_i \in X$ and $v_{i+1} \notin X$; and hence (q, q) is cut-admissible.

Remove one occurrence of each distinct C_i from L, forming L_0. If $n(G, q) < n(G, p)$ then by induction there is a list of circuits L' using each edge $e \in E$ $q(e)$ times; and then $L_0 \cup L'$ is a list of circuits of G using each edge e $p(e)$ times, as required. We may therefore assume that $n(G, q) = n(G, p)$, and hence that $q \equiv p$ and L_0 is null. So L is $C_1, \ldots, C_k$, and the C_i's are all distinct.

Now if $C \in L_1$ and $v_i, v_j \in V(C)$ with $j > i$, then $v_i \to v_j$ and so

$$x = v_0 \to v_1 \to \cdots \to v_i \to v_j \to v_{j+1} \to \cdots \to v_k = y;$$

thus from the minimality of k, $j = i + 1$. It follows that for $1 \le i \le k - 1$, there are at most two circuits in L_1 passing through v_i; namely, C_i and C_{i+1}. But there are only $p(e_0) - 2$ circuits in L_2 altogether, and so

$$q(\partial(\{v_i\})) \le 2(p(e_0) - 2) + 2 \cdot 2 = 2p(e_0).$$

On the other hand, by the assumed nonapplicability of (2), v_i is incident with some edge e with $p(e) > 1$; and by choice of e_0, $p(e) \ge p(e_0)$. But

$$p(e) \le p(\partial(\{v_i\}) - \{e\})$$

since (p, p) is cut-admissible, and so

$$p(\partial(\{v_i\})) \ge 2p(e_0).$$

We know that $p \equiv q$; and so we have equality throughout, and hence C_i and C_{i+1} are both in L_1, for $1 \le i \le k - 1$. It follows that L_2 is null, and that $p(e_0) = 2$.

Now suppose that $1 \le i < j \le k$ and that C_i and C_j have a vertex v in common. Then $v_{i-1} \to v$ and $v \to v_j$, and so from the minimality of k, we deduce that $j = i + 1$. Thus for each $v \in V$ there are at most two members of L passing through v; it follows that $q(e) \le 2$ for each $e \in E$. But $p \equiv q$, and so $p \le 2$.

Let F be the set of edges $e \in E$ such that $p(e) = 1$. Since $p(B)$ is even for each cut B, it follows that $|F \cap B|$ is even for each B, and hence that F is a cycle. (A *cycle* is an edge-disjoint union of circuits.) Moreover, G is bridgeless since (p, p) is cut-admissible, and so by a version of the four colour theorem [1, 8] there are three cycles F_1, F_2, F_3 of G such that each edge is in precisely two of them. $F_i \triangle F$ ($\triangle$ denotes symmetric difference) is also a cycle for $i = 1, 2, 3$; and the list of circuits obtained by combining these three cycles uses each edge e $p(e)$ times, as required. ∎

4. Attempts at Extension and Variation

The truth of (3.5) for planar graphs, and its falsity in general, might lead one to suspect that we should really be thinking about sums of cuts rather than sums of circuits, and then perhaps we would obtain a theorem implying (3.5), valid for all graphs. Certainly a result like Hoffman's theorem (1.3) is true (in integers) for sums of directed cuts; indeed (1.3) can be generalized to regular matroids (see [3]). But this does not work for (3.5); and perhaps more surprisingly, (1.4) itself has no analogue for cuts. For a counterexample, let G be K_5, and let G' be a $K_{2,3}$ subgraph; and define $p(e) = 1$ ($e \in E(G')$), and $p(e) = 2$, otherwise. Then (p, p) has the property corresponding to "cut-admissible," that is, for each circuit C and $e \in E(C)$,

$$p(e) \le p(E(C) - \{e\})$$

and yet there is no map $\alpha: \mathscr{B} \to \mathbb{Q}^+$ such that $\sum_{B \in \mathscr{B}} \alpha(B) f_B = p$. [Here $\mathscr{B}$ is the collection of all cuts (or minimal cuts).]

The failure of the analogue of (1.4) for sums of cuts is really the same phenomenon as the failure of the obvious necessary conditions for the existence of a multi-commodity flow to be sufficient. The truth of (1.4) itself tells us that the problem which is dual (in the circuit-cut sense) to the multicommodity flow problem does have a nice solution. We study this in the next section.

One might ask which matroids have a property like (1.4)? In view of the note after (2.1), this is equivalent to the problem

(4.1) *For which matroids* $\mathbf{M}$ *with collection of circuits* $\mathscr{C}$ *is it true that for each map* $p: E(\mathbf{M}) \to \mathbb{Q}^+$ *the following are equivalent:*

(i) *there exists* $\alpha: \mathscr{C} \to \mathbb{Q}^+$ *such that* $\sum_{C \in \mathscr{C}} \alpha(C) f_C = p$;
(ii) *for each cocircuit* B *and* $e \in B$, $p(e) \le p(B - \{e\})$?

It is straightforward to check that the class of matroids satisfying (4.1) is closed under taking minors; moreover, $U_4{}^2$ does not have the property, and so all such matroids are binary. Nor can they have minors isomorphic to the binary affine geometry on eight points, or to the bond matroid of K_5. But I have no idea whether the exclusion of these minors is also sufficient.

The following is another matroid problem, which (believe it or not) was the starting point for the research in this paper.

(4.2) *For which binary matroids* M *is it true that for each map* $p: E(M) \to \mathbb{Z}^+$, *the following are equivalent*:
   (i) *there exists* $\alpha: \mathscr{C} \to \mathbb{Z}^+$ *such that* $\sum_{C \in \mathscr{C}} \alpha(C) f_C = p$;
   (ii) *for each cocircuit* B, $p(B)$ *is even, and for each* $e \in B$, $p(e) \leq p(B - \{e\})$?

Again, this property is closed under taking minors, and so no such matroid has a minor isomorphic to the cycle matroid of the Petersen graph, the bond matroid of K_5, or the dual of the Fano matroid (for the last, take $p \equiv 2$). I suspect that exclusion of these three minors is also sufficient (see Tutte [8]). It was proved in [7] that binary matroids with no Fano dual minor have an integral max-flow min-cut property, but evidently this is not the same thing.

One might even consider extending (1.4) beyond (4.1) to clutters. [A *clutter* L is a collection of subsets of $E(L)$, no one included in another.] The max-flow min-cut property itself is quite interesting in a clutter formulation (see [6, 7]). But here there is a forceful negative result, which says that we get no further than (4.1) this way.

(4.3) *If* L *is a clutter such that for every* $p: E(L) \to \mathbb{Q}^+$ *the following are equivalent*:
   (i) *there exists* $\alpha: L \to \mathbb{Q}^+$ *such that* $\sum_{A \in L} \alpha(A) f_A = p$;
   (ii) *for each* $X \subseteq E(L)$ *and* $x \in X$, *if* $p(x) > p(X - \{x\})$ *then* $X \cap A = \{x\}$ *for some* $A \in L$,

then L *is the collection of circuits of a matroid.*

Proof Suppose $A_1, A_2 \in L$ are distinct and $z \in A_1 \cap A_2$. Define $p: E(L) \to \mathbb{Q}^+$ by $p(x) = 1$ $(x \in A_1 \cup A_2)$ and $p(x) = 0$, otherwise. Then p satisfies (ii) and so by hypothesis satisfies (i); it follows that there exists $A \in L$ with $A \subseteq A_1 \cup A_2$ and $z \notin A$. Hence L is the collection of circuits of a matroid, by the weak circuit axiom. ∎

Another problem which suggests itself is the following. What if we permit α to take negative values? But then the problems seem to become much less subtle. For instance, we have the following two theorems.

(4.4) *Let* M *be a matroid, and* $\mathscr{C}$ *its collection of circuits. Given* $p: E(M) \to \mathbb{Q}$, *the following are equivalent*:
   (i) *there exists* $\alpha: \mathscr{C} \to \mathbb{Q}$ *such that* $\sum_{C \in \mathscr{C}} \alpha(C) f_C = p$;
   (ii) *if* $\{e\}$ *is a cocircuit*, $p(e) = 0$; *and if* $\{e_1, e_2\}$ *is a cocircuit*, $p(e_1) = p(e_2)$.

(4.5) *Let* $G = (V, E)$ *be a graph, with collection of circuits* $\mathscr{C}$. *Given* $p: E(G) \to \mathbb{Z}$, *the following are equivalent*:
   (i) *there exists* $\alpha: \mathscr{C} \to \mathbb{Z}$ *such that* $\sum_{C \in \mathscr{C}} \alpha(C) f_C = p$;
   (ii) *if* $\{e\}$ *is a cut*, $p(e) = 0$; *if* $\{e_1, e_2\}$ *is a cut*, $p(e_1) = p(e_2)$; *and for any cut* B, $p(B)$ *is even.*

The proofs of these theorems are easy, and we omit them. [Hint for (4.5): If e is an edge of a graph and is not a bridge, then there are two circuits C_1, C_2 which intersect in e and edges coparallel with e only.] Note, however, that (4.5) does not extend to all binary matroids (take $p \equiv 2$ in the Fano dual). (4.4) implies the (much-proved) result that no matroid has more points than hyperplanes. (Consider the rank of the cocircuit/point incidence matrix.)

5. Multicommodity Flows and Cuts

The following problem is well known; it is called the multi-commodity network flow problem.

(5.1) *Let $G = (V, E)$ be a graph, and let $E' \subseteq E$ and $c: E \to \mathbb{Q}^+$ be given. Let $\mathcal{R}$ be the collection of circuits of G which use precisely one edge of E'. When is there a map $\alpha: \mathcal{R} \to \mathbb{Q}^+$ such that*

$$\sum_{C \in \mathcal{R}} \alpha(C) f_C(e) \leq c(e) \qquad \text{if} \quad e \notin E'$$

and

$$\sum_{C \in \mathcal{R}} \alpha(C) f_C(e) \geq c(e) \qquad \text{if} \quad e \in E'?$$

(We have changed the normal formulation of the problem slightly by adding the dummy edges E'; this is just to clarify the connection with (5.2).]

In [5], Hu says it was a popular conjecture that such an α exists if and only if for each cut B, $c(B \cap E') \leq c(B - E')$. This is true for $|E'| = 1$ (the max-flow min-cut theorem) and for $|E'| = 2$ (this is Hu's 2-commodity flow theorem) but false in general for $|E'| \geq 3$ (see [5]).

In this section we solve the problem dual to (5.1) (in the circuit-cut sense), with the following theorem.

(5.2) *Let $G = (V, E)$ be a graph, and let $E' \subseteq E$ and $c: E \to \mathbb{Q}^+$ be given. Let $\mathcal{R}$ be the collection of cuts of G which use precisely one edge of E'. Then the following are equivalent:*

(i) *there exists $\alpha: \mathcal{R} \to \mathbb{Q}^+$ such that*

$$\sum_{C \in \mathcal{R}} \alpha(C) f_C(e) \leq c(e) \qquad \text{if} \quad e \notin E',$$

$$\sum_{C \in \mathcal{R}} \alpha(C) f_C(e) \geq c(e) \qquad \text{if} \quad e \in E';$$

(ii) *for each circuit C of G,*

$$c(E(C) \cap E') \leq c(E(C) - E').$$

Proof Certainly (i) implies (ii); we sketch a proof of the converse. So suppose that c satisfies (ii). Define $d: E \to \mathbb{Q}^+$ by

$$d(e) = \begin{cases} c(e) & (e \notin E'), \\ -c(e) & (e \in E'). \end{cases}$$

Then for every circuit C, $d(E(C)) \geq 0$. It follows from (1.4) [or its special case (2.5)] that for $w: E \to \mathbb{Q}^+$, if $w(e) \leq w(B - \{e\})$ for every cut B and $e \in B$, then $\sum_{e \in E} w(e)d(e) \geq 0$.

Let $\mathscr{F}$ be the set of all functions $f: E \to \mathbb{Q}$ of the form

$$f(e_1) = -1 \qquad f(e) = \begin{cases} 1 & (e \in B - \{e_1\}), \\ 0 & (e \in E - B), \end{cases}$$

where B is a cut and $e_1 \in B$. Then we have that if $w: E \to \mathbb{Q}^+$ satisfies $\sum w(e)f(e) \geq 0$ for each $f \in \mathscr{F}$, then $\sum w(e)d(e) \geq 0$. Well-known linear programming results imply that there exists $\beta: \mathscr{F} \to \mathbb{Q}^+$ such that $d \geq \sum_{f \in \mathscr{F}} \beta(f)f$. Moreover, we can choose β so that for $f \in \mathscr{F}$, if $\beta(f) \neq 0$ and f corresponds to B, e, then $B \cap E' = \{e\}$. (This follows easily from the fact that if B_1, B_2 are cuts, and $e_1 \in B_1 \cap B_2$ and $e_2 \in B_2 - B_1$ then there is a third cut $B_3 \subseteq B_1 \cup B_2$ with $e_2 \in B_3$ and $e_1 \notin B_3$.) Then define $\alpha: \mathscr{R} \to \mathbb{Q}^+$ in terms of β in the obvious way, and the theorem is proved. ∎

As a corollary, we have (from planar duality) a solution to (5.1) when the graph G is planar.

Theorem (5.2) is actually equivalent to (2.5), because the argument may be reversed. We see that the fact that (5.1) has no nice solution in general corresponds to the fact that (1.4) and (2.5) do not work for sums of cuts.

ACKNOWLEDGMENTS

I have had valuable discussions with several people which have contributed to this paper; but in particular I am grateful to Rick Giles for his help with (2.2), and to Alan Hoffman, who drew my attention to the approach in Section 5.

REFERENCES

1. K. Appel and W. Haken, Every planar map is four colorable, *Illinois J. Math.* **21** (1977), 429–567.
2. L. R. Ford, Jr. and D. R. Fulkerson, "Flows in Networks," Princeton Univ. Press, Princeton, New Jersey, 1962.
3. D. R. Fulkerson, Networks, frames, blocking systems, *in* "Mathematics of the Decision Science," Lectures in Applied Mathematics, pp. 303–334. Amer. Math. Soc., Providence, Rhode Island, 1968.

4. A. J. Hoffman, Some recent applications of the theory of linear inequalities to extremal combinatorial analysis, *Proc. Symp. Appl. Math.* **10** (1960), 113–127.
5. T. C. Hu, Multi-commodity network flows, *Operations Res.* **11** (1963) 344–360.
6. A. Lehman, On the length-width inequality, (mimeo. 1965).
7. P. D. Seymour, The matroids with the max-flow min-cut property, *J. Combinatorial Theory Ser. B*, **23** (1977) 189–222.
8. W. T. Tutte, On the algebraic theory of graph coloring, *J. Combinatorial Theory* **1** (1966) 15–50.
9. A. Kotzig, Ein Beitrag zur Theorie der endlichen Graphen mit linearen Faktoren II, *Mat. Fyz. Casopis* **9** (1959) 136–159.
10. P. D. Seymour, On odd cuts and plane multicommodity flows, *Proc. London Math. Soc.*, to appear.
11. P. D. Seymour, Matroids and multicommodity flows, in preparation.

AMS 05C99

MERTON COLLEGE
OXFORD, ENGLAND

Notes added in proof, 10 November 1978:

(i) Jack Edmonds points out that (1.4) may be derived from matching theory.
(ii) The author has now solved (4.1) [11]. The conjecture after (4.2) is wrong.
(iii) A stronger (half-integral) form of (5.2) will appear in [10].

Unsolved Problems

A CONJECTURE ON PLANAR GRAPHS

MICHAEL O. ALBERTSON

and

DAVID BERMAN

Let G be a graph with vertex set $V(G)$ and suppose $f(G)$ is defined as follows:

$$f(G) = \max_{X \subseteq V(G)} \{|X|/|V(G)|: X \text{ induces a forest}\}.$$

Results on acyclic colorings imply that if G is a planar graph then $f(G)$ is at least $\frac{2}{5}$. We conjecture that if G is a planar graph then $f(G)$ is at least $\frac{1}{2}$. The truth of this conjecture would settle the Erdös–Vising problem (see [1, Problem 36, p. 251]), presumably without the use of the four color theorem. Lower bounds for $f(G)$ on other classes of graphs would also be of interest.

REFERENCE

1. J. A. Bondy and U. S. R. Murty, "Graph Theory with Applications." Macmillan, New York, 1976.

Michael O. Albertson
SMITH COLLEGE
NORTHAMPTON, MASSACHUSETTS

David Berman
UNIVERSITY OF NEW ORLEANS
NEW ORLEANS, LOUISIANA

SELECTED OPEN PROBLEMS IN GRAPH THEORY

ANTON KOTZIG

1. Graphs with a Regular Path-Connectedness

In "Graph Theory with Applications" by J. A. Bondy and U. S. R. Murty, one can find the following conjecture of mine (see p. 246, Problem 4): There exists no graph with the property that every pair of vertices is connected by a unique path of length k ($k > 2$). Let us call a graph with this property a $P(k)$-graph.

If a $P(k)$-graph (say, G) exists for some $k > 2$, then G is clearly connected. Furthermore, because each edge of G belongs to exactly one $(k + 1)$-cycle, G is uniquely edge-decomposable into $(k + 1)$-cycles, and hence is eulerian. One can also prove that

(i) G contains a $2n$-cycle with $3 \le n \le k - 4$ (which implies immediately that $k \ge 7$).

(ii) G contains no $2n$-cycle with $n \in \{2, k - 3, k - 2, k - 1, k\}$.

(iii)   any two $(k + 1)$-cycles of G have at least six, and at most k, vertices in common.

By using the preceding ideas, one can deduce that there is no $P(k)$-graph with $2 < k < 9$. Much stronger results could be obtained if any of the following weaker conjectures (all for $k > 2$) were verified:

Conjecture 1   *A $P(k)$-graph contains no vertex of degree two.*

Conjecture 2   *A $P(k)$-graph contains no triangle.*

Conjecture 3   *Every $P(k)$-graph contains at least three $(k + 1)$-cycles.*

Problem 1   Prove or disprove the preceding conjectures.

2. Regular and Strongly Regular Self-Complementary Graphs

A graph G is said to be *self-complementary* if it is isomorphic with its complement $\bar{G}$. The fundamental properties of self-complementary graphs were discovered and described nearly simultaneously in the papers by H. Sachs [Über Selbstcomplementare Graphen, *Publ. Math. Debrecen* **9** (1962) 270–288], G. Ringel [Selbst-complementare Graphen, *Arch. Math. (Basel)* **14** (1963) 354–358] and R. C. Read [On the number of self-complementary graphs and digraphs, *J. London Math. Soc.* **38** (1963) 99–104].

It is well known that if G is self-complementary, then $|V(G)| \equiv 0$ or 1 (mod 4) and the diameter of G is either two or three. If, moreover, G is regular, then $|V(G)| \equiv 1$ (mod 4), the diameter of G is two, and the number of triangles in G equals $\frac{1}{3}(4k + 1)k(k - 1)$, where $|V(G)| = 4k + 1$. There exist self-complementary graphs on $4k$ and $4k + 1$ vertices for every natural number k.

Let G be a self-complementary graph and let P be an *isomorphism-permutation* of G [that is, a permutation of $V(G)$ describing an isomorphism between G and $\bar{G}$]. It is known that P has at most one cycle of length one and that the length of every other cycle of P is divisible by four. Thus P has a fixed point if and only if $|V(G)| \equiv 1$ (mod 4). I think a deeper study of the properties of the set of all the isomorphism-permutations of a given self-complementary graph may yield a lot of new and possibly surprising results, particularly if the graph is, in addition, assumed to be regular or strongly regular. (A regular graph is said to be *strongly regular* if every edge of G belongs to the same number of triangles and every pair of nonadjacent vertices has the same number of neighbors in common.) Several of the problems in this area appear to be hopeful:

Problem 2 Is it true that, for every regular self-complementary graph G, there is at least one isomorphism-permutation P such that, except for the cycle of length one, every cycle of P is of length exactly four?

Problem 3 Characterize the subset $F(G)$ of $V(G)$ consisting of the vertices which are fixed points of at least one isomorphism-permutation of the regular self-complementary graph G. [Obviously, if $v \in F(G)$ and $|V(G)| = 4k + 1$ then v belongs to exactly $k(k - 1)$ triangles of G. But is it true that whenever $|V(G)| = 4k + 1$ and v belongs to exactly $k(k - 1)$ triangles then $v \in F(G)$?]

Problem 4 Characterize the subset $N(G)$ of $E(G)$ consisting of the edges (u, v) of G with the property that there exists at least one isomorphism-permutation P such that u and v are neighbors in a cycle of P.

Problem 5 Is it true that a regular self-complementary graph G is strongly regular if and only if $F(G) = V(G)$ and $N(G) = E(G)$?

It is known that there exists a strongly regular self-complementary graph with $4k + 1$ vertices for $k \in \{1, 2, \ldots, 7\}, k \neq 5$. These graphs are known to be unique for $k \in \{1, 2, 3\}$ (see Fig. 1).

Problem 6 Is it true that a strongly regular self-complementary graph with $4k + 1$ vertices exists for every natural number k, where $4k + 1 = x^2 + y^2$ and x and y are integers? What is the smallest integer k with the property that there exist at least two nonisomorphic strongly regular self-complementary graphs on $4k + 1$ vertices?

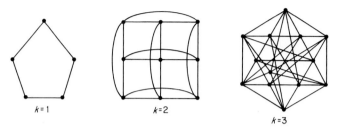

Figure 1

Let R_k denote the set of all regular self-complementary graphs with $4k + 1$ vertices. For $G \in R_k$, denote by $\mu(G)$ the maximal number of mutually edge-disjoint hamiltonian cycles of G, and define the function μ on the set of all natural numbers as follows:

$$\mu(k) = \min_{G \in R_k} \{\mu(G)\}.$$

Then, clearly, $\mu(k) \le k$. Recently W. Jackson [Edge-disjoint Hamiltonian cycles in regular graphs of large degree, *J. London Math. Soc.*, submitted for publication] has shown that $\mu(k) \ge [k/3]$. The construction of $G \in R_k$ with the property that $\mu(G) = k$ for every natural number k is also known. Moreover one can very easily prove that $\mu(1) = 1$ and $\mu(2) = 2$. The next value $\mu(3)$ is unknown and no example of a $G \in R_k$ is known for which $\mu(G) < k$.

Problem 7 Find the exact values of $\mu(k)$ for small $k > 2$.

3. Regular S-Graphs

Let G be a connected graph, let v be a vertex of G, and let $M(v)$ denote the set of vertices of G whose distance from v is locally maximal [in other words, $x \in M(v)$ if and only if $d_G(v, x) \ge d_G(v, y)$ for every neighbour y of x]. G is said to be an *S-graph* if G is bipartite and $|M(v)| = 1$ for every $v \in V(G)$. The unique element of $M(v)$ is called the *opposite* vertex of v in G and is denoted by $\bar{v}$.

It can be proved that, in any S-graph G with diameter $\Delta(G)$,

(i) the function which maps each vertex onto its opposite is an involutory isomorphism;

(ii) $d_G(v, x) + d_G(x, \bar{v}) = d_G(v, \bar{v}) = \Delta(G)$ for every pair $\{v, x\} \subset V(G)$;

(iii) the cartesian product $F \times G$ of two graphs F and G is an S-graph if and only if both F and G are S-graphs.

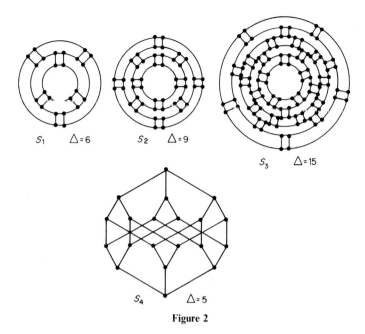

Figure 2

An S-graph G is called *primitive* if it is not a cartesian product of two S-graphs. In view of (iii), in order to characterize all S-graphs it suffices to characterize those that are primitive. Except for some results which can be obtained easily by simple observations, very little is known about such graphs (even regular ones).

If an S-graph G contains a vertex of degree one, then clearly $G = K_2$; and if a regular S-graph G contains a vertex of degree two, then all the vertices of G are of degree two and G is a $2n$-cycle, for every $n \geq 2$ (but G is primitive only if $n > 2$). The problems begin with regular S-graphs of degree three. In fact, only four primitive cubic S-graphs (shown in Fig. 2) have so far been found.

Problem 8 What is the set D of possible values of the diameter in a primitive cubic S-graph? [For the graphs shown in Fig. 2, we have $\Delta(S_1) = 6$, $\Delta(S_2) = 9$, $\Delta(S_3) = 15$, $\Delta(S_4) = 5$; hence, $\{5, 6, 9, 15\} \subset D$.]

Problem 9 What is the set K of possible values of the girth in a primitive cubic S-graph? [Clearly this girth g must be even, because an S-graph is a bipartite graph. For the graphs shown in Fig. 2, we have $g(S_1) = g(S_2) = g(S_3) = 4$, $g(S_4) = 6$; hence, $\{4, 6\} \subset K$.]

For every natural number n, there exists an S-graph of diameter n; the n-dimensional cube $K_2 \times K_2 \times \cdots \times K_2$ (n times) is an example of such a graph. It is an S-graph by (iii), and has diameter n since $\Delta(K_2) = 1$ and

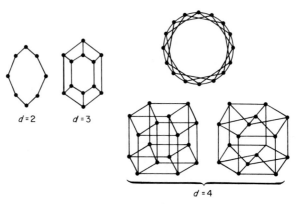

$d=2$ $d=3$

$d=4$

Figure 3

$\Delta(F \times G) = \Delta(F) + \Delta(G)$. A description has been given of all S-graphs with diameter less than four (all are regular), and a simple method of constructing all S-graphs with diameter four is also known (four is the smallest diameter for which there exist nonregular S-graphs). But the following problem remains open:

Problem 10 Let d be a natural number. What is the number $\rho(d)$ of nonisomorphic d-regular S-graphs with diameter four? [Known results $\rho(1) = 0$, $\rho(2) = \rho(3) = 1$, $\rho(4) = 3$; see Fig. 3.]

No method of construction of regular primitive S-graph with degree greater than two and diameter greater than four is known, and there are many open problems concerning nonregular S-graphs. For example, it is known that the degree of a vertex in a given S-graph with diameter four can have at most two distinct values; but what is the greatest number of distinct values in an S-graph with diameter five, six, ... ?

REFERENCES

A. Kotzig, Centrally symmetric graphs (in Russian), *Czechoslovak Math. J.* **18** (93) (1968) 606–615; *Math. Reviews* **38**, No. 5664.

F. Glivjak, A. Kotzig, and J. Plesnik, Remark on the graphs with a central symmetry, *Monatsh. Math.* **74** (1970) 302–307; *Math. Reviews* **44**, No. 105.

A. Berman and A. Kotzig, Some problems in the theory of graphs with a central symmetry, *Publ. CRM.* Sept. (1973) 329.

4. On 1-Factorizations of Cartesian Products of Regular Graphs

Let $G_1, G_2, \ldots, G_n$ be regular graphs and let $H = G_1 \times G_2 \times \cdots \times G_n$. I have proved the following:

Theorem 4.1 *If either*

(i) *some G_i is 1-factorizable, or*
(ii) *there exist G_i and G_j ($i \neq j$) each containing at least one 1-factor then H is factorizable.*

Although both (i) and (ii) are sufficient conditions for the existence of a 1-factorization of H, neither of them is necessary. This is a consequence of

Theorem 4.2 *Let F be a cycle of length greater than three and let G be an arbitrary cubic graph. Then the cartesian product $H = F \times G$ is 1-factorizable.*

Thus if F is a pentagon and G is isomorphic to the graph of Fig. 4, then $H = F \times G$ is 1-factorizable. However, neither F nor G has a 1-factor.

Figure 4 G.

Problem 11 Let G be a bridgeless cubic graph and let $H = G \times K_3$. Is it true that H must have at least one 1-factorization?

Remark If G contains a bridge, then H contains an edge-cut E of cardinality three, and then the 5-regular graph $H = K_3 \times G$ cannot be 1-factorizable. (This is a consequence of the well-known theorem of Tutte: E must contain an odd number of edges, and hence at least one edge, of each 1-factor of G; since $|E| = 3$, it follows that H has at most three disjoint 1-factors.) Thus the problem is open only for a bridgeless G.

Problem 12 Can one find a pair of cubic graphs F and G such that $F \times G$ has no 1-factorization?

5. Decompositions of Cartesian Products of Regular Graphs into Hamiltonian Cycles

In 1963, I made the following conjecture:

Conjecture 4 *Let r and s be natural numbers and let $G_1, G_2, \ldots, G_r$ be graphs each decomposable into (exactly) s hamiltonian cycles. Then the cartesian product $H = G_1 \times G_2 \times \cdots \times G_r$ is decomposable into rs hamiltonian cycles.*

This conjecture was verified for $r = 2$, with no restrictions on s, and the following theorem was proved:

Theorem 5.1 *If the conjecture is true for a given r and $s = 1$, then it is true for the same r and every natural s.*

Thus the considerations may be restricted to the case where the graphs $G_i (i = 1, \ldots, r)$ are cycles.

Theorem 5.2 *If the conjecture is true with $r = m$ and also with $r = n$, then it is true with $r = mn$.*

This implies that the conjecture holds for $r = 2^n$, where n is an arbitrary natural number, and that it is sufficient to verify the conjecture for prime numbers $r > 2$. Thus the natural question is:

Problem 13 For what prime numbers r ($r > 2$) is the following conjecture true: Every cartesian product of r cycles is decomposable into r hamiltonian cycles?

I think that the very strong property of the graphs G_i in our Conjecture 4 (to be decomposable into s hamiltonian cycles) is not necessary, and the following problem (the simplest possible generalization of the previously mentioned problem) may be interesting:

Problem 14 Is it true that the cartesian product of two d-regular d-vertex-connected (d-edge-connected) graphs is decomposable into hamiltonian cycles?

REFERENCE

A. Kotzig, Every cartesian product of two circuits is decomposable into hamiltonian circuits, *Publ. CRM* Sept. (1973) 323.

6. Four Open Problems for the Graph of a d-Dimensional Cube

(a) Decompositions of complete graphs.

The graph of a d-dimensional cube will be denoted by W_d. Thus $W_d = K_2 \times K_2 \times \cdots \times K_2$ (d times). I have obtained the following results on decompositions of complete graphs into d-dimensional cubes:

Theorem 6.1 *Let d and m be natural numbers such that $d > 1$ and $m \equiv 1 \pmod{d \cdot 2^d}$. Then K_m is decomposable into graphs all isomorphic with W_d.*

Theorem 6.2 *If d is even, then K_m is decomposable into graphs all isomorphic with W_d if and only if $m \equiv 1 \pmod{d \cdot 2^d}$.*

Theorem 6.3 *Let d be odd and let K_m be decomposable into graphs all isomorphic with W_d and $m \not\equiv 1 \pmod{d \cdot 2^d}$. Then $m \equiv 1 \pmod{d}$ and $m \equiv 0 \pmod{2^d}$.*

As an illustration of Theorem 6.3 a decomposition of K_{16} into five cubic factors each consisting of two components isomorphic with W_3 is depicted in Fig. 5.

Problem 15 Let d be a prime number, $d > 2$. Is it true that K_m is decomposable into graphs all isomorphic with W_d if and only if $m \equiv 1$ or $(d + 1)2^{d-1} \pmod{d \cdot 2^d}$?

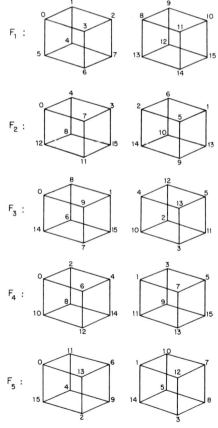

Figure 5 $K_{16} = F_1 \bigcup^i F_2 \bigcup^i F_3 \bigcup^i F_4 \bigcup^i F_5$.

Problem 16 Let d be odd and let P_d be the set of all natural numbers m with $m \equiv 1 \pmod{d \cdot 2^d}$ and with the property that K_m is decomposable into d factors each of whose components is isomorphic with W_d. What can one say about P_d ($d = 3, 5, 7, \ldots$)? As Fig. 5 illustrates, $16 \in P_3$.

(b) A special edge-colouring of W_d.

A *τ-colouring* of W_d is a d-edge-colouring in which the edges of each 4-cycle of W_d receive distinct colours. Obviously a τ-colouring of W_d can exist only if $d \geq 4$.

I have proved two theorems concerning τ-colourings:

Theorem 6.4 *A τ-colouring of W_d exists for every d divisible by four.*

Theorem 6.5 *If there exists a τ-colouring of W_d then there exists a τ-colouring of W_{d+4n} for every natural number n.*

However, the following problem (now more than fourteen years old) remains open:

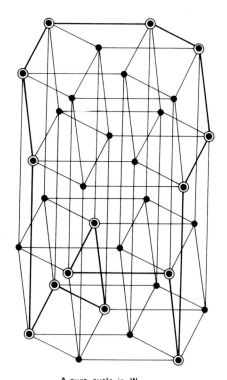

A pure cycle in W_5.

Figure 6 A pure cycle in W_5.

Problem 17 What is the smallest integer $r \equiv 1$ ($s \equiv 2$, respectively, $t \equiv 3$, respectively) (mod 4) with the property that W_r (W_s, respectively; W_t, respectively) has a τ-colouring?

REFERENCE

A. Kotzig, Linear factors in lattice graphs, *Mat.-Fyz. Čas. SAV* **14** (1964) 104–133 (Theorem 7); *Math. Reviews* **30**, No. 1504.

(c) Pure cycles in W_d.

A cycle C in W_d is said to be a *pure cycle* of W_d if every edge joining two vertices of C belongs to C. Clearly every 4-cycle of W_d ($d > 1$) is a pure cycle of W_d and four is the smallest length of a pure cycle in W_d. Denote by $\lambda(d)$ the largest possible length of a pure cycle in W_d. Then one can find very easily that $\lambda(1) = 0$ and $\lambda(i) = 2i$ for $1 < i < 5$. Figure 6 shows that $\lambda(5) \geq 14$ [it is not difficult to prove that $\lambda(5) = 14$] and nearly nothing is known about the exact values $\lambda(n)$ for $n > 5$.

Problem 18 Find the exact values (or at least good estimates) of $\lambda(n)$ for small n's.

CENTRE DE RECHERCHES MATHEMATIQUES
UNIVERSITÉ DE MONTRÉAL
MONTRÉAL, QUÉBEC

SOME UNSOLVED PROBLEMS ON ONE-FACTORIZATIONS OF GRAPHS

P. D. SEYMOUR

Let $G = (V, E)$ be a graph, possibly with multiple edges but without loops, and with $|V| \geq 3$. Define $\chi'(G)$ = edge-chromatic number, $\Delta(G)$ = maximum vertex-degree, and

$$\Gamma(G) = \max\left(\frac{|\langle X \rangle|}{\frac{1}{2}(|X| - 1)} : X \subseteq V \text{ with } |X| \geq 3 \text{ and odd}\right)$$

(here $\langle X \rangle$ is the set of edges of G with both ends in X). It is easy to see that $\chi'(G) \geq \Delta(G), \Gamma(G)$, and so

$$\chi'(G) \geq \max(\Delta(G), \{\Gamma(G)\})$$

where { } denotes upper integer part.

Problem 1 Is it true that $\chi'(G) \le \max(\Delta(G) + 1, \{\Gamma(G)\})$?

(This is true if the right hand side ≤ 6; also if G is simple, by Vizing's theorem.)

Problem 2 Is it true that if G is planar then,
$$\chi'(G) = \max(\Delta(G), \{\Gamma(G)\})?$$

(This of course contains the four colour theorem, and various other difficult conjectures; so the search is really for a counterexample.)

If G is regular of degree r and $\Gamma(G) \le r$, we say that G is an *r-graph*. An equivalent condition is: G is r-regular and, for every $X \subseteq V$ with $|X|$ odd, there are at least r edges with just one end in X.

Problem 3 Is it true that if $r \ge 4$ then every r-graph has a 1-factor whose deletion gives an $(r - 1)$-graph?

A gross, but still open, weakening of this is the following.

Problem 4 Is there any integer $r \ge 3$ such that every r-connected r-regular graph with an even number of vertices has two disjoint 1-factors?

I conjecture that the answer in each case is "yes."

MERTON COLLEGE
OXFORD, ENGLAND

IMBALANCE OF TREES

B. SIMEONE

If V is a set with n elements and $\pi \equiv \{V_1, \ldots, V_p\}$ is a *p-partition* of V into nonempty subsets, the *imbalance* of π is defined to be $\sum_{i=1}^{p} |n_i - (n/p)|$, where n_i is the number of elements of V_i.

Assuming that $V = V(G)$ is the vertex-set of a connected graph G, define π to be *connected* if, for each $1 \le i \le p$, the subgraph $G[V_i]$ induced by V_i is connected.

Consider the following problem.

Problem Given a connected graph G with n vertices, and an integer p, $1 \le p \le n$, find a connected p-partition of $V(G)$ with minimum imbalance.

Denote such minimum imbalance by $u_p(G)$. It can be (nonconstructively) shown [1] that any connected graph G has a spanning tree T^* such that

$u_p(T^*) = u_p(G)$. [Of course one has $u_p(T) \geq u_p(G)$ for any spanning tree T of G.]

Question   Is there any algorithm, polynomial in n and p, for solving the problem when G is a tree?

REFERENCE

1. E. L. Aparo and B. Simeone, Equipartizione su un grafo: Un'applicazione all'informatica medica, in "Applicazioni del Calcolo" (I. Galligani, ed.), pp. 77–86. Consiglio Nazionale delle Ricerche, Rome, 1975. (In Italian.)

UNIVERSITY OF WATERLOO
WATERLOO, ONTARIO

THE FOUR COLOR PROBLEM FOR LOCALLY PLANAR GRAPHS

WALTER STROMQUIST

A graph drawn on a surface of high genus may have a chromatic number much greater than four. But what if the graph is locally planar? Since the most important obstacles to coloring seem to be local, should not most such graphs be four-colorable?

Several more-or-less equivalent conditions can be used to formalize the idea of a locally planar graph G on a surface S. For example, we might assume that any circuit of G which is not contractible in S contains at least k vertices. Or, given a fixed surface S with a metric, we might assume that every edge of G has length less than ε.

Michael Albertson and I think we have proved that given a strong enough local-planarity condition, any graph on any surface can be five-colored [1]. The proof uses the four color theorem for planar graphs. Steve Fisk discovered that if a triangulation of any surface has exactly two odd vertices and they are adjacent, then the triangulation cannot be four-colored [2]. Using this fact, graphs can be constructed on any compact surface (except the sphere) which satisfy any local-planarity condition but which require five colors.

Are Fisk's examples the only ones, or are there other global obstacles to four-coloring?

Conjecture Let G be imbedded in a surface S, and assume that G has at least four odd vertices and no 3-circuits which are not faces. If G satisfies a sufficiently strong local-planarity condition, then G is four-colorable.

REFERENCES

1. M. Albertson and W. Stromquist, in preparation.
2. S. Fisk, Combinatorial structures on triangulations I, *Advances in Math.* **11**, No. 3 (1973) 326–338.

BROOKHAVEN NATIONAL LABORATORY
UPTON, NEW YORK

A DISSECTION PROBLEM AND AN INTERTWINING PROBLEM

PETER UNGAR

The first problem arose in a conversation with Ian C. Percival when we were students in London, c. 1951.

Problem Is every cyclically 4-connected cubic graph with more than two vertices the union of two edge-disjoint trees?

A famous theorem of Whitney [2] states that in any cyclically 4-connected cubic plane graph G there is a curve W which passes through all the regions into which G divides the plane exactly once. The curve W cuts G into two pieces and by assigning the edges cut by W to one or the other of these pieces in a suitable manner one gets a decomposition of G into two edge-disjoint trees. Conversely, one can easily obtain a Whitney circuit W from a 2-tree decomposition of G. Thus, for planar cyclically 4-connected cubic graphs the two-tree decomposition property is equivalent to Whitney's theorem, and our problem is, can one drop the assumption of planarity?

The proof of Whitney's theorem is laborious and utilizes the Jordan curve theorem at every turn. For this reason we thought it very unlikely at first that the planarity assumption could simply be dropped. But later I found that randomly drawn graphs with up to 80 nodes as well as such particularly high genus graphs as the 12-cage and Balaban's graph of girth ten [1] could all easily be decomposed into two edge-disjoint trees. Moreover, there appeared to be a great deal of freedom in making the decompositions in all cases, the highly nonplanar as well as the planar.

The second problem I wish to present originated in conversations with Martin Milgram about the missing link in proving an analog of Kuratowski's theorem for graphs not representable on a given surface. L. Lovasz told us that he had also made this conjecture.

Milgram defines a graph G with no nodes of degree two to be an *intertwining* of two graphs H and K if G has subgraphs homeomorphic to H and to K but no proper subgraph of G has two such subgraphs.

Conjecture *Two graphs H, K have only finitely many different intertwinings.*

As a trial run, Milgram proved this when H is $K_{3,3}$ and K is K_5. I then showed that if a graph G has eight nodes A_i, B_i, C_i, D_i ($i = 1, 2$), with a pair of disjoint paths connecting A_1 and A_2, and B_1 and B_2, and also a pair of disjoint paths connecting C_1 and C_2, and D_1 and D_2, and if no proper subgraph of G has such paths between these eight nodes, then G has at most 60 nodes (of degree other than 2).

REFERENCES

1. A. T. Balaban, A trivalent graph of girth ten, *J. Combinatorial Theory Ser. B* **12** (1972) 1–5.
2. H. Whitney, A theorem on graphs, *Ann. of Math.* **32** (1931) 378–390.

NEW YORK UNIVERSITY
NEW YORK, NEW YORK

Note added in Proof It has now been proved that a similar statement with two sets of k disjoint paths would imply the intertwining conjecture. (Oral communication from L. Lovász.)